Algorithms and Computation in Mathematics · Volume 1

Editors

E. Becker   M. Bronstein   H. Cohen
D. Eisenbud   R. Gilman

Springer
Berlin
Heidelberg
New York
Barcelona
Budapest
Hong Kong
London
Milan
Paris
Santa Clara
Singapore
Tokyo

Manuel Bronstein

Symbolic Integration I

Transcendental Functions

With 3 Figures

Springer WITHDRAWN

Manuel Bronstein

ETH Zürich
Institute for Scientific Computation
ETH Zentrum IFW
CH-8092 Zürich, Switzerland

e-mail: Manuel.Bronstein@inf.ethz.ch

Library of Congress Cataloging-in-Publication Data

Bronstein, Manuel.
  Symbolic integration I : transcendential functions / Manuel
Bronstein.
      p.   cm. -- (Algorithms and computation in mathematics ; v. 1)
  Includes bibliographical references and index.
  ISBN 3-540-60521-5 (hardcover : alk. paper)
  1. Integrals--Data processing.   I. Title.   II. Series.
QA308.B76   1997
515.43--dc20
                                                              96-41956
                                                                   CIP

Mathematics Subject Classification (1991):
68Q40, 12H05, 12Y05, 33B10, 28-04

ISSN 1431-1550

ISBN 3-540-60521-5 Springer-Verlag Berlin Heidelberg New York

© Springer-Verlag Berlin Heidelberg 1997
Printed in Germany

The use of general descriptive names, registered names, trademarks, etc. in this publication does not imply, even in the absence of a specific statement, that such names are exempt from the relevant protective laws and regulations and therefore free for general use.

Typesetting: Camera-ready copy produced from the author's output file using a Springer TeX macro package

SPIN 10521634   41/3143 – 5 4 3 2 1 0 – Printed on acid-free paper

Foreword

This book brings together two streams of research in mathematics and computing that were begun in the nineteenth century and made possible through results brought to fruition in the twentieth century.

Methods for indefinite integration have been important ever since the invention of the calculus in the 1700s. In the 1800s Abel and Liouville began the earliest mathematical research on algorithmic methods on integration in finite terms leading to what might be considered today as an early mathematical vision of a complete algorithmic solution for integrating elementary functions. In an 1842 publication Lady Ada Augusta, Countess of Lovelace, describing the capabilities of Babbage's analytical engine put forth the vision that computational devices could do algebraic as well as numerical calculations when she said that "[Babbage's Analytical Engine] can arrange and combine its numerical quantities exactly as if they were *letters* or any other *general* symbols; and in fact it might bring out its results in algebraical *notation* were provisions made accordingly." Thus these two visions set the stage for a century and a half of research that partially culminates in this book.

Progress in the mathematical realm continued through out the nineteenth and twentieth centuries. The Russian mathematician Mordukhai-Boltovskoi wrote the first two books on this subject in 1910 and 1913[1].

With the invention of electronic computers in the late 1930s and early 1940s, a new impetus was given to both the mathematical and computational streams of work. In the meantime in the mathematical world important progress had been made on algebraic methods of research. Ritt began to apply the new algebraic techniques to the problem of integration in finite terms, an approach that has proven crucially important. In 1948 he published the results of his research in a little book, *Integration in Finite Terms*. The use of these algebraic ideas were brought to further fruition by Kolchin, Rosenlicht, and, particularly for problems of symbolic integration, by three of Rosenlicht's Ph.D. students — Risch, Singer, and Bronstein[2].

---

[1] *On the Integration in Finite Terms of Linear Differential Equations*. Warsaw, 1910 (in Russian) and *On the Integration of Transcendental Functions*. Warsaw, 1913 (in Russian).

[2] Let me hasten to add that there have been important contributions by many others and it is not my intention to give a complete history of the field in this short paragraph, but to indicate some of main streams of work that have led to the current book.

On the computational side, matters rested until 1953 when two early programs were written, one by Kahrimanian at Temple University and another by Nolan at Massachusetts Institute of Technology, to do analytic differentiation — the inverse of indefinite integration. There was active research in the late 1950s and early 1960s on list processing packages and languages that laid the implementation foundations for today's computer algebra systems. Slagle's 1961 thesis was an early effort to write a program, in LISP, to do symbolic integration. With the advent of general computer algebra systems, some kind of symbolic integration facility was implemented in most. These integration capabilities opened the eyes of many early users of symbolic mathematical computation to the amazing potential of this form of computation. But yet none of the systems had a complete implementation of the full algorithm that Risch had announced in barest outline in 1970. There were a number of reasons for this. First and foremost, no one had worked out the many aspects of the problem that Risch's announcement left incomplete.

Starting with his Ph.D. dissertation and continuing in a series of beautiful and important papers, Bronstein set out to fill in the missing components of Risch's 1970 announcement. Meanwhile working at the IBM T. J. Watson Research Center, he carried out an almost complete implementation of the integration algorithms for elementary functions. It is the most complete implementation of symbolic integration algorithms to date.

In this book, Bronstein brings these mathematical and computational streams of research together in a highly effective manner. He presents the algorithmic details in pseudo-code that is easy to implement in most of the general computer algebra systems. Indeed, my students and I have implemented and tested many of the algorithms in MAPLE and MACSYMA. Bronstein's style and appropriate level of detail makes this a straightforward task, and I expect this book to be the standard starting place for future implementers of symbolic integration algorithms. Along with the algorithms, he presents the mathematics necessary to show that the algorithms work correctly. This is a very interesting story in its own right and Bronstein tells it well. Nonetheless, for those primarily interested in the algorithms, much of the mathematics can be skipped at least in a first study. But the full beauty of the subject is to be most appreciated by studying both aspects.

The full treatment of the subject is a long one and it is not finished in this volume. The longer and more difficult part involving the integration of algebraic functions must await a second volume. This volume serves as a good foundation to the topic of symbolic integration and as a nice introduction to the literature for integration of algebraic functions and for other aspects such as integration involving non-elementary functions. Study, learn, implement, and enjoy!

B. F. Caviness

Preface

The integration problem, which is as old as calculus and differentiation, can be informally stated very concisely: given a formula for a function $f(x)$, determine whether there is a formula for a differentiable function $F(x)$ satisfying

$$\frac{dF}{dx} = f(x)$$

and compute such an $F(x)$, which is called an antiderivative of $f(x)$ and is denoted

$$F(x) = \int f(x)dx$$

if it exists. Yet, while symbolic differentiation is a rather simple mechanical process, suitable as an exercise in a first course in analysis or computer programming, the inverse problem has been challenging scientists since the time of Leibniz and Newton, and is still a challenge for mathematicians and computer scientists today. Despite the many great strides made since the 19th century in showing that integration is in essence a mechanical process, although quite more complicated than differentiation, most calculus and analysis textbooks give students the impression that integration is at best a mixture of art and science, with flair in choosing the right change of variable or approach being an essential ingredient, as well as a comprehensive table of integrals.

The goal of this book is to show that computing symbolic antiderivatives is in fact an algorithmic process, and that the integration procedure for transcendental functions can be carried out by anyone with some familiarity with polynomial arithmetic. The integration procedure we describe is also capable of deciding when antiderivatives are not elementary, and proving it as a byproduct of its calculations. For example the following classical nonelementary integrals

$$\int e^{x^2} dx \,, \qquad \int \frac{dx}{\log(x)} \,, \qquad \int \frac{\sin(x)dx}{x} \,,$$

can be proven nonelementary with minimal calculations.

The algorithmic approach, pioneered by Abel and Liouville in the past century, eventually succeeded in producing a mechanical procedure for deciding whether an elementary function has an elementary antiderivative, and for computing one if so. This procedure, which Risch described in a series of reports [58, 59, 60, 61], unfortunately not all of them published, forms the basis of most of the symbolic integration algorithms of the past 20 years, all of them loosely grouped under the appellation *Risch algorithm*. The procedure which we describe in this book also has its roots in the original Risch algorithm [60] and its improvements, our main sources besides Risch being [11, 12, 68, 74].

We have tried to keep the presentation as elementary as possible, with the minimal background for understanding the algorithm being an introductory course in algebra, where the topics rings and fields, polynomial greatest common divisors, irreducible polynomials and resultants are covered[3]. Some additional background in field theory, essentially algebraic and transcendental extensions, is occasionally used in the proofs associated with the algorithm. The reader willing to accept the algorithm without proof can skip those sections while learning the algorithm.

We have also generalized and extended the original Risch algorithm to a wider class of functions, thereby offering the following features, some of them new, to the reader already familiar with symbolic integration:

– The algorithms in this book use only rational operations, avoiding factorization of polynomials into irreducibles.
– Extensions by tangents and arc-tangents are treated directly, thereby real trigonometric functions are integrated without introducing complex exponentials and logarithms in the computations.
– Antiderivatives in elementary extensions can still be computed when arbitrary primitives are allowed in the integrand, *e.g.* $\mathrm{Erf}(x)$, rather than logarithms.
– Several subalgorithms are applicable to a large class of non-Liouvillian extensions, thereby allowing integrals to be computed for such functions.

The material in this book has been used in several courses for advanced undergraduates in mathematics or computer science at the Swiss Federal Institute of Technology in Zurich:

– In a one-semester course on symbolic integration, emphasizing the algorithmic and implementation aspects. This course covers Chap. 2 in depth, Chap. 3 and 4 superficially, then concentrates on Chap. 5, 6, 7 and 8.
– In the first part of a one-semester course on differential algebra. This course covers Chap. 3, 4 and 5 in depth, turning after Liouville's Theorem to other topics (e.g. differential Galois theory).
– In the last part of a one-semester introductory course in computer algebra, where some algorithms from Chap. 2 and 5 are presented, usually without proofs.

In all those courses, the material of Chap. 1 is covered as and when needed, depending on the background of the students. Chap. 9 contains complete

---

[3] Those topics are reviewed in Chap. 1.

proofs of several structure theorems and can be presented independently of the rest of this book.

By presenting the algorithm in pseudocode in various "algorithm boxes" throughout the text, we also hope to make this book useful for programmers implementing symbolic integrators: by following the pseudocode, they should be able to write an integrator without studying in detail the associated theory.

The reader will notice that several topics in symbolic integration are missing from this book, the main one being the integration of algebraic functions. Including algorithms for integrating algebraic and mixed algebraic-transcendental functions would however easily double the size of this book, as well as increase the mathematical prerequisites, since those algorithms require prior familiarity with algebraic curves and functions. We have thus decided to cover algebraic functions in a second volume, which will hopefully appear in the near future. In the meantime, this book is an adequate preparation to the extensive literature on the integration of algebraic functions [8, 9, 13, 14, 26, 58, 59, 61, 76]. Another related topic is integration in nonelementary terms, *i.e.* with new special functions allowed in the antiderivatives. Here also, the reader should have no difficulty moving on to the research literature [5, 6, 20, 21, 38, 39, 55, 79] after completing this book.

Acknowledgements

I am thankful to several colleagues and students who have read and corrected many early drafts of this book. I am particularly grateful to Bob Caviness, Thom Mulders and Paul Zimmermann, who corrected many errors in the final text and suggested several improvements. Sergei Abramov, Cedric Bächler, Johannes Grabmeier, David Stoutemyer, Jacques-Arthur Weyl and Clifton Williamson have also helped a great deal with their corrections and suggestions. Of course, I am fully responsible for any error that may remain.

Finally, I wish to thank Dr. Martin Peters and his staff at Springer-Verlag for their great patience with this project.

M. Bronstein

Table of Contents

1. Algebraic Preliminaries

We review in this chapter the basic algebraic structures and algorithms that will be used throughout this book. This chapter is not intended to be a replacement for an introductory course in abstract algebra, and we expect the reader to have already encountered the definitions and fundamental properties of rings, fields and polynomials. We only recall those definitions here and describe some algorithms on polynomials that are not always covered in introductory algebra courses. Since they are well-known algorithms in computer algebra, we do not reprove their correctness here, but give references instead. For a comprehensive introduction to constructive algebra and algebraic algorithms, including more efficient alternatives for computing greatest common divisors of polynomials, we recommend consulting introductory computer algebra textbooks [2, 29, 31, 50, 82]. Readers with some background in algebra can skip this chapter and come back to it later as needed.

1.1 Groups, Rings and Fields

An algebraic structure is usually a set together with one or more operations on it, operations that satisfy some computation rules called axioms. In order not to always list all the satisfied axioms for a given structure, short names have been given to the most common structures. Groups, rings and fields are such structures, and we recall their definitions in this section.

Definition 1.1.1. *A* group $(G, \circ)$ *is a nonempty set* G, *together with an operation* $\circ : G \times G \to G$ *satisfying the following axioms:*

(i)   *(Associativity)* $\forall a, b, c \in G, a \circ (b \circ c) = (a \circ b) \circ c$.
(ii) *(Identity element)* $\exists e \in G$ *such that* $\forall a \in G, e \circ a = a \circ e = a$.
(iii) *(Inverses)* $\forall a \in G, \exists a^{-1} \in G$ *such that* $a \circ a^{-1} = a^{-1} \circ a = e$.

In addition, $\circ$ *is called* commutative *(or* Abelian*) if* $a \circ b = b \circ a$ *for all* $a, b \in G$, *and* $(G, \circ)$ *is called a* commutative group *(or* Abelian group*) if it is a group and* $\circ$ *is commutative.*

Example 1.1.1. Let $G = GL(\mathbb{Q}, 2)$ be the set of all the 2 by 2 matrices with rational number coefficients and nonzero determinant, and let $\circ$ denote the usual matrix multiplication. $(G, \circ)$ is then a group: associativity can easily be checked, the identity element is the identity matrix, and the inverse of a matrix in G is given by

$$\begin{pmatrix} a & b \\ c & d \end{pmatrix}^{-1} = \frac{1}{ad - bc} \begin{pmatrix} d & -b \\ -c & a \end{pmatrix}$$

which is in G since the determinant of any element of G is nonzero. Note that $(G, \circ)$ is not a commutative group since

$$\begin{pmatrix} 1 & 1 \\ 0 & 1 \end{pmatrix} \circ \begin{pmatrix} 0 & 1 \\ 1 & 0 \end{pmatrix} = \begin{pmatrix} 1 & 1 \\ 1 & 0 \end{pmatrix}$$

and

$$\begin{pmatrix} 0 & 1 \\ 1 & 0 \end{pmatrix} \circ \begin{pmatrix} 1 & 1 \\ 0 & 1 \end{pmatrix} = \begin{pmatrix} 0 & 1 \\ 1 & 1 \end{pmatrix}.$$

Example 1.1.2. Let $G = \mathcal{M}_{2,2}(\mathbb{Q})$ be the set of all the 2 by 2 matrices with rational number coefficients, and let $\circ$ denote the usual matrix addition. It can easily be checked that $(G, \circ)$ is a commutative group with the zero matrix as identity element.

Definition 1.1.2. *A ring $(R, +, \cdot)$ is a set R, together with two operations $+ : R \times R \to R$ and $\cdot : R \times R \to R$ such that:*

(i) $(R, +)$ is a commutative group.
(ii) (Associativity) $\forall a, b, c \in R, a \cdot (b \cdot c) = (a \cdot b) \cdot c$.
(iii) (Multiplicative identity) $\exists i \in R$ such that $\forall a \in R, i \cdot a = a \cdot i = a$.
(iv) (Distributivity)

$$\forall a, b, c \in R, a \cdot (b + c) = (a \cdot b) + (a \cdot c) \text{ and } (a + b) \cdot c = (a \cdot c) + (b \cdot c).$$

$(R, +, \cdot)$ *is called a* commutative ring *if it is a ring and $\cdot$ is commutative. In addition, we define the* characteristic of R *to be 0 if $ni \neq e$ for any positive integer n, the smallest positive integer m such that $mi = e$ otherwise. Let R and S be rings. A map $\phi : R \to S$ is a* ring–homomorphism *if $\phi(e_R) = e_S$, $\phi(i_R) = i_S$, and $\phi(a + b) = \phi(a) + \phi(b)$ and $\phi(ab) = \phi(a) \cdot \phi(b)$ for any $a, b \in R$. A* ring–isomorphism *is a bijective ring–homomorphism.*

In the rest of this book, whenever $(R, +, \cdot)$ is a ring, we write 0 for the identity element of R with respect to $+$, 1 for the identity element of R with respect to $\cdot$, and for $a, b \in R$, we write ab instead of $a \cdot b$.

Example 1.1.3. Let $R = \mathcal{M}_{2,2}(\mathbb{Q})$ be the set of all the 2 by 2 matrices with rational number coefficients, and let $+$ denote matrix addition and $\cdot$ denote matrix multiplication. $(R, +, \cdot)$ is then a ring, but not a commutative ring (see example 1.1.1). Since

$$n \, i = n \begin{pmatrix} 1 & 0 \\ 0 & 1 \end{pmatrix} = \begin{pmatrix} n & 0 \\ 0 & n \end{pmatrix}$$

is nonzero for any positive integer n, R has characteristic 0.

Example 1.1.4. Let $R = \mathbb{Z}_6$ (the integers modulo 6) with $+$ and $\cdot$ being the addition and multiplication of integers modulo 6. $(R, +, \cdot)$ is then a commutative ring, and the map $\phi : \mathbb{Z} \to \mathbb{Z}_6$ defined by $\phi(n) = n \pmod 6$ is a ring–homomorphism. Since $1 + 1 + 1 + 1 + 1 + 1 = 0$ in $\mathbb{Z}_6$, and $n1 \neq 0$ for $0 < n < 6$, $\mathbb{Z}_6$ has characteristic 6. Note that $2 \cdot 3 = 0$ in $\mathbb{Z}_6$, while $2 \neq 0$ and $3 \neq 0$, so we cannot in general deduce from an equation $ab = 0$ that either a or b must be 0. Commutative rings where we can make this simplification are very useful and common, so they receive a special name.

Definition 1.1.3. *An* integral domain $(R, +, \cdot)$ *is a commutative ring where* $0 \neq 1$ *and*
$$\forall a, b \in R, a \cdot b = 0 \Longrightarrow a = 0 \text{ or } b = 0.$$

Example 1.1.5. Let $R = \mathbb{Z}[\sqrt{-5}] = \{a + b\sqrt{-5}; a, b \in \mathbb{Z}\}$ with $+$ and $\cdot$ denoting complex addition and multiplication. $(R, +, \cdot)$ is then an integral domain.

We now come to the problem of factoring, *i.e.* writing elements of an integral domain as a product of other elements.

Definition 1.1.4. *Let $(R, +, \cdot)$ be an integral domain, and $x, y \in R$. We say that x divides y, and write $x \mid y$, if $y = xt$ for some $t \in R$. An element $x \in R$ is called a* unit *if $x \mid 1$. The set of all the units of R is written R^*. We say that $z \in R$ is a* greatest common divisor *(gcd) of $x_1, \ldots, x_n$ and write $z = \gcd(x_1, \ldots, x_n)$ if:*

(i) $z \mid x_i$ for $1 \leq i \leq n$,
(ii) $\forall t \in R, t \mid x_i$ for $1 \leq i \leq n \Longrightarrow t \mid z$.

In addition, we say that x and y are coprime *if there exists a unit $u \in R^*$, which is a gcd of x and y.*

Example 1.1.6. Let $R = \mathbb{Z}\left[\sqrt{-5}\right]$ as in example 1.1.5, $x = 6$ and $y = 2 + 2\sqrt{-5}$. A norm argument shows that x and y have no gcd in R. Let $N : R \to \mathbb{Z}$ be the map given by $N(a + b\sqrt{-5}) = a^2 + 5b^2$ for $a, b \in \mathbb{Z}$. It can easily be checked that $N(uv) = N(u)N(v)$ for any $u, v \in R$, so $u \mid v$ in R implies that $N(u) \mid N(v)$ in $\mathbb{Z}$. Suppose that $z \in R$ is a greatest common divisor of x and y, and let $n = N(z) \geq 0$. Then, $n \mid N(x) = 36$ and $n \mid N(y) = 24$, so $n \mid 12$ in $\mathbb{Z}$. We have $2 \mid x$ and $2 \mid y$ in R, so $4 = N(2) \mid n$ in $\mathbb{Z}$. In addition, $1 + \sqrt{-5} \mid y$ in R, and

$$6 = 2 \cdot 3 = (1 + \sqrt{-5})(1 - \sqrt{-5}) \tag{1.1}$$

so $1 + \sqrt{-5} \mid x$ in R, hence $6 = N(1 + \sqrt{-5}) \mid n$ in $\mathbb{Z}$. Thus, $12 \mid n$ in $\mathbb{Z}$, so $n = 12$. Writing $z = a + b\sqrt{-5}$ for some $a, b \in \mathbb{Z}$, this implies that $N(z) = a^2 + 5b^2 = 12$, hence that $a^2 \equiv 2 \pmod 5$. But the squares in $\mathbb{Z}_5$ are $0, 1$ and 4, so this equation has no solution, implying that x and y have no gcd in R.

Although gcd's do not always exist, whenever they exist, they are unique up to multiplication by units.

Theorem 1.1.1. *Let* $(R, +, \cdot)$ *be an integral domain, and* $x, y \in R$. *If* z *and* t *are both gcd's of* x *and* y, *then* $z = ut$ *and* $t = vz$ *for some* $u, v \in R^*$.

Proof. Suppose that both z and t are gcd's of x and y. Then, $t \mid z$ since $t \mid x$, $t \mid y$, and $z = \gcd(x, y)$. Thus, $z = ut$ for some $u \in R$. Similarly, $z \mid t$, so $t = vz$ for some $v \in R$. Hence $z = ut = uvz$, so $(1 - uv)z = 0$. If $z \neq 0$, then $1 = uv$, so $u, v \in R^*$. If $z = 0$, then $t = vz = 0$, so $z = 1t$ and $t = 1z$.   □

Definition 1.1.5. *Let* R *be an integral domain. A nonzero element* $p \in R \setminus R^*$ *is called* prime *if for any* $a, b \in R$, $p \mid ab \Longrightarrow p \mid a$ *or* $p \mid b$. *A nonzero element* $p \in R \setminus R^*$ *is called* irreducible *if for any* $a, b \in R$, $p = ab \Longrightarrow a \in R^*$ *or* $b \in R^*$.

Example 1.1.7. Let $R = \mathbb{Z}\left[\sqrt{-5}\right]$ as in example 1.1.5, and check that $2, 3, 1 + \sqrt{-5}$ and $1 - \sqrt{-5}$ are all irreducible elements of R. Equation (1.1) then shows that the same element can have several different factorizations into irreducibles. Therefore, integral domains where such a factorization is unique receive a special name.

Definition 1.1.6. *A* unique factorization domain (UFD) $(R, +, \cdot)$ *is an integral domain where for any nonzero* $x \in R \setminus R^*$, *there are* $u \in R^*$, *coprime irreducibles* $p_1, \ldots, p_n \in R$ *and positive integers* $e_1, \ldots, e_n$ *such that* $x = u p_1^{e_1} \cdots p_n^{e_n}$. *Furthermore, this factorization is unique up to multiplication of* u *and the* p_i's *by units and up to permutation of the indices.*

Example 1.1.8. Let $R = \mathbb{Q}[X, Y]$ be the set of all the polynomials in the variables X and Y and with rational number coefficients. It is a classical result ([40] Chap. V §6, [77] §5.4) that $(R, +, \cdot)$ is a unique factorization domain where $+$ and $\cdot$ denote polynomial addition and multiplication respectively.

In any integral domain, a prime is always irreducible. The converse is not always true, but it holds in unique factorization domains. Thus, we can use interchangeably "prime" or "irreducible" whenever we are in a unique factorization domain, so, "the prime factorization of x" and "the irreducible factorization of x" have the same meaning.

Theorem 1.1.2 ([40] Chap. II §4). *Let* $(R, +, \cdot)$ *be an integral domain. Then every prime* $p \in R$ *is irreducible. If* R *is a unique factorization domain, then every irreducible* $p \in R$ *is prime.*

In addition, gcd's always exist in UFD's, and can be obtained from the irreducible factorizations.

Theorem 1.1.3. *If* R *is a UFD, then any* $x, y \in R$ *have a gcd in* R.

Proof. Let $x, y \in R$, and suppose first that $x = 0$. Then $y \mid y$, $y \mid 0$, and any $t \in R$ that divides x and y must divide y, so y is a gcd of x and y. Similarly, x is a gcd of x and y if $y = 0$, so suppose now that $x \neq 0$ and $y \neq 0$, and let $x = u \prod_{p \in \mathcal{X}} p^{n_p}$ and $y = v \prod_{p \in \mathcal{Y}} p^{m_p}$ be the irreducible factorizations of x and y, where $\mathcal{X}$ and $\mathcal{Y}$ are finite sets of irreducibles. We choose the units u and v so that any irreducible dividing both x and y is in $\mathcal{X} \cap \mathcal{Y}$. Let then

$$z = \prod_{p \in \mathcal{X} \cap \mathcal{Y}} p^{\min(n_p, m_p)} \in R. \qquad (1.2)$$

We have

$$x = z u \prod_{p \in \mathcal{X} \cap \mathcal{Y}} p^{n_p - \min(n_p, m_p)} \prod_{p \in \mathcal{X} \setminus \mathcal{Y}} p^{n_p}$$

so $z \mid x$. A similar formula shows that $z \mid y$. Suppose that $t \mid x$ and $t \mid y$ for some $t \in R$, and let $t = w \prod_{p \in \mathcal{T}} p^{e_p}$ be its irreducible factorization where $\mathcal{T}$ is a finite set of irreducibles. For $p \in \mathcal{T}$, we have $x = tb = p^{e_p} ab$ for some $a, b \in R$, so $sp \in \mathcal{X}$ for some $s \in R^*$. Replacing w by ws^{-e_p}, we can assume that $p \in \mathcal{X}$, and $e_p \leq n_p$ by the unicity of the irreducible factorization. Similarly, we get $p \in \mathcal{Y}$ and $e_p \leq m_p$ since $t \mid y$. Hence, $\mathcal{T} \subseteq \mathcal{X} \cap \mathcal{Y}$ and $e_p \leq \min(n_p, m_p)$ for any $p \in \mathcal{T}$. Thus,

$$z = t w^{-1} \prod_{p \in \mathcal{T}} p^{\min(n_p, m_p) - e_p} \prod_{p \in (\mathcal{X} \cap \mathcal{Y}) \setminus \mathcal{T}} p^{\min(n_p, m_p)}$$

which means that $t \mid z$, hence that $z = \gcd(x, y)$.   □

It is a classical result due to Gauss that polynomials can be factored uniquely into irreducibles.

Theorem 1.1.4 ([40] Chap. V §6, [77] §5.4). *If R is a UFD, then the polynomial ring $R[X_1, \ldots, X_n]$ is a UFD.*

Definition 1.1.7. *Let $(G, \circ)$ be a group with identity element e. We say that $H \subseteq G$ is a subgroup of $(G, \circ)$ if:*

(i) $e \in H$.
(ii) $\forall a, b \in H, a \circ b \in H$.
(iii) $\forall a \in H, a^{-1} \in H$.

In practice, given a subset H of a group G, it is equivalent to check the above properties (i), (ii) and (iii), or that H is not empty and that $a \circ b^{-1} \in H$ for any $a, b \in H$.

Example 1.1.9. Let $G = GL(\mathbb{Q}, 2)$ as in example 1.1.1 with $\circ$ denoting matrix multiplication, and let $H = SL(\mathbb{Q}, 2)$ be the subset of G consisting of all the matrices whose determinant is equal to 1. The identity matrix is in H, so H is not empty, and for any $a, b \in H$, the determinant of $a \circ b^{-1}$ is the quotient of the determinant of a by the determinant of b, which is 1, so H is a subgroup of G.

Definition 1.1.8. *Let $(R, +, \cdot)$ be a commutative ring. A subset I of R is called an* ideal *if $(I, +)$ is a subgroup of $(R, +)$ and $xa \in I$ for any x in R and a in I. Let $x_1, \ldots, x_n \in R$. The* ideal generated by $\{x_1, \ldots, x_n\}$ *is the smallest ideal of R containing $\{x_1, \ldots, x_n\}$, and is denoted $(x_1, \ldots, x_n)$. An ideal $I \subseteq R$ is called* principal *if $I = (x)$ for some $x \in R$.*

In fact, the ideal generated by $\{x_1, \ldots, x_n\}$ is just the set of all the linear combinations of the x_i's with coefficients in R.

Theorem 1.1.5. *Let $(R, +, \cdot)$ be a commutative ring, and $x_1, \ldots, x_n \in R$. Then,*

$$(x_1, \ldots, x_n) = \{a_1 x_1 + \cdots + a_n x_n; a_1, \ldots, a_n \in R\}.$$

Proof. Let $I = \{a_1 x_1 + \cdots + a_n x_n, a_1, \ldots, a_n \in R\}$. Then $x_i \in I$ for any i. Let $a = \sum_{i=1}^{n} a_i x_i \in I$ and $b = \sum_{i=1}^{n} b_i x_i \in I$. We have $a - b = \sum_{i=1}^{n} (a_i - b_i) x_i \in I$, so $(I, +)$ is a subgroup of $(R, +)$. For any $x \in R$, we have $xa = \sum_{i=1}^{n} (xa_i) x_i \in I$, so I is an ideal of R containing $\{x_1, \ldots, x_n\}$. Let now J be any ideal of R containing $\{x_1, \ldots, x_n\}$, and let $a = \sum_{i=1}^{n} a_i x_i \in I$. For each i, $x_i \in J$, so $a_i x_i \in J$ since $RJ \subseteq J$, so $a \in J$ since $(J, +)$ is a group. Hence $I \subseteq J$, so $I = (x_1, \ldots, x_n)$. $\square$

Example 1.1.10. Let $R = \mathbb{Q}[X, Y]$ as in example 1.1.8, and let $I = (X, Y)$. It can be checked that I is not principal, hence that not every ideal of R is principal. Naturally, this means that integral domains where every ideal is principal receive a special name.

Definition 1.1.9. *A* principal ideal domain (PID) $(R, +, \cdot)$ *is an integral domain where any ideal is principal.*

Example 1.1.11. Let $R = \mathbb{Q}[X]$ be the set of all the univariate polynomials in X with rational number coefficients. $(R, +, \cdot)$ is then a principal ideal domain ([40] Chap. V §4, [77] §3.7) where $+$ and $\cdot$ denote polynomial addition and multiplication respectively.

The last, and most useful, type of ring that we use in this book, is an integral domain in which Euclidean division can be carried out.

Definition 1.1.10. *A* Euclidean domain $(R, +, \cdot)$ *is an integral domain together with a map $\nu : R \setminus \{0\} \to \mathbb{N}$ such that:*

(i) $\forall a, b \in R \setminus \{0\}, \nu(ab) \geq \nu(a)$.

(ii) (Euclidean division) For any $a, b \in R$, $b \neq 0$, there are $q, r \in R$ such that $a = bq + r$ and either $r = 0$ or $\nu(r) < \nu(b)$.

The map ν is called the size function *of the Euclidean domain.*

Example 1.1.12. The ring $(\mathbb{Z}, +, \cdot)$ of the integers with the usual addition and multiplication is a Euclidean domain with $\nu(a) = |a|$, a fact that was known to Euclid, and which is the origin of the name.

Even though the notions of principal ideal domains and Euclidean domains are defined for an arbitrary integral domain, there is in fact a linear hierarchy of integral domains.

Theorem 1.1.6 ([77] §3.7). *Every Euclidean domain is a PID.*

Theorem 1.1.7 ([40] Chap. II §4, [77] §3.8). *Every PID is a UFD.*

Since every PID is a UFD, and gcd's always exist in UFD's by Theorem 1.1.3, then gcd's always exist in PID's. We show that in PID's, the gcd of two elements generates the same ideal than them.

Theorem 1.1.8. *If R is a PID, then $(x, y) = (\gcd(x, y))$ for any $x, y \in R$.*

Proof. Let $x, y \in R$ and $z \in R$ be a generator of the ideal (x, y), i.e. $(z) = (x, y)$. Then, $x \in (z)$, so $x = zu$ for some $u \in R$, which means that $z \mid x$. Similarly, $y \in (z)$, so $z \mid y$. In addition, $z \in (x, y)$, so $z = ax + by$ for some $a, b \in R$. Let $t \in R$ be such that $t \mid x$ and $t \mid y$. Then $x = ct$ and $y = dt$ for some $c, d \in R$. Hence, $z = act + bdt = (ac + bd)t$ so $t \mid z$, which implies that $z = \gcd(u, v)$.   □

We finally recall some important definitions and results about fields.

Definition 1.1.11. *A field $(F, +, \cdot)$ is a commutative ring where $(F \setminus \{0\}, \cdot)$ is a group, i.e. every nonzero element is a unit $(F^* = F \setminus \{0\})$.*

Example 1.1.13. Let $F = \mathbb{Z}_5$ (the integers modulo 5) with $+$ and $\cdot$ being the addition and multiplication of integers modulo 5. $(F, +, \cdot)$ is then a field.

Example 1.1.14. Let R be an integral domain and define the relation $\sim$ on $R \times R \setminus \{0\}$ by $(a, b) \sim (c, d)$ if $ad = bc$. It can easily be checked that $\sim$ is an equivalence relation on $R \times R \setminus \{0\}$ and that the set of equivalence classes is a field with the usual operations

$$\frac{a}{b} + \frac{c}{d} = \frac{ad + bc}{bd} \quad \text{and} \quad \frac{a}{b}\frac{c}{d} = \frac{ac}{bd}$$

where a/b denotes the equivalence class of (a, b). This field is called the *quotient field* of R. For example, the quotient field of $\mathbb{Z}$ is $\mathbb{Q}$ and the quotient field of the polynomial ring $D[x]$ is the rational function field $D(x)$ when D is an integral domain.

Definition 1.1.12. *Let $F \subseteq E$ be fields. An element $\alpha \in E$ is called* algebraic *over F if $p(\alpha) = 0$ for some nonzero polynomial $p \in F[X]$, transcendental over F otherwise. E is called an* algebraic extension *of F if all the elements of E are algebraic over F.*

Definition 1.1.13. *A field F is called* algebraically closed *if for every polynomial $p \in K[X] \setminus K$ there exists $\alpha \in F$ such that $p(\alpha) = 0$. A field E is called an* algebraic closure *of F if E is an algebraically closed algebraic extension of F.*

Note that if F is algebraically closed, then any $p \in K[X] \setminus K$ factors linearly as $p = c \prod_{i=1}^{n} (X - \alpha_i)^{e_i}$ over F: p must have one root α in F by definition, and $p/(X - \alpha)$ factors linearly over F by induction. The fundamental result about algebraic closures is a result of E. Steinitz which states that they exist and are essentially unique.

Theorem 1.1.9 ([40] Chap. VII §2, [77] §10.1). *Every field F has an algebraic closure, and any two algebraic closures of F are isomorphic.*

In view of the above theorem, we can refer to *the* algebraic closure of a field F, and we denote it $\overline{F}$. The last result we mention in this section is Hilbert's Nullstellensatz, which is not needed in the algorithm, but is needed in order to eliminate the possibility of new transcendental constants appearing in antiderivatives. We present it here in both its classical forms.

Theorem 1.1.10 (Weak Nullstellensatz, [77] §16.5). *Let F be an algebraically closed field, I an ideal of the polynomial ring $F[X_1, \ldots, X_n]$ and $V(I)$ be the subset of F^n given by*

$$V(I) = \{(x_1, \ldots, x_n) \in F^n \text{ s.t. } p(x_1, \ldots, x_n) = 0 \text{ for all } p \in I\}. \quad (1.3)$$

Then, $V(I) = \emptyset \iff 1 \in I$.

Theorem 1.1.11 (Nullstellensatz, [40] Chap. X §2, [77] §16.5). *Let F be an algebraically closed field, I an ideal of the polynomial ring $F[X_1, \ldots, X_n]$ and $V(I)$ be given by (1.3). For any $p \in K[X_1, \ldots, X_n]$, if $p(x_1, \ldots, x_n) = 0$ for every $(x_1, \ldots, x_n) \in V(I)$, then $p^m \in I$ for some integer $m > 0$.*

1.2 Euclidean Division and Pseudo-Division

Let K be a field and x be an indeterminate over K. We first describe the classical polynomial division algorithm ([77] §3.4), which, given $A, B \in K[x]$, $B \neq 0$, produces unique $Q, R \in K[x]$ such that $A = BQ + R$ and either $R = 0$ or $\deg(R) < \deg(B)$. This shows that the polynomial ring $K[x]$ is a Euclidean domain with the degree for size function when K is field. Q and R are called the *quotient of A by B*, and the *remainder of A modulo B* respectively.

PolyDivide(A, B) (* Euclidean Polynomial Division *)

(* Given a field K and $A, B \in K[x]$ with $B \neq 0$, return $Q, R \in K[x]$ such that $A = BQ + R$ and either $R = 0$ or $\deg(R) < \deg(B)$. *)

$Q \leftarrow 0, R \leftarrow A$
while $R \neq 0$ and $\delta \leftarrow \deg(R) - \deg(B) \geq 0$ **do**
$\quad T \leftarrow \frac{\mathrm{lc}(R)}{\mathrm{lc}(B)} x^\delta, Q \leftarrow Q + T, R \leftarrow R - BT$
return(Q, R)

Example 1.2.1. Here is the Euclidean division of $A = 3x^3 + x^2 + x + 5$ by $B = 5x^2 - 3x + 1$ in $\mathbb{Q}[x]$:

Q	R	δ	T
0	$3x^3 + x^2 + x + 5$	1	$\frac{3}{5}x$
$\frac{3}{5}x$	$\frac{14}{5}x^2 + \frac{2}{5}x + 5$	0	$\frac{14}{25}$
$\frac{3}{5}x + \frac{14}{25}$	$\frac{52}{25}x + \frac{111}{25}$	-1	

Thus,

$$A = B\left(\frac{3}{5}x + \frac{14}{25}\right) + \left(\frac{52}{25}x + \frac{111}{25}\right).$$

This algorithm requires the coefficients to be from a field because it makes the quotient in K of the two leading coefficients. If K is an integral domain, the leading coefficient of B does not always divide exactly the leading coefficient of A, so Euclidean division is not always possible. For example it is not possible in the above example to do a Euclidean division of A by B in $\mathbb{Z}[x]$. But it is possible to apply **PolyDivide** to $25A$ and B in $\mathbb{Z}[x]$ since all the divisions in $\mathbb{Z}$ will then be exact. In general, given an integral domain D and $A, B \in D[x]$, applying **PolyDivide** to $b^{\delta+1}A$ and B where $b = \mathrm{lc}(B)$ and $\delta = \max(-1, \deg(A) - \deg(B))$ only generates exact divisions in D, and the Q and R it returns are respectively called the *pseudo-quotient of A by B* and *pseudo-remainder of A modulo B*. They satisfy $b^{\delta+1}A = BQ + R$ and either $R = 0$ or $\deg(R) < \deg(B)$. We write $\mathrm{pquo}(A, B)$ and $\mathrm{prem}(A, B)$ for the pseudo-quotient and pseudo-remainder of A and B. It is more efficient in practice to multiply A by b iteratively, as is done in the algorithm below, rather than once by $b^{\delta+1}$.

PolyPseudoDivide(A, B) (* Euclidean Polynomial Pseudo-Division *)

(* Given an integral domain D and $A, B \in D[x]$ with $B \neq 0$, return $\mathrm{pquo}(A, B)$ and $\mathrm{prem}(A, B)$. *)

$b \leftarrow \mathrm{lc}(B)$, $N \leftarrow \deg(A) - \deg(B) + 1$, $Q \leftarrow 0$, $R \leftarrow A$
while $R \neq 0$ **and** $\delta \leftarrow \deg(R) - \deg(B) \geq 0$ **do**
    $T \leftarrow \mathrm{lc}(R)x^\delta$, $N \leftarrow N - 1$, $Q \leftarrow bQ + T$, $R \leftarrow bR - TB$
return$(b^N Q, b^N R)$

Example 1.2.2. With the A and B of example 1.2.1, we get $b = 5$, $N = 2$, and

Q	R	δ	T	N
0	$3x^3 + x^2 + x + 5$	1	$3x$	1
$3x$	$14x^2 + 2x + 25$	0	14	0
$15x + 14$	$52x + 111$	-1		

so $25A = B(15x + 14) + (52x + 111)$.

1.3 The Euclidean Algorithm

Let D be a Euclidean domain and $\nu : D \setminus \{0\} \to \mathbb{N}$ its size function. The Euclidean division in D can be used to compute the greatest common divisor of any two elements of D. The basic idea, which goes back to Euclid who used it to compute the gcd of two integers, is that if $a = bq + r$, then $\gcd(a, b) = \gcd(b, r)$. Since $\gcd(x, 0) = x$ for any $x \in D$, the last nonzero element in the sequence $(a_i)_{i \geq 0}$ defined by

$$a_0 = a, \quad a_1 = b, \quad \text{and} \quad (q_i, a_i) = \textbf{PolyDivide}(a_{i-2}, a_{i-1}) \text{ for } i \geq 2$$

is then a gcd of a and b. Since for $a_i \neq 0$ and $i \geq 1$, either $a_{i+1} = 0$ or $\nu(a_{i+1}) < \nu(a_i)$, that sequence can only have a finite number of nonzero elements. This yields an algorithm for computing $\gcd(a, b)$ by repeated Euclidean divisions.

Euclidean(a, b) (* Euclidean algorithm *)

    (* Given a Euclidean domain D and $a, b \in D$, return $\gcd(a, b)$. *)

    while $b \neq 0$ **do**
        $(q, r) \leftarrow$ **PolyDivide**(a, b) (* $a = bq + r$ *)
        $a \leftarrow b$
        $b \leftarrow r$
    return a

Example 1.3.1. Applying the Euclidean algorithm to

$$a = x^4 - 2x^3 - 6x^2 + 12x + 15 \quad \text{and} \quad b = x^3 + x^2 - 4x - 4$$

in $D = \mathbb{Q}[x]$ gives:

a	b	q	r
$x^4 - 2x^3 - 6x^2 + 12x + 15$	$x^3 + x^2 - 4x - 4$	$x - 3$	$x^2 + 4x + 3$
$x^3 + x^2 - 4x - 4$	$x^2 + 4x + 3$	$x - 3$	$5x + 5$
$x^2 + 4x + 3$	$5x + 5$	$\frac{1}{5}x + \frac{3}{5}$	0
$5x + 5$	0		

so $5x + 5$ is a gcd of a and b in $\mathbb{Q}[x]$.

    The Euclidean algorithm can be easily extended to return not only a gcd of a and b, but also elements s and t in D such that $sa + tb = \gcd(a, b)$. Such elements always exist since $\gcd(a, b)$ belongs to the ideal generated by a and b by Theorem 1.1.8.

ExtendedEuclidean(a, b) (* Extended Euclidean algorithm *)

(* Given a Euclidean domain D and $a, b \in D$, return $s, t, g \in D$ such that $g = \gcd(a, b)$ and $sa + tb = g$. *)

$a_1 \leftarrow 1, a_2 \leftarrow 0, b_1 \leftarrow 0, b_2 \leftarrow 1$
while $b \neq 0$ **do**
   $(q, r) \leftarrow$ **PolyDivide**(a, b) (* $a = bq + r$ *)
   $a \leftarrow b, b \leftarrow r$
   $r_1 \leftarrow a_1 - q\,b_1, r_2 \leftarrow a_2 - q b_2$
   $a_1 \leftarrow b_1, a_2 \leftarrow b_2, b_1 \leftarrow r_1, b_2 \leftarrow r_2$
return(a_1, a_2, a)

Example 1.3.2. Using the same a and b as in example 1.3.1:

a	b	q	r
$x^4 - 2x^3 - 6x^2 + 12x + 15$	$x^3 + x^2 - 4x - 4$	$x - 3$	$x^2 + 4x + 3$
$x^3 + x^2 - 4x - 4$	$x^2 + 4x + 3$	$x - 3$	$5x + 5$
$x^2 + 4x + 3$	$5x + 5$	$\frac{1}{5}x + \frac{3}{5}$	0
$5x + 5$	0		

a_1	a_2	b_1	b_2
1	0	0	1
0	1	1	$-x + 3$
1	$-x + 3$	$-x + 3$	$x^2 - 6x + 10$
$-x + 3$	$x^2 - 6x + 10$	$\frac{1}{5}x^2 - \frac{4}{5}$	$-\frac{1}{5}x^3 + \frac{3}{5}x^2 + \frac{3}{5}x - 3$

Thus, $5x + 5$ is a gcd of a and b in $\mathbb{Q}[x]$, and

$$(-x + 3)a + (x^2 - 6x + 10)b = 5x + 5. \tag{1.4}$$

If only one of the coefficients s or t is needed, a variant of the extended Euclidean algorithm that computes only that coefficient can be used:

HalfExtendedEuclidean(a, b) (* Half extended Euclidean algorithm *)

(* Given a Euclidean domain D and $a, b \in D$, return $s, g \in D$ such that $g = \gcd(a, b)$ and $sa \equiv g \pmod{b}$. *)

$a_1 \leftarrow 1, b_1 \leftarrow 0$
while $b \neq 0$ **do**
   $(q, r) \leftarrow$ **PolyDivide**(a, b) (* $a = bq + r$ *)
   $a \leftarrow b, b \leftarrow r$
   $r_1 \leftarrow a_1 - q\,b_1, a_1 \leftarrow b_1, b_1 \leftarrow r_1$
return(a_1, a)

This "half" variant of the algorithm is also used as a more efficient alternative to the extended Euclidean algorithm, since the second coefficient can be obtained from the first via

$$t = \frac{g - sa}{b}$$

where the division is always exact.

ExtendedEuclidean(a, b)
(* Extended Euclidean algorithm – "half/full" version *)

    (* Given a Euclidean domain D and $a, b \in D$, return $s, t, g \in D$ such that $g = \gcd(a, b)$ and $sa + tb = g$. *)

    $(s, g) \leftarrow$ **HalfExtendedEuclidean**(a, b) (* $sa \equiv g \pmod{b}$ *)
    $(t, r) \leftarrow$ **PolyDivide**$(g - sa, b)$ (* r must be 0 *)
    return(s, t, g)

Example 1.3.3. Recomputing the extended gcd of the a and b of example 1.3.1, we get:

1. $(s, g) =$ **HalfExtendedEuclidean**$(a, b) = (-x + 3, 5x + 5)$
2. $g - sa = x^5 - 5x^4 + 30x^2 - 16x$
3. $(t, r) =$ **PolyDivide**$(g - sa, b) = (x^2 - 6x + 10, 0)$

so we recover (1.4).

The extended Euclidean algorithm can also be used to solve the diophantine equation

$$sa + tb = c \tag{1.5}$$

where $a, b, c \in D$ are given and $s, t \in D$ are the unknowns. For (1.5) to have a solution, it is necessary and sufficient that c be in the ideal generated by a and b, i.e. that c be a multiple of $\gcd(a, b)$ in D. The extended Euclidean algorithm first solves the equation $sa + tb = \gcd(a, b)$, and there remains only to multiply the solutions by $c/\gcd(a, b)$ to get a solution of (1.5). It should be noted that when c is in the ideal generated by a and b, then (1.5) has as many solutions as the number of elements of D (when a and b are nonzero), since $sa + tb = (s + bd)a + (t - ad)b$ for any $d \in D$. Since there can be no confusion with the previous extended Euclidean algorithm, which has only two parameters, we also call this algorithm the "extended Euclidean algorithm". As before, a half-extended version exists when only one of the coefficients is needed. We remark that the versions of the algorithm that we present here, and use extensively in the sequel, all return a solution s or (s, t) such that either $s = 0$ or $\nu(s) < \nu(b)$. An important consequence of this in polynomial rings (where $\nu(p) = \deg(p)$) is that if $\deg(c) < \deg(a) + \deg(b)$, then we also get either $t = 0$ or $\deg(t) < \deg(a)$. Indeed, if we had $\deg(s) < \deg(b)$ and $\deg(t) \geq \deg(a)$, then we would have $\deg(c) = \deg(sa + tb) = \deg(tb) = \deg(t) + \deg(b) \geq \deg(a) + \deg(b)$.

ExtendedEuclidean(a, b, c)
(* Extended Euclidean algorithm – diophantine version *)

(* Given a Euclidean domain D and $a, b, c \in D$ with $c \in (a, b)$, return $s, t \in D$ such that $sa + tb = c$ and either $s = 0$ or $\nu(s) < \nu(b)$. *)

$(s, t, g) \leftarrow$ **ExtendedEuclidean**(a, b) (* $g = sa + tb$ *)
$(q, r) \leftarrow$ **PolyDivide**(c, g) (* $c = gq + r$ *)
if $r \neq 0$ **then error** "c is not in the ideal generated by a and b"
$s \leftarrow qs, \ t \leftarrow qt$
if $s \neq 0$ and $\nu(s) \geq \nu(b)$ **then**
    $(q, r) \leftarrow$ **PolyDivide**(s, b) (* $s = bq + r$ *)
    $s \leftarrow r, \ t \leftarrow t + qa$
return(s, t)

Example 1.3.4. Suppose that we want to solve $sa + tb = x^2 - 1$ in $\mathbb{Q}[x]$ with the a and b of example 1.3.1. Applying **ExtendedEuclidean** we get:

1. $(s, t, g) = $ **ExtendedEuclidean**$(a, b) = (-x + 3, x^2 - 6x + 10, 5x + 5)$
2. $(q, r) = $ **PolyDivide**$(x^2 - 1, 5x + 5) = ((x - 1)/5, 0)$
3. $s \leftarrow qs = (-x^2 + 4x - 3)/5$
4. $t \leftarrow qt = (x^3 - 7x^2 + 16x - 10)/5$

So we get the following solution:

$$\left(\frac{-x^2 + 4x - 3}{5} \right) a + \left(\frac{x^3 - 7x^2 + 16x - 10}{5} \right) b = x^2 - 1. \qquad (1.6)$$

HalfExtendedEuclidean(a, b, c)
(* Half extended Euclidean algorithm – diophantine version *)

(* Given a Euclidean domain D and $a, b, c \in D$ with $c \in (a, b)$, return $s \in D$ such that $sa \equiv c \pmod{b}$ and either $s = 0$ or $\nu(s) < \nu(b)$. *)

$(s, g) \leftarrow$ **HalfExtendedEuclidean**(a, b) (* $sa \equiv g \pmod{b}$ *)
$(q, r) \leftarrow$ **PolyDivide**(c, g) (* $c = gq + r$ *)
if $r \neq 0$ **then error** "c is not in the ideal generated by a and b"
$s \leftarrow qs$
if $s \neq 0$ and $\nu(s) \geq \nu(b)$ **then**
    $(q, r) \leftarrow$ **PolyDivide**(s, b) (* $s = bq + r$ *)
    $s \leftarrow r$
return s

As earlier, the "half" variant yields a more efficient alternative to the extended diophantine version, since the second coefficient can be obtained via

$$t = \frac{c - sa}{b}$$

where the division is always exact.

ExtendedEuclidean(a, b, c)
(* Extended Euclidean algorithm – "half/full" diophantine version *)

    (* Given a Euclidean domain D and $a, b, c \in D$ with $c \in (a, b)$, return $s, t \in D$ such that $sa + tb = c$ and either $s = 0$ or $\nu(s) < \nu(b)$. *)

    $s \leftarrow$ **HalfExtendedEuclidean**(a, b, c) (* $sa \equiv c \pmod b$ *)
    $(t, r) \leftarrow$ **PolyDivide**$(c - sa, b)$ (* r must be 0 *)
    return(s, t)

Example 1.3.5. Solving $sa + tb = x^2 - 1$ in $\mathbb{Q}[x]$ with the a and b of example 1.3.1, we get

1. $s =$ **HalfExtendedEuclidean**$(a, b, x^2 - 1) = (-x^2 + 4x - 3)/5$
2. $c - sa = x^2 - 1 - sa = (x^6 - 6x^5 + 5x^4 + 30x^3 - 46x^2 - 24x + 40)/5$
3. $(t, r) =$ **PolyDivide**$(c - sa, b) = ((x^3 - 7x^2 + 16x - 10)/5, 0)$

so we recover (1.6).

Since the extended Euclidean algorithm can be used to solve diophantine equations, it is also useful for computing partial fraction decompositions. Let $d \in D \setminus \{0\}$ and let $d = d_1 \cdots d_n$ be any factorization of d (not necessarily into irreducibles) where $\gcd(d_i, d_j) = 1$ for $i \neq j$. Then, for any $a \in D \setminus \{0\}$, there are $a_0, a_1, \ldots, a_n$ in D such that either $a_i = 0$ or $\nu(a_i) < \nu(d_i)$ for $i \geq 1$, and

$$\frac{a}{d} = \frac{a}{\prod_{i=1}^{n} d_i} = a_0 + \sum_{i=1}^{n} \frac{a_i}{d_i}.$$

Such a decomposition is called the *partial fraction decomposition of a/d with respect to the factorization $d = \prod_{i=1}^{n} d_i$*, and computing it reduces to solving equations of the form (1.5), so to the extended Euclidean algorithm. Indeed, write first $a = da_0 + r$ by the Euclidean division, where either $r = 0$ or $\nu(r) < \nu(d)$. If $n = 1$, then $a/d = a_0 + r/d$ is already in the desired form. Otherwise, since $\gcd(d_i, d_j) = 1$ for $i \neq j$, we have $\gcd(d_1, d_2 \cdots d_n) = 1$, so by the extended Euclidean algorithm, we can find a_1 and b in D such that

$$r = a_1 (d_2 \cdots d_n) + b d_1 \tag{1.7}$$

and either $a_1 = 0$ or $\nu(a_1) < \nu(d_1)$. We can recursively find $b_0, a_2, \ldots, a_n \in D$ such that either $a_i = 0$ or $\nu(a_i) < \nu(d_i)$, and

$$\frac{b}{d_2 \cdots d_n} = b_0 + \sum_{i=2}^{n} \frac{a_i}{d_i}.$$

Dividing (1.7) by d and adding a_0, we get

$$\frac{a}{d} = a_0 + \frac{r}{d} = a_0 + \frac{a_1}{d_1} + \frac{b}{d_2 \cdots d_n} = (a_0 + b_0) + \sum_{i=1}^{n} \frac{a_i}{d_i}.$$

We note that in the case of polynomial rings, since $\deg(r) < \deg(d) = \deg(d_1) + \deg(d_2 \cdots d_n)$ and $\deg(a_1) < \deg(d_1)$ in (1.7), then $\deg(b) < \deg(d_2 \cdots d_n)$, so $b_0 = 0$.

PartialFraction$(a, d_1, \ldots, d_n)$ (* Partial fraction decomposition *)

(* Given a Euclidean domain D, a positive integer n and $a, d_1, \ldots, d_n \in D \setminus \{0\}$ with $\gcd(d_i, d_j) = 1$ for $i \neq j$, return $a_0, a_1, \ldots, a_n \in D$ such that

$$\frac{a}{d_1 \cdots d_n} = a_0 + \sum_{i=1}^{n} \frac{a_i}{d_i}$$

and either $a_i = 0$ or $\nu(a_i) < \nu(d_i)$ for $i \geq 1$. *)

$(a_0, r) \leftarrow$ **PolyDivide**$(a, d_1 \cdots d_n)$ (* $a = (d_1 \cdots d_n)a_0 + r$ *)
if $n = 1$ then return(a_0, r)
$(a_1, t) \leftarrow$ **ExtendedEuclidean**$(d_2 \cdots d_n, d_1, r)$ (* $\nu(a_1) < \nu(d_1)$ *)
$(b_0, a_2, \ldots, a_n) \leftarrow$ **PartialFraction**$(t, d_2, \ldots, d_n)$
return$(a_0 + b_0, a_1, a_2, \ldots, a_n)$

Example 1.3.6. We compute the partial fraction decomposition of

$$f = \frac{a}{d} = \frac{x^2 + 3x}{x^3 - x^2 - x + 1} \in \mathbb{Q}(x)$$

with respect to the factorization $d = (x + 1)(x^2 - 2x + 1) = d_1 d_2$. Applying **PartialFraction** to a, d_1 and d_2, we get:

1. $(a_0, r) =$ **PolyDivide**$(a, d) = (0, x^2 + 3x)$
2.

$$(a_1, t) = \textbf{ExtendedEuclidean}(x^2 - 2x + 1, x + 1, x^2 + 3x)$$
$$= \left(-\frac{1}{2}, \frac{3x + 1}{2}\right)$$

3. $(b_0, a_2) =$ **PartialFraction**$((3x + 1)/2, x^2 - 2x + 1) = (0, (3x + 1)/2)$

so the partial fraction decomposition of f is

$$\frac{x^2 + 3x}{x^3 - x^2 - x + 1} = \frac{-1/2}{x + 1} + \frac{(3x + 1)/2}{x^2 - 2x + 1}.$$

We can combine this with the Euclidean division to get a refinement of the partial fraction decomposition: let $m \geq 1$ and $d \in D \setminus \{0\}$. Then, for any $a \in D \setminus \{0\}$, there are $a_0, a_1, \ldots, a_m \in D$ such that either $a_j = 0$ or $\nu(a_j) < \nu(d)$ for $j \geq 1$, and

$$\frac{a}{d^m} = a_0 + \sum_{j=1}^{m} \frac{a_j}{d^j}.$$

Such a decomposition is called the *d-adic expansion of* a/d^m. Write $a = dq + a_m$ by the Euclidean division, where either $a_m = 0$ or $\nu(a_m) < \nu(d)$. Then,

$$\frac{a}{d^m} = \frac{dq + a_m}{d^m} = \frac{q}{d^{m-1}} + \frac{a_m}{d^m}.$$

If $m = 1$, then the above is in the desired form with $a_0 = q$. Otherwise, we recursively find $a_0, a_1, \ldots, a_{m-1} \in D$ such that either $a_j = 0$ or $\nu(a_j) < \nu(d)$ for $j \geq 1$, and

$$\frac{q}{d^{m-1}} = a_0 + \sum_{j=1}^{m-1} \frac{a_j}{d^j}.$$

Thus,

$$\frac{a}{d^m} = \frac{q}{d^{m-1}} + \frac{a_m}{d^m} = a_0 + \sum_{j=1}^{m} \frac{a_j}{d^j}.$$

Let now $d \in D \setminus \{0\}$ and let $d = d_1^{e_1} \cdots d_n^{e_n}$ be any factorization of d, not necessarily into irreducibles, where $\gcd(d_i, d_j) = 1$ for $i \neq j$, and the e_i's are positive integers. Then, for any $a \in D \setminus \{0\}$, we can first compute the partial fraction decomposition of a/d with respect to $d = b_1 \cdots b_n$ where $b_i = d_i^{e_i}$:

$$\frac{a}{d} = a_0 + \sum_{i=1}^{n} \frac{a_i}{b_i} = a_0 + \sum_{i=1}^{n} \frac{a_i}{d_i^{e_i}}$$

and then compute the d_i-adic expansion of each summand to get

$$\frac{a}{d} = \frac{a}{\prod_{i=1}^{n} d_i^{e_i}} = \tilde{a} + \sum_{i=1}^{n} \sum_{j=1}^{e_i} \frac{a_{ij}}{d_i^j}$$

where $\tilde{a} \in D$ and either $a_{ij} = 0$ or $\nu(a_{ij}) < \nu(d_i)$ for each i and j. This decomposition is called the *complete partial fraction decomposition of* a/d *with respect to the factorization* $d = \prod_{i=1}^{n} d_i^{e_i}$, or simply the complete partial fraction decomposition of a/d when the factorization of d into irreducibles[1] is used.

---

[1] We show in Sect. 2.7 how to compute that decomposition for linear factors without factoring d.

PartialFraction$(a, d_1, \ldots, d_n, e_1, \ldots, e_n)$
(* Complete partial fraction decomposition *)

(* Given a Euclidean domain D, positive integers $n, e_1, \ldots, e_n$ and
$a, d_1, \ldots, d_n \in D \setminus \{0\}$ with $\gcd(d_i, d_j) = 1$ for $i \neq j$, return
$a_0, a_{1,1}, \ldots, a_{1,e_1}, \ldots, a_{n,1}, \ldots, a_{n,e_n} \in D$ such that

$$\frac{a}{d_1^{e_1} \cdots d_n^{e_n}} = a_0 + \sum_{i=1}^{n} \sum_{j=1}^{e_i} \frac{a_{ij}}{d_i^j}$$

and either $a_{ij} = 0$ or $\nu(a_{ij}) < \nu(d_i)$. *)

$(a_0, a_1, \ldots, a_n) \leftarrow$ **PartialFraction**$(a, d_1^{e_1}, \ldots, d_n^{e_n})$
for $i \leftarrow 1$ **to** n **do**
   for $j \leftarrow e_i$ **to** 1 **step** -1 **do**
      $(q, a_{ij}) \leftarrow$ **PolyDivide**(a_i, d_i) (* $a_i = d_i q + a_{ij}$ *)
      $a_i \leftarrow q$
   $a_0 \leftarrow a_0 + a_i$
return$(a_0, a_{1,1}, \ldots, a_{1,e_1}, \ldots, a_{n,1}, \ldots, a_{n,e_n})$

Example 1.3.7. We compute the complete partial fraction fraction decomposition of

$$f = \frac{a}{d} = \frac{x^2 + 3x}{x^3 - x^2 - x + 1} \in \mathbb{Q}(x)$$

with respect to the factorization $d = (x+1)(x-1)^2 = d_1 d_2^2$. Applying **PartialFraction** to a, d_1, d_2, and the exponents 1 and 2, we get:

$$(a_0, a_1, \ldots, a_n) = \textbf{PartialFraction}(x^2 + 3x, x+1, (x-1)^2) = (0, -\frac{1}{2}, \frac{3x+1}{2})$$

and then:

i	j	a_i	d_i	q	a_{ij}	a_0
1	1	$-1/2$	$x+1$	0	$-1/2$	0
2	2	$(3x+1)/2$	$x-1$	$3/2$	2	0
2	1	$3/2$	$x-1$	0	$3/2$	0

so the complete partial fraction decomposition of f is

$$\frac{x^2+3x}{x^3-x^2-x+1} = \frac{-1/2}{x+1} + \frac{2}{(x-1)^2} + \frac{3/2}{x-1}.$$

The algorithm for computing partial fraction decompositions that we presented here dates back to Hermite in the 19th century. There are alternative and faster approaches for rational functions that we do not detail here. See [1, 36] for other approaches and their complexities.

1.4 Resultants and Subresultants

We describe in this section the fundamental properties of the resultant of two polynomials. Although they originate from 19th-century work on solving systems of nonlinear equations, resultants play a crucial role in integration. Throughout this section, let R be a commutative ring and x be an indeterminate over R.

Definition 1.4.1. Let $A, B \in R[x] \setminus \{0\}$. Write $A = a_n x^n + \cdots + a_1 x + a_0$ and $B = b_m x^m + \cdots + b_1 x + b_0$ where $a_n \neq 0$, $b_m \neq 0$ and at least one of n or m is nonzero. The Sylvester matrix of A and B is the $n + m$ by $n + m$ matrix defined by

$$
S(A,B) = \left. \left(\begin{array}{cccccccc}
a_n & \cdots & \cdots & \cdots & a_1 & a_0 & & \\
 & & \ddots & & & & & \\
 & & a_n & \cdots & \cdots & \cdots & a_1 & a_0 \\
b_m & \cdots & b_1 & b_0 & & & & \\
 & \ddots & & & & & & \\
 & & \ddots & & & & & \\
 & & & \ddots & & & & \\
 & & & & b_m & \cdots & b_1 & b_0
\end{array} \right) \right\} \begin{array}{l} \\ \\ m \ rows \\ \\ \\ \\ n \ rows \\ \\ \end{array}
$$

where the A-rows are repeated m times and the B-rows are repeated n times. The resultant of A and B is the determinant of $S(A, B)$.

Example 1.4.1. Let $R = \mathbb{Z}[t]$, $A = 3tx^2 - t^3 - 4 \in R[x]$, and $B = x^2 + t^3 x - 9 \in R[x]$. The Sylvester matrix of A and B is

$$
S(A,B) = \begin{pmatrix}
3t & 0 & -t^3 - 4 & 0 \\
0 & 3t & 0 & -t^3 - 4 \\
1 & t^3 & -9 & 0 \\
0 & 1 & t^3 & -9
\end{pmatrix}
$$

and the resultant of A and B is

$$
\det(S(A,B)) = -3t^{10} - 12t^7 + t^6 - 54t^4 + 8t^3 + 729t^2 - 216t + 16 .
$$

The first useful property of the resultant of two polynomials is that it can be expressed in terms of their roots.

Theorem 1.4.1 ([40] Chap. V §10, [77] §5.9). *Let* $\alpha_1, \ldots, \alpha_n, \beta_1, \ldots, \beta_m, a$ *and* b *be in* R *with* $a \neq 0$, $b \neq 0$, $A = a(x - \alpha_1) \cdots (x - \alpha_n)$ *and* $B = b(x - \beta_1) \cdots (x - \beta_m)$. *Then,*

$$
\begin{aligned}
\mathrm{resultant}(A, B) &= a^m b^n \prod_{i=1}^{n} \prod_{j=1}^{m} (\alpha_i - \beta_j) = a^m \prod_{i=1}^{n} B(\alpha_i) \\
&= (-1)^{nm} b^n \prod_{j=1}^{m} A(\beta_j) = (-1)^{mn} \mathrm{resultant}(B, A) .
\end{aligned}
$$

As a consequence, the resultant of two polynomials over an integral domain R is 0 if and only if they have a common zero in the algebraic closure of the quotient field of R.

Corollary 1.4.1 ([40] Chap. V §10, [77] §5.8). *Suppose that R is an integral domain. Let K be the quotient field of R and $\overline{K}$ the algebraic closure of K Then, for any $A, B \in R[x] \setminus \{0\}$,*

$$\text{resultant}(A, B) = 0 \iff \exists \gamma \in \overline{K} \text{ such that } A(\gamma) = B(\gamma) = 0.$$

Proof. Let $A, B \in R[x] \setminus \{0\}$, and let

$$A = a \prod_{i=1}^{n} (x - \alpha_i)^{e_i} \quad \text{and} \quad B = b \prod_{j=1}^{m} (x - \beta_j)^{f_j}$$

be the prime factorizations of A and B in $\overline{K}[x]$. We have $a \neq 0 \neq b$ since $A \neq 0 \neq B$, so by Theorem 1.4.1 we get

$$C = \text{resultant}(A, B) = a^M b^N \prod_{i=1}^{n} \prod_{j=1}^{m} (\alpha_i - \beta_j)^{e_i f_j}$$

where $M = \sum_{j=1}^{m} f_j$ and $N = \sum_{i=1}^{n} e_i$. Since $\overline{K}$ is a field, if $C = 0$ then $\alpha_{i_0} - \beta_{j_0} = 0$ for some i_0 and j_0. But then $A(\gamma) = B(\gamma) = 0$ where $\gamma = \alpha_{i_0} = \beta_{j_0} \in \overline{K}$. Conversely, if $A(\gamma) = B(\gamma) = 0$ for some $\gamma \in \overline{K}$, then, since $\overline{K}$ is a field, $\gamma = \alpha_{i_0} = \beta_{j_0}$ for some i_0 and j_0, so $\alpha_{i_0} - \beta_{j_0} = 0$, so $C = 0$. □

Another property is that the resultant of two polynomials is in the ideal that they generate.

Theorem 1.4.2 ([40] Chap. V §10, [77] §5.8). *For any $A, B \in R[x] \setminus \{0\}$, there are $S, T \in R[x]$ such that* $\text{resultant}(A, B) = SA + TB$.

As a consequence, the resultant of two polynomials over a unique factorization domain is 0 if and only if they have a non-trivial common factor.

Corollary 1.4.2 ([77] §5.8). *Suppose that R is a unique factorization domain. Then, for any $A, B \in R[x] \setminus \{0\}$,*

$$\text{resultant}(A, B) = 0 \iff \deg(\gcd(A, B)) > 0.$$

Subresultants are polynomials obtained from submatrices of the Sylvester matrix.

Definition 1.4.2. *Let $A, B \in R[x] \setminus \{0\}$, $n = \deg(A), m = \deg(B)$, S be the Sylvester matrix of A and B, and j be an integer such that $0 \leq j < \min(n, m)$. Let $_jS$ be the $n + m - 2j$ by $n + m$ matrix obtained by deleting from S:*

(i)   rows $m - j + 1$ to m (i.e. the last j rows corresponding to A),
(ii) rows $m + n - j + 1$ to $m + n$ (i.e. the last j rows corresponding to B).

Furthermore, for $0 \leq i \leq j$, let $_jS_i$ be the square matrix obtained by deleting columns $m + n - 2j$ to $m + n$ (i.e. the last $2j + 1$ columns) of $_jS$ except for column $m + n - i - j$. The jth subresultant of A and B is then

$$S_j(A, B) = \sum_{i=0}^{j} \det(_jS_i)x^i \in R[x].$$

It is clear from the definition that $\deg(S_j(A, B)) \leq j$ for each j. Following the standard terminology [46], we call $S_j(A, B)$ *defective* if $\deg(S_j(A, B)) < j$, *regular* otherwise. In addition, $_0S_0 = S$, so $S_0(A, B) = \text{resultant}(A, B)$.

Example 1.4.2. Let $A = x^2 + 1$ and $B = x^2 - 1$ in $\mathbb{Z}[x]$. The Sylvester matrix of A and B is

$$S(A, B) = \begin{pmatrix} 1 & 0 & 1 & 0 \\ 0 & 1 & 0 & 1 \\ 1 & 0 & -1 & 0 \\ 0 & 1 & 0 & -1 \end{pmatrix}$$

and the submatrices of Definition 1.4.2 are $_0S = _0S_0 = S(A, B)$,

$$_1S = \begin{pmatrix} 1 & 0 & 1 & 0 \\ 1 & 0 & -1 & 0 \end{pmatrix}, \quad _1S_0 = \begin{pmatrix} 1 & 1 \\ 1 & -1 \end{pmatrix} \quad \text{and} \quad _1S_1 = \begin{pmatrix} 1 & 0 \\ 1 & 0 \end{pmatrix}$$

so the subresultants of A and B are $S_0 = \det(_0S_0) = 4 = \text{resultant}(A, B)$ and $S_1 = \det(_1S_0) + \det(_1S_1)x = -2$, which is defective.

Another useful property of subresultants is that they commute with ring homomorphisms when the degrees do not decrease, and that they specialize in a predictable way when only one degree decreases: any ring homomorphism $\sigma : R \to S$ induces the homomorphism of polynomial rings $\overline{\sigma} : R[x] \to S[x]$ given by

$$\overline{\sigma}\left(\sum a_j x^j\right) = \sum \sigma(a_j)x^j. \tag{1.8}$$

The following theorem describes how $S_j(\overline{\sigma}(A), \overline{\sigma}(B))$ can be computed from $S_j(A, B)$, when at least one of the leading coefficients of A or B is not taken to 0 by σ.

Theorem 1.4.3 ([50] §7.8). *Let $\sigma : R \to S$ be a ring homomorphism, $\overline{\sigma} : R[x] \to S[x]$ be given by (1.8), and $A, B \in R[x] \setminus \{0\}$. If $\deg(\overline{\sigma}(A)) = \deg(A)$ then*

$$\overline{\sigma}(S_j(A, B)) = \sigma(\text{lc}(A))^{\deg(B) - \deg(\overline{\sigma}(B))} S_j(\overline{\sigma}(A), \overline{\sigma}(B))$$

for $0 \leq j < \min(\deg(A), \deg(\overline{\sigma}(B)))$.

Note in particular that $\overline{\sigma}(S_j(A, B)) = S_j(\overline{\sigma}(A), \overline{\sigma}(B))$ when either A or B is monic, or when $\deg(\overline{\sigma}(A)) = \deg(A)$ and $\deg(\overline{\sigma}(B)) = \deg(B)$.

Theorem 1.4.3 will be used for specialization homomorphisms, when R is of the form $R = D[t_1, \ldots, t_n]$ where the t_i's are independent indeterminates, S is a ring containing D, $\alpha_1, \ldots, \alpha_n$ are given elements of S, and $\sigma : R \to S$

is the ring homomorphism that is the identity on D and that takes each t_i to α_i. In this case, Theorem 1.4.3 states that under certain circumstances, evaluating a subresultant for given values of the parameters t_i yields the corresponding subresultant of the two initial polynomials evaluated with the same values.

Example 1.4.3. Let $A = 3tx^2 - t^3 - 4$ and $B = x^2 + t^3x - 9$ in $\mathbb{Z}[t][x]$. The Sylvester matrix of A and B is

$$S(A,B) = \begin{pmatrix} 3t & 0 & -t^3 - 4 & 0 \\ 0 & 3t & 0 & -t^3 - 4 \\ 1 & t^3 & -9 & 0 \\ 0 & 1 & t^3 & -9 \end{pmatrix}$$

and the submatrices of Definition 1.4.2 are $_0S = {}_0S_0 = S(A,B)$,

$$_1S = \begin{pmatrix} 3t & 0 & -t^3 - 4 & 0 \\ 1 & t^3 & -9 & 0 \end{pmatrix}, \quad {}_1S_0 = \begin{pmatrix} 3t & -t^3 - 4 \\ 1 & -9 \end{pmatrix} \text{ and } {}_1S_1 = \begin{pmatrix} 3t & 0 \\ 1 & t^3 \end{pmatrix}$$

so the subresultants of A and B are

$$\begin{aligned} S_0(A,B) &= \text{resultant}_x(A,B) = \det({}_0S_0) = \\ &= -3t^{10} - 12t^7 + t^6 - 54t^4 + 8t^3 + 729t^2 - 216t + 16 \, , \\ S_1(A,B) &= \det({}_1S_1)x + \det({}_1S_0) = 3t^4x + t^3 - 27t + 4 \, . \end{aligned}$$

Consider now the evaluation map $t \to 1$, *i.e.* the homomorphism $\sigma : \mathbb{Z}[t] \to \mathbb{Z}$ given by $\sigma(t) = 1$ and $\sigma(n) = n$ for $n \in \mathbb{Z}$. We have $\overline{\sigma}(A) = 3x^2 - 5$, and $\overline{\sigma}(B) = x^2 + x - 9$, so Theorem 1.4.3 implies that

$$\begin{aligned} S_0(\overline{\sigma}(A), \overline{\sigma}(B)) &= \text{resultant}_x(3x^2 - 5, x^2 + x - 9) = \overline{\sigma}(S_0(A,B)) = 469 \, , \\ S_1(\overline{\sigma}(A), \overline{\sigma}(B)) &= \overline{\sigma}(3t^4x + t^3 - 27t + 4) = 3x - 22 \, . \end{aligned}$$

1.5 Polynomial Remainder Sequences

We now introduce polynomial remainder sequences, which are generalizations of the Euclidean algorithm for computing gcd's and resultants. Let D be an integral domain and x be an indeterminate over D throughout this section.

Definition 1.5.1. Let $A, B \in D[x]$ with $B \neq 0$ and $\deg(A) \geq \deg(B)$. A Polynomial Remainder Sequence (PRS) for A and B is a sequence $(R_i)_{i \geq 0}$ in $D[x]$ satisfying

(i) $R_0 = A$, $R_1 = B$,
(ii) For $i \geq 1$,
$$\beta_i R_{i+1} = \begin{cases} 0 & \text{if } R_i = 0 \\ \text{prem}(R_{i-1}, R_i) & \text{if } R_i \neq 0 \end{cases}$$
where $(\beta_i)_{i \geq 1}$ is a sequence of nonzero elements of D.

It is clear from the definition that either $R_{i+1} = 0$ or $\deg(R_{i+1}) < \deg(R_i)$ for $i \geq 1$, hence,

(i) A PRS has finitely many non-zero elements.
(ii) If $R_i \neq 0$, $R_j \neq 0$, $\deg(R_i) = \deg(R_j)$ and $i, j \geq 1$, then $i = j$ (*i.e.* only R_0 and R_1 may have the same degree).

Definition 1.5.2. *Let $A, B \in D[x]$. A is similar to B if there are $a, b \in D \setminus \{0\}$ such that $aA = bB$.*

From the definition of a PRS, we see that various choices for the β_i's yield different types of PRS. For example, the PRS obtained with $\beta_i = 1$ is just the sequence of the successive pseudo-remainders of A and B, and is called the *Euclidean PRS* of A and B. The PRS obtained with β_i set to the gcd in D of the coefficients of $\operatorname{prem}(R_{i-1}, R_i)$ is called the *primitive PRS* of A and B. An important fact is that if D is a unique factorization domain, then the last nonzero element of a PRS is similar to a gcd of A and B.

Theorem 1.5.1. *Suppose that D is a unique factorization domain, and let $A, B \in D[x]$ with $B \neq 0$ and $\deg(A) \geq \deg(B)$. Let $(R_0, R_1, \ldots, R_k, 0, \ldots)$ be any PRS of A and B with $R_k \neq 0$. Then $\gcd(R_i, R_{i+1})$ is similar to $\gcd(R_j, R_{j+1})$ for $0 \leq i, j \leq k$. In particular ($i = 0, j = k$), R_k is similar to $\gcd(A, B)$.*

Proof. Let i be such that $0 \leq i < k$, $G = \gcd(R_i, R_{i+1})$ and $H = \gcd(R_{i+1}, R_{i+2})$. Since $i < k$, $R_{i+1} \neq 0$, so, from the definitions of a PRS and of a pseudo-remainder, there are $\alpha, \beta \in D \setminus \{0\}$ and $Q \in D[x]$ such that

$$\alpha R_i = R_{i+1} Q + \beta R_{i+2}\,.$$

Hence $H \mid \alpha R_i$, but $H \mid \alpha R_{i+1}$ so $H \mid \alpha G$ since αG is a gcd of αR_i and αR_{i+1}. From the above equation we also get $G \mid \beta R_{i+2}$. But $G \mid \beta R_{i+1}$ so $G \mid \beta H$. So there are $Q_1, Q_2 \in D[x]$ such that $\alpha G = H Q_1$ and $\beta H = G Q_2$. This implies that $\alpha \beta G = G Q_1 Q_2$, hence that $Q_1, Q_2 \in D$, so G is similar to H. Thus, the theorem holds for $j = i+1$. Since similarity is transitive, it holds for $0 \leq i < j \leq k$. Since similarity is symmetric, it holds for $0 \leq i \neq j \leq k$. It is trivial for $i = j$, so it holds for $0 \leq i, j \leq k$. □

Thus, any PRS of A and B contains $\gcd(A, B)$. In addition, all the nonzero subresultants of A and B are similar to some element in the PRS. The following fundamental theorem of PRS gives explicit formulas for the similarity coefficients.

Theorem 1.5.2 (Fundamental PRS Theorem,[31] Chap. 7,[46]). *Let A and $B \neq 0$ be in $D[x]$ with $\deg(A) \geq \deg(B)$, and let $(R_0, R_1, \ldots, R_k, 0, \ldots)$ be any PRS of A and B with $R_k \neq 0$. For $i = 1, \ldots, k$, let $n_i = \deg(R_i)$ and r_i be the leading coefficient of R_i. Then, for any j in $\{0, \ldots, \deg(B) - 1\}$,*

$$S_j(A,B) = \begin{cases} \eta_i R_i & \text{if } j = n_{i-1} - 1 \\ \tau_i R_i & \text{if } j = n_i \\ 0 & \text{otherwise} \end{cases}$$

where

$$\eta_i = (-1)^{\phi_i} r_{i-1}^{1-n_{i-1}+n_i} \prod_{j=1}^{i-1} \left[\left(\frac{\beta_j}{r_j^{1+n_{j-1}-n_j}} \right)^{1+n_j-n_{i-1}} r_j^{n_{j-1}-n_{j+1}} \right]$$

$$\tau_i = (-1)^{\sigma_i} r_i^{n_{i-1}-n_i-1} \prod_{j=1}^{i-1} \left[\left(\frac{\beta_j}{r_j^{1+n_{j-1}-n_j}} \right)^{n_j-n_i} r_j^{n_{j-1}-n_{j+1}} \right] \qquad (1.9)$$

and

$$\phi_i = \sum_{j=1}^{i-1} (n_j - n_{i-1} + 1)(n_{j-1} - n_{i-1} + 1), \qquad \sigma_i = \sum_{j=1}^{i-1} (n_{j-1} - n_i)(n_j - n_i).$$

$$(1.10)$$

The *Subresultant PRS* of A and B is a particular PRS, introduced by Collins [22] and Brown [16], for which $\eta_i = 1$ in Theorem 1.5.2. It is obtained with the following recursion for β_i:

$$R_0 = A, \quad R_1 = B, \quad \gamma_1 = -1, \qquad \beta_1 = (-1)^{\delta_1+1}$$

and

$$\begin{cases} \gamma_{i+1} &= (-\text{lc}(R_i))^{\delta_i} \gamma_i^{1-\delta_i} \\ \beta_{i+1} &= -\text{lc}(R_i)\gamma_{i+1}^{\delta_i+1} \end{cases}$$

for $i \geq 1$ where $\delta_i = \deg(R_{i-1}) - \deg(R_i)$. Its key property is given by the following theorem.

Theorem 1.5.3 ([16] §7, [22, 46]). *Let A and B be in $D[x]$ with $\deg(A) \geq \deg(B)$, $(R_0, R_1, R_2, \ldots, R_k, 0, \ldots)$ be the subresultant PRS of A and B with $R_k \neq 0$, and $n_i = \deg(R_i)$ for $i = 1, \ldots, k$. Then, for any j in $\{0, \ldots, \deg(B) - 1\}$,*

$$S_j(A,B) = \begin{cases} R_i & \text{if } j = n_{i-1} - 1 \\ \tau_i R_i & \text{if } j = n_i \\ 0 & \text{otherwise} \end{cases}$$

where τ_i is given by formula (1.9).

This theorem yields the so-called *subresultant algorithm* for computing the resultant of A and B: if $\deg(A) \geq \deg(B)$, then $\text{resultant}(A,B) = S_0(A,B)$ by definition, so we compute the subresultant PRS of A and B. If $\deg(R_k) > 0$, then A and B have a common factor, so $\text{resultant}(A,B) = 0$. Otherwise Theorem 1.5.3 implies that $S_0(A,B)$ is equal to either R_k if $\deg(R_{k-1}) = 1$, or $\tau_k R_k$ if $\deg(R_{k-1}) > 1$. In that last case, the computation of τ_k can be simplified since $n_k = 0$: (1.10) becomes $\sigma_k = \sum_{j=1}^{k-1} n_{j-1} n_j$, so $(-1)^{\sigma_k} = \prod_{j=1}^{k-1} (-1)^{n_{j-1} n_j}$. A factor of -1 appears in this product whenever both n_{j-1} and n_j are odd. Furthermore, since $\deg(R_k) = 0$, $r_k = R_k$ and (1.9) becomes

$$\tau_k = (-1)^{\sigma_k} R_k^{n_{k-1}-1} \prod_{j=1}^{k-1} \left[\left(\frac{\beta_j}{r_j^{1+n_{j-1}-n_j}} \right)^{n_j} r_j^{n_{j-1}-n_{j+1}} \right].$$

If $\deg(A) < \deg(B)$, we compute the subresultant PRS of B and A, and resultant$(A, B) = (-1)^{\deg(A) \deg(B)}$resultant$(B, A)$ by Theorem 1.4.1.

SubResultant(A, B) (* Subresultant algorithm *)

(* Given an integral domain D and $A, B \in D[x]$ with $B \neq 0$ and $\deg(A) \geq \deg(B)$, return resultant(A, B) and the subresultant PRS $(R_0, R_1, \ldots, R_k, 0)$ of A and B. *)

$R_0 \leftarrow A$, $R_1 \leftarrow B$
$i \leftarrow 1$, $\gamma_1 \leftarrow -1$
$\delta_1 \leftarrow \deg(A) - \deg(B)$
$\beta_1 \leftarrow (-1)^{\delta_1+1}$
while $R_i \neq 0$ **do**
    $r_i \leftarrow \mathrm{lc}(R_i)$
    $(Q, R) \leftarrow$ **PolyPseudoDivide**(R_{i-1}, R_i)
    $R_{i+1} \leftarrow R/\beta_i$ (* this division is always exact *)
    $i \leftarrow i + 1$
    $\gamma_i \leftarrow (-r_{i-1})^{\delta_{i-1}} \gamma_{i-1}^{1-\delta_{i-1}}$
    $\delta_i \leftarrow \deg(R_{i-1}) - \deg(R_i)$
    $\beta_i \leftarrow -r_{i-1} \gamma_i^{\delta_i}$
$k \leftarrow i - 1$
if $\deg(R_k) > 0$ **then** **return**$(0, (R_0, R_1, \ldots, R_k, 0))$
if $\deg(R_{k-1}) = 1$ **then** **return**$(R_k, (R_0, R_1, \ldots, R_k, 0))$
$s \leftarrow 1$, $c \leftarrow 1$ (* s will be $(-1)^{\sigma_k}$, $s R_k^{\deg(R_{k-1})-1} c$ will be τ_k *)
for $j \leftarrow 1$ **to** $k - 1$ **do**   (* compute $\tau_k R_k$ *)
    if $\deg(R_{j-1})$ is odd and $\deg(R_j)$ is odd **then** $s \leftarrow -s$
    $c \leftarrow c (\beta_j/r_j^{1+\delta_j})^{\deg(R_j)} r_j^{\deg(R_{j-1})-\deg(R_{j+1})}$ (* exact division *)
return$(s c R_k^{\deg(R_{k-1})}, (R_0, R_1, \ldots, R_k, 0))$

Example 1.5.1. Here is the subresultant algorithm for $A = x^2 + 1$ and $B = x^2 - 1$ in $\mathbb{Z}[x]$:

i	R_i	γ_i	δ_i	β_i	r_i	$r_i^{1+\delta_i}$
0	$x^2 + 1$				1	
1	$x^2 - 1$	-1	0	-1	1	1
2	-2	-1	2	-1	-2	-8
3	0					

We get $k = 2$, $\deg(R_2) = 0$ and $\deg(R_1) = 2$, so we compute s and c:

j	$\deg(R_{j-1})$	$\deg(R_j)$	s	c
1	2	2	1	1
2	2	0	1	1

so $R = s\,c\,R_2^2 = 4 = \text{resultant}(x^2 + 1, x^2 - 1)$.

Example 1.5.2. Here is the subresultant algorithm for $A = 3tx^2 - t^3 - 4$ and $B = x^2 + t^3 x - 9$ in $D[x]$ where $D = \mathbb{Z}[t]$:

i	R_i	γ_i	δ_i	β_i	r_i	$r_i^{1+\delta_i}$
0	A				$3t$	
1	B	-1	0	-1	1	1
2	$3t^4 x + t^3 - 27t + 4$	-1	1	1	$3t^4$	$9t^8$
3	R	$-3t^4$	1	$9t^8$	R	R^2
4	0					

where

$$R = -3t^{10} - 12t^7 + t^6 - 54t^4 + 8t^3 + 729t^2 - 216t + 16 \in D.$$

We get $k = 3$, $\deg(R_3) = 0$ and $\deg(R_2) = 1$, so $R = \text{resultant}_x(A, B)$, as in example 1.4.3.

1.6 Primitive Polynomials

Let D be a unique factorization domain, and x be an indeterminate over D. Then, gcd's always exist in D by Theorem 1.1.3, and we study in this section the properties of the gcd of the coefficients of an element of $D[x]$.

Definition 1.6.1. Let $A = \sum_{i=0}^{n} a_i x^i \in D[x] \setminus \{0\}$. The content of A is

$$\text{content}(A) = \gcd(a_0, \ldots, a_n) \in D.$$

We also say that A is primitive if $\text{content}(A) \in D^*$. Finally, the primitive part of A is given by

$$\text{pp}(A) = \frac{A}{\text{content}(A)} \in D[x].$$

By convention, $\text{content}(0) = \text{pp}(0) = 0$ and 0 is not primitive.

Note that the content and primitive part are defined, like the gcd, up to multiplication by a unit. We make the convention throughout this book however that the choice of unit is made consistently in order that $A = \text{content}(A)\text{pp}(A)$ for any $A \in D[x]$. In addition, primitivity depends on the ring D, and nonprimitive polynomials can become primitive when D is embedded into a larger UFD: $4x + 6$ is not primitive as an element of $\mathbb{Z}[x]$, but becomes primitive as an element of $\mathbb{Q}[x]$. In fact, if D is a field, then every nonzero polynomial is primitive. Let $P \in D[x] \setminus D$ be irreducible. Since $P = \text{content}(P)\text{pp}(P)$ and $\text{pp}(P)$ is not a unit, it follows that $\text{content}(P)$ must be a unit, hence that P is primitive.

The main property of contents is that they are multiplicative. This classical result is due to Gauss and is known as Gauss' Lemma:

Lemma 1.6.1 ([40], Chap. V, §6, [77], §5.4).

$$\text{content}(AB) = \text{content}(A)\,\text{content}(B) \quad \textit{for any } A, B \in D[x]\,.$$

As a result, a product of primitive polynomials is itself primitive. This has an effect on the leading coefficients of prime factorizations in $D[x]$: let $A \in D[x]$ be nonzero, and $A = u \prod_{j=1}^{m} p_j^{d_j} \prod_{i=1}^{n} P_i^{e_i}$ be its prime factorization where $u \in D^*$, each p_j is an irreducible of D, and each P_i is an irreducible of $D[X] \setminus D$. Each P_i is primitive as noted above, hence $\prod_{i=1}^{n} P_i^{e_i}$ is primitive, so $\text{content}(A) = uv \prod_{j=1}^{m} p_j^{d_j}$ for some $v \in D^*$ by Lemma 1.6.1. If A is primitive, the unicity of the prime factorization over D implies that $m = 0$, so we can choose an appropriate unit for the content so that the prime factorization of $\text{pp}(A)$ is of the form $\text{pp}(A) = \prod_{i=1}^{n} P_i^{e_i}$ where the P_i are coprime and $\deg(P_i) > 0$. We use this fact in the following definition, as well as whenever we use primitive parts in the integration algorithm.

Definition 1.6.2. *Let $A \in D[x]$ and $\text{pp}(A) = \prod_{i=1}^{n} P_i^{e_i}$ be the prime factorization of its primitive part where $e_i \geq 1$ for each i. We define the* squarefree part *of A to be*

$$A^* = \prod_{i=1}^{n} P_i$$

and for $k \in \mathbb{Z}$, $k \geq 0$, the k-deflation of A to be

$$A^{-k} = \prod_{i=1}^{n} P_i^{\max(0,e_i-k)} = \prod_{i\,|\,e_i>k} P_i^{e_i-k}\,.$$

Note that $A^{-0} = \text{pp}(A)$. For convenience we call A^{-1} simply the *deflation* of A, and denote if by A^-, *i.e.*

$$A^- = A^{-1} = \prod_{i=1}^{n} P_i^{e_i-1}\,.$$

As consequences of the definition we have the following useful relations:

$$A^* A^- = \mathrm{pp}(A),\tag{1.11}$$

$$A^{-k} = A^{-i} {}^{-j} \text{ where } i, j \geq 0 \text{ and } i + j = k.$$

A special case of the above relation is

$$A^{-k+1} = A^{-k} {}^-\tag{1.12}$$

which together with (1.11) implies that

$$A^{-k+1} = \frac{A^{-k}}{A^{-k} {}^*} \qquad \text{for } k \geq 0.\tag{1.13}$$

Although the squarefree part and deflations are defined in terms of the prime factorization, it turns out that they can be computed by gcd computations in $D[x]$. The basic idea is that a prime factor of A divides dA/dx once less than A.

Theorem 1.6.1. *Let $A, P \in D[x] \setminus D$ and $n > 0$ be an integer. Then,*

(i) $P^{n+1} \mid A \Longrightarrow P^n \mid \gcd(A, dA/dx)$,
(ii) if P is prime and $\mathrm{char}(D) = 0$, then $P^n \mid \gcd(A, dA/dx) \Longrightarrow P^{n+1} \mid A$.

Proof. (i) Suppose that $P^{n+1} \mid A$, then there exists $B \in D[x]$ such that $A = P^{n+1} B$. Hence,

$$\frac{dA}{dx} = P^{n+1} \frac{dB}{dx} + (n+1) P^n B \frac{dP}{dx}$$

so $P^n \mid dA/dx$, which implies that $P^n \mid \gcd(A, dA/dx)$.
(ii) Suppose that D has characteristic 0, P is prime, and $P^n \mid \gcd(A, dA/dx)$. Let $m > 0$ be the unique integer such that $P^m \mid A$ and $P^{m+1} \nmid A$. Then, there exists $B \in D[x]$ such that $A = P^m B$ and $P \nmid B$. As in part (i), we have

$$\frac{dA}{dx} = P^m \frac{dB}{dx} + m P^{m-1} B \frac{dP}{dx}.$$

We have $m \geq n$ since $P^n \mid A$. Suppose that $m = n$. Then,

$$\frac{dA}{dx} - P^n \frac{dB}{dx} = n P^{n-1} B \frac{dP}{dx}.$$

We have $P^n \mid dA/dx$, so $P^n \mid n P^{n-1} B (dP/dx)$, hence $P \mid nB(dP/dx)$. But P is prime and $P \nmid B$, so $P \mid n(dP/dx)$. In characteristic 0, $n(dP/dx)$ is nonzero and has a smaller degree than P, so $P \nmid n(dP/dx)$. Hence $m \neq n$, so $m > n$, which implies that $P^{n+1} \mid A$. $\qquad\square$

An immediate consequence of Theorem 1.6.1 is that when D has characteristic 0, then

$$A^- = \gcd\left(A, \frac{dA}{dx}\right)\tag{1.14}$$

for any primitive A, and A^* can then be computed by (1.11). The further deflations of A can be computed recursively with (1.12). Squarefree parts and deflations are thus easier to compute than prime factorizations. We use this fact in the next section where we introduce the notion of squarefree factorization.

1.7 Squarefree Factorization

Let D be a unique factorization domain, and x be an indeterminate over D. $D[x]$ is then a unique factorization domain, so every $A \in D[x]$ has a factorization into irreducibles. Such a factorization is usually difficult to compute in general, but there are other factorizations that are easier to compute and that can be used instead for many purposes. We introduce in this section the *squarefree factorization*, which is the one primarily used by integration algorithms.

Definition 1.7.1. $A \in D[x]$ *is* squarefree *if there exists no* $B \in D[x] \setminus D$ *such that* $B^2 \mid A$ *in* $D[x]$.

Equivalently, A is squarefree if $e_i = 1$ for $i = 1, \ldots, n$ in any prime factorization of A over D.

Definition 1.7.2. *Let* $A \in D[x]$. *A* squarefree factorization *of* A *is a factorization of the form* $A = \prod_{i=1}^m A_i^i$ *where each* A_i *is squarefree and* $\gcd(A_i, A_j) \in D$ *for* $i \neq j$.

Note that there is no need to require a separate leading coefficient in D^* and prime factors in D as in the prime factorization, since the elements of D are automatically squarefree by our definition. In addition, if we have a squarefree factorization of the primitive part of A of the form $\mathrm{pp}(A) = \prod_{i=1}^m A_i^i$, then

$$A = (\mathrm{content}(A)\, A_1) \prod_{i=2}^m A_i^i$$

is a squarefree factorization of A, so it is sufficient to compute squarefree factorizations of primitive parts. In addition, we assume that D has characteristic 0 (see [31, 32, 82] for squarefree factorizations in positive characteristic). In characteristic 0, a squarefree factorization of A separates the zeroes of A by equal multiplicities, since a zero of A must be a zero of exactly one A_i, and its multiplicity in A is then i. We use this fact in order to express the A_i's in terms of the deflations of A and vice-versa.

Lemma 1.7.1. *Let* $A \in D[x] \setminus D$, $\mathrm{pp}(A) = \prod_{i=1}^n P_i^{e_i}$ *be a prime factorization of* $\mathrm{pp}(A)$, $m = \max(e_1, \ldots, e_m)$ *and* $A_i = \prod_{j \mid e_j = i} P_j$ *for* $1 \leq i \leq m$. *Then,*

(i) $A^{-k} = \prod_{i=k+1}^m A_i^{i-k} = A_{k+1} A_{k+2}^2 \cdots A_m^{m-k}$ *for any integer* $k \geq 0$.
(ii)

$$A_i = \frac{A^{-i-1^*}}{A^{-i^*}} \quad \text{for } 1 \leq i \leq m. \tag{1.15}$$

(iii) $\mathrm{pp}(A) = \prod_{i=1}^m A_i^i$ *is a squarefree factorization of* $\mathrm{pp}(A)$.

Proof. (i) We have

$$\prod_{i=k+1}^{m} A_i^{i-k} = \prod_{i=k+1}^{m} \prod_{j|e_j=i} P_j^{i-k} = \prod_{j|e_j>k} P_j^{e_j-k} = A^{-k}.$$

(ii) From (i) we have

$$A_i = \frac{A_i A_{i+1} \cdots A_m}{A_{i+1} \cdots A_m} = \frac{A^{-i-1^*}}{A^{-i^*}}.$$

(iii) Each A_i is squarefree, since it is a product of coprime irreducibles. In addition, $\gcd(A_i, A_j) \in D$ for $i \neq j$ since each prime factor of $pp(A)$ appears in its factorization with a unique exponent. Finally, using $k = 0$ in (i), we get $pp(A) = A^{-0} = \prod_{i=1}^{m} A_i^i$, which is a squarefree factorization of $pp(A)$. □

Since deflations and squarefree parts can be computed by gcd's as explained in the previous section, we get the following squarefree factorization algorithm for a primitive A: by (1.14), we have $A^{-1} = A^- = \gcd(A, dA/dx)$, which gives us $A^{-0^*} = A^* = pp(A)/A^-$. Once we have A^{-k^*} and A^{-k+1} for $k \geq 0$, the sequence can be continued by

$$\gcd\left(A^{-k^*}, A^{-k+1}\right) = \gcd\left(A_{k+1} \cdots A_m, A_{k+2} A_{k+3}^2 \cdots A_m^{m-k-1}\right) = A^{-k+1^*},$$

and A_{k+1} and A^{-k+2} are obtained by (1.15) and (1.13) respectively. We continue this sequence until $A^{-k+1} \in D$, which implies that A^{-k} is squarefree, in which case $k = m - 1$, and $A_m = A^{-k}$. This squarefree factorization algorithm uses only rational operations plus gcd computations in $D[x]$.

Squarefree(A)   (* Musser's squarefree factorization *)

(* Given a unique factorization domain D of characteristic 0 and $A \in D[x]$, return $A_1, \ldots, A_m \in D[x]$ such that $A = \prod_{k=1}^{m} A_k^k$ is a squarefree factorization of A. *)

$c \leftarrow \text{content}(A), S \leftarrow A/c$ (* $S = pp(A)$ *)
$S^- \leftarrow \gcd(S, dS/dx)$
$S^* \leftarrow S/S^-$
$k \leftarrow 1$
while $\deg(S^-) > 0$ **do**   (* $S^- = A^{-k}, S^* = A^{-k-1^*}$ *)
    $Y \leftarrow \gcd\left(S^*, S^-\right)$ (* $Y = A^{-k^*}$ *)
    $A_k \leftarrow S^*/Y$ (* $A_k = A^{-k-1^*}/A^{-k^*}$ *)
    $S^* \leftarrow Y$ (* $S^* = A^{-k^*}$ *)
    $S^- \leftarrow S^-/Y$ (* $S^- = A^{-k+1}$ *)
    $k \leftarrow k + 1$
$A_k \leftarrow S^*$
return$\left((cS^-) A_1, \ldots, A_k\right)$

Example 1.7.1. Applying **Squarefree** to $A = x^8 + 6x^6 + 12x^4 + 8x^2 \in \mathbb{Q}[x]$, we get $c = 1$, $S = A$, $dS/dx = 8x^7 + 36x^5 + 48x^3 + 16x$,

$$S^- = \gcd\left(S, \frac{dS}{dx}\right) = x^5 + 4x^3 + 4x$$

and $S^* = S/S^- = x^3 + 2x$. Then,

k	S^*	S^-	Y	A_j
1	$x^3 + 2x$	$x^5 + 4x^3 + 4x$	$x^3 + 2x$	1
2	$x^3 + 2x$	$x^2 + 2$	$x^2 + 2$	x
3	$x^2 + 2$	1		$x^2 + 2$

Hence,

$$A = x^8 + 6x^6 + 12x^4 + 8x^2 = x^2(x^2 + 2)^3.$$

Yun [80] showed that it is possible to be more efficient than the above algorithm by reducing the degree of the polynomials appearing in the gcd inside the loop. His idea is to consider the following sequence of polynomials:

$$
\begin{aligned}
Y_k &= \sum_{i=k}^{m}(i - k + 1)\frac{dA_i}{dx}\frac{A^{-k-1\,*}}{A_i} \\
&= \sum_{i=k}^{m}(i - k + 1)A_k \cdots A_{i-1}\frac{dA_i}{dx}A_{i+1}\cdots A_m \quad \text{for } k \geq 1 \quad (1.16)
\end{aligned}
$$

whose properties are summarized in the following lemma.

Lemma 1.7.2. *With the above notation,*

$$\gcd(A^{-i-1\,*}, Y_i) \in D,$$

$$\frac{dA^{-i-1}}{dx} = A^{-i}Y_i, \tag{1.17}$$

and with A_i as defined in Lemma 1.7.1,

$$Y_i - \frac{dA^{-i-1\,*}}{dx} = A_iY_{i+1} \tag{1.18}$$

for $1 \leq i \leq m$.

Proof. Let $1 \leq i \leq j \leq m$. Then,

$$\gcd(A_j, A_i \cdots A_{j-1}\frac{dA_j}{dx}A_{j+1}\cdots A_m) \in D$$

since A_j is squarefree and the A_i are pairwise relatively prime. Since

$$A_j \mid A_i \cdots A_{k-1}\frac{dA_k}{dx}A_{k+1}\cdots A_m \quad \text{for } j \neq k,$$

this implies that $\gcd(A_j, Y_i) \in D$, hence that $\gcd(A^{-i-1*}, Y_i) \in D$. Let $1 \leq i \leq m$. Using Lemma 1.7.1 and (1.11) we get

$$
\begin{aligned}
\frac{dA^{-i-1}}{dx} &= \frac{d}{dx}\left(\prod_{j=i}^{m} A_j^{j-i+1}\right) = \sum_{j=i}^{m}(j-i+1)\frac{dA_j}{dx}\frac{A^{-i-1}}{A_j} \\
&= \sum_{j=i}^{m}(j-i+1)\frac{dA_j}{dx}\frac{A^{-i-1^-}A^{-i-1^*}}{A_j} \\
&= A^{-i-1^-}\sum_{j=i}^{m}(j-i+1)\frac{dA_j}{dx}\frac{A^{-i-1^*}}{A_j} = A^{-i-1^-}Y_i\,.
\end{aligned}
$$

From

$$
\frac{dA^{-k-1^*}}{dx} = \frac{d}{dx}\left(\prod_{j=k}^{m} A_j\right) = \sum_{j=k}^{m}\frac{dA_j}{dx}\frac{A^{-k-1^*}}{A_j}
$$

we get

$$
\begin{aligned}
Y_i - \frac{dA^{-i-1^*}}{dx} &= \sum_{j=i}^{m}(j-i+1)\frac{dA_j}{dx}\frac{A^{-i-1^*}}{A_j} - \sum_{j=i}^{m}\frac{dA_j}{dx}\frac{A^{-i-1^*}}{A_j} \\
&= \sum_{j=i+1}^{m}(j-i)\frac{dA_j}{dx}\frac{A^{-i-1^*}}{A_j} \\
&= A_i\sum_{j=i+1}^{m}(j-i)\frac{dA_j}{dx}\frac{A^{-i^*}}{A_j} = A_iY_{i+1}\,.
\end{aligned}
$$

$\square$

Since $A^{-i-1^*} = A_iA^{-i^*}$ and $\gcd(A^{-i^*}, Y_{i+1}) \in D$, we conclude from (1.18) that

$$
\gcd\left(A^{-i-1^*}, Y_i - \frac{dA^{-i-1^*}}{dx}\right) = A_i \tag{1.19}
$$

which yields Yun's squarefree factorization algorithm: assuming as before that A is primitive, we have $A^- = \gcd(A, dA/dx)$, which gives us

$$
A^{-0^*} = A^* = \mathrm{pp}(A)/A^- \quad \text{and} \quad Y_1 = \frac{dA/dx}{A^-} \text{ by (1.17)}\,.
$$

Once we have A^{-k-1^*} and Y_k, A_k is computed by (1.19), and Y_{k+1} and A^{-k^*} are obtained by (1.18) and (1.15) respectively. We continue this sequence until $Y_k = dA^{-k-1^*}/dx$, which implies that A^{-k-1} is squarefree, in which case $k = m$, and $A_k = A^{-k-1} = A^{-k-1^*}$. The difference between this squarefree factorization algorithm and the previous one, is that $Y_k - dA^{-k-1^*}/dx$ appears in the main gcd computation instead of A^{-k}.

Squarefree(A)   (* Yun's squarefree factorization *)

(* Given a unique factorization domain D of characteristic 0 and $A \in D[x]$, return $A_1, \ldots, A_m \in D[x]$ such that $A = \prod_{k=1}^{m} A_k^k$ is a squarefree factorization of A. *)

$c \leftarrow \text{content}(A), S \leftarrow A/c$ $(* \ S = \text{pp}(A) \ *)$
$S' \leftarrow dS/dx$
$S^- \leftarrow \gcd(S, S')$
$S^* \leftarrow S/S^-$
$Y \leftarrow S'/S^-$
$k \leftarrow 1$
while $(Z \leftarrow Y - dS^*/dx) \neq 0$ **do** $(* \ S^* = A^{-k-1^*}, Y \leftarrow Y_k \ *)$
    $A_k \leftarrow \gcd(S^*, Z)$ $(* \ (1.19) \ *)$
    $S^* \leftarrow S^*/A_k$ $(* \ S^* = A^{-k^*} \ *)$
    $Y \leftarrow Z/A_k$ $(* \ Y = Y_{k+1} \ *)$
    $k \leftarrow k + 1$
$A_k \leftarrow S^*$
return$(c \ A_1, \ldots, A_k)$

Example 1.7.2. Here is a step-by-step execution of Yun's algorithm on the A of example 1.7.1. We get $c = 1$, $S = A$, $S' = dS/dx = 8x^7 + 36x^5 + 48x^3 + 16x$, and $S^- = \gcd(S, S') = x^5 + 4x^3 + 4x$. Then,

k	S^*	Y	Z	A_j
1	$x^3 + 2x$	$8x^2 + 4$	$5x^2 + 2$	1
2	$x^3 + 2x$	$5x^2 + 2$	$2x^2$	x
3	$x^2 + 2$	$2x$	0	$x^2 + 2$

Hence,

$$A = x^8 + 6x^6 + 12x^4 + 8x^2 = x^2(x^2 + 2)^3.$$

The second arguments to the repeated gcd computations inside the loop are in the Z-column, and their degrees are smaller than in the corresponding S^--column of example 1.7.1.

Exercises

Exercise 1.1. Use the Euclidean Algorithm to compute the gcd of 217 and 413 in $\mathbb{Z}$.

Exercise 1.2. Find integers x, y such, that
(a) $12x + 19y = 1$.
(b) $3x + 2y = 5$.

Exercise 1.3. Find the inverse of 14 in $\mathbb{Z}_{37}$.

Exercise 1.4. Compute the gcd of $2x^3 - \frac{19}{5}x^2 - x + \frac{6}{5}$ and $x^2 + \frac{1}{3}x - \frac{14}{3}$ in $\mathbb{Q}[x]$.

Exercise 1.5. Compute the pseudo-quotient and pseudo-remainder of $x^4 - 7x + 7$ by $3x^2 - 7$ in $\mathbb{Z}[x]$.

Exercise 1.6. Compute the quotient and remainder (or pseudo-quotient and pseudo-remainder) of $7x^5 + 4x^3 + 2x + 1$ by $2x^3 + 3$ in $\mathbb{Z}_5[x]$, $\mathbb{Z}_{11}[x]$, $\mathbb{Z}[x]$ and $\mathbb{Q}[x]$. In each case determine over which kind of algebraic structure you are computing.

Exercise 1.7. Compute the primitive PRS and the subresultant PRS of $x^4 + x^3 - t$ and $x^3 + 2x^2 + 3tx - t + 1$ in $\mathbb{Z}[t][x]$.

Exercise 1.8. Compute the gcd of $4x^4 + 13x^3 + 15x^2 + 7x + 1$ and $2x^3 + x^2 - 4x - 3$ in a) $\mathbb{Q}[x]$ and b) $\mathbb{Z}[x]$.

Exercise 1.9. Compute a squarefree factorization of $x^8 - 5x^6 + 6x^4 + 4x^2 - 8$.

Exercise 1.10. Prove that 2 is irreducible but not prime in $\mathbb{Z}\left[\sqrt{-5}\right]$.

Exercise 1.11. Prove that similarity as defined in Definition 1.5.2 is an equivalence relation.

Exercise 1.12. Prove that if a, b are in a Euclidean domain D, and $a = qb + r$ for some $q, r \in D$, then $\gcd(a, b) = \gcd(b, r)$.

Exercise 1.13. Use the Extended Euclidean algorithm and Theorem 1.4.1 to prove Theorem 1.4.2.

Exercise 1.14. Use a loop invariant to prove that the Extended Euclidean algorithm is correct.

2. Integration of Rational Functions

We describe in this chapter algorithms for the integration of rational functions. This case, which is the simplest since rational functions always have elementary integrals, is important because the algorithms for integrating more complicated functions are essentially generalizations of the techniques used for rational functions. Since the algorithms and theorems of this chapter are special cases of the Risch algorithm, we postpone the proof of their correctness until the later chapters on integrating transcendental functions. Throughout this chapter, let K be a field[1] of characteristic 0, x an indeterminate over K, and $'$ denote the derivation d/dx on $K(x)$, so x is the integration variable. By a *rational function w.r.t.* x, we mean a quotient of two polynomials in x, allowing other expressions provided they do not involve x. For example, $\log(y)/(x - e - \pi)$ is a rational function w.r.t. x, where $K = \mathbb{Q}(\log(y), e, \pi)$. We see from this example that computing in the algebraic closure $\overline{K}$ of K, while possible in theory, is in general ineffective or impractical. Thus, modern algorithms try to avoid computing in extensions of K as long as possible.

Introduction

The problem of integrating rational functions seems to be as old as differentiation. According to Ostrogradsky [53], both Newton and Leibniz attempted to compute antiderivatives of rational functions, neither of them obtaining a complete algorithm. Leibniz' approach was to compute an irreducible factorization of the denominator over the reals, then a partial fraction decomposition where the denominators have degree 1 or 2 in x, and then to integrate each summand separately. However, he could not completely handle the case of a quadratic denominator. In the early 18th century, Johan Bernoulli perfected the partial fraction decomposition method and completed Leibniz' method (*Acta Eruditorum*, 1703), giving what seems to be the oldest integration algorithm on record ([47] Chap. IX p. 353). Amazingly, it remains the method found in today's calculus textbooks and taught to high-school and university students in introductory analysis courses. The major computational problem with this method is of course computing the complete factorization

---

[1] The reader unfamiliar with algebra can think of K throughout this chapter as either the set of rational, real or complex numbers.

of a polynomial over the reals. This problem was already an active research area in the 19th century, and as early as 1845, the Russian mathematician M. W. Ostrogradsky [53] presented a new algorithm that computes the rational part of the integral without factoring whatsoever. Although his method was taught to Russian students, and appears in older Russian analysis textbooks ([30] Chap. VIII §2), it was not widely taught in the rest of the world, where competing or similar methods were independently discovered[2]. Thus, Hermite [34] published in 1872 a different algorithm that achieved the same goal, namely computing the rational part of the integral without factoring. And more recently, E. Horowitz independently discovered essentially Ostrogradsky's method and presented it with a detailed complexity analysis [36]. The problem of computing the transcendental part of the integral without factoring remained open for over a century, and was finally solved in recent papers [42, 68, 74].

2.1 The Bernoulli Algorithm

This approach, both the oldest and simplest, is not often used in practice because of the cost of factoring in $\mathbb{R}[x]$, but it is important since it provides the theoretical foundations for all the subsequent algorithms. Let $f \in \mathbb{R}(x)$ be our integrand, and write $f = P + A/D$ where $P, A, D \in \mathbb{R}[x]$, $\gcd(A, D) = 1$, and $\deg(A) < \deg(D)$. Let

$$D = c \prod_{i=1}^{n} (x - a_i)^{e_i} \prod_{j=1}^{m} (x^2 + b_j x + c_j)^{f_j}$$

be the irreducible factorization of D over $\mathbb{R}$, where c, the a_i's, b_j's and c_j's are in $\mathbb{R}$ and the e_i's and f_j's are positive integers. Computing the partial fraction decomposition of f, we get

$$f = P + \sum_{i=1}^{n} \sum_{k=1}^{e_i} \frac{A_{ik}}{(x - a_i)^k} + \sum_{j=1}^{m} \sum_{k=1}^{f_j} \frac{B_{jk} x + C_{jk}}{(x^2 + b_j x + c_j)^k}$$

where the A_{ik}'s, B_{jk}'s and C_{jk}'s are in $\mathbb{R}$. Hence,

$$\int f = \int P + \sum_{i=1}^{n} \sum_{k=1}^{e_i} \int \frac{A_{ik}}{(x - a_i)^k} + \sum_{j=1}^{m} \sum_{k=1}^{f_j} \int \frac{B_{jk} x + C_{jk}}{(x^2 + b_j x + c_j)^k} .$$

Computing $\int P$ poses no problem (it will for any other class of functions), and for the other terms we have

---

[2] I would like to thank Prof. S. A. Abramov, Moscow State University, for bringing this point to my attention.

$$\int \frac{A_{ik}}{(x-a_i)^k} = \begin{cases} A_{ik}(x-a_i)^{1-k}/(1-k) & \text{if } k > 1 \\ A_{i1}\log(x-a_i) & \text{if } k = 1 \end{cases} \tag{2.1}$$

and, noting that $b_j^2 - 4c_j < 0$ since $x^2 + b_j x + c_j$ is irreducible in $\mathbb{R}[x]$,

$$\int \frac{B_{j1}x + C_{j1}}{(x^2 + b_j x + c_j)} = \frac{B_{j1}}{2}\log(x^2 + b_j x + c_j)$$

$$+ \frac{2C_{j1} - b_j B_{j1}}{\sqrt{4c_j - b_j^2}}\arctan\left(\frac{2x + b_j}{\sqrt{4c_j - b_j^2}}\right)$$

and for $k > 1$,

$$\int \frac{B_{jk}x + C_{jk}}{(x^2 + b_j x + c_j)^k} = \frac{(2C_{jk} - b_j B_{jk})x + b_j C_{jk} - 2c_j B_{jk}}{(k-1)(4c_j - b_j^2)(x^2 + b_j x + c_j)^{k-1}}$$

$$+ \int \frac{(2k-3)(2C_{jk} - b_j B_{jk})}{(k-1)(4c_j - b_j^2)(x^2 + b_j x + c_j)^{k-1}}.$$

This last formula can be used recursively until $k = 1$, thus producing the complete integral.

Example 2.1.1. Consider $f = 1/(x^3 + x) \in \mathbb{Q}(x)$. The denominator of f factors over $\mathbb{R}$ as $x^3 + x = x(x^2 + 1)$, and the partial fraction decomposition of f is

$$\frac{1}{x^3 + x} = \frac{1}{x} - \frac{x}{x^2 + 1}.$$

So from the above formulas we get

$$\int \frac{dx}{x^3 + x} = \log(x) - \frac{1}{2}\log(x^2 + 1). \tag{2.2}$$

Example 2.1.2. Consider $f = 1/(x^2 + 1)^2 \in \mathbb{Q}(x)$. The denominator of f factors over $\mathbb{R}$ as $(x^2 + 1)^2$, and the partial fraction decomposition of f is $1/(x^2 + 1)^2$, so from the above formulas with $j = 1$, $k = 2$, $b_1 = B_{12} = 0$ and $c_1 = C_{12} = 1$, we get

$$\int \frac{dx}{(x^2 + 1)^2} = \frac{2x}{4(x^2 + 1)} + \int \frac{2dx}{4(x^2 + 1)} = \frac{x}{2(x^2 + 1)} + \frac{1}{2}\arctan(x).$$

A variant of Bernoulli's algorithm that works over an arbitrary field K of characteristic 0, is to factor D linearly over the algebraic closure of K, $D = \prod_{i=1}^q (x - \alpha_i)^{e_i}$, and then use (2.1) on each term of the following partial fraction decomposition of f:

$$f = P + \sum_{i=1}^q \sum_{j=1}^{e_i} \frac{A_{ij}}{(x - \alpha_i)^j}. \tag{2.3}$$

We note that this approach is then equivalent to expanding f into its Laurent series at all its finite poles, since at $x = \alpha_i$, the Laurent series is

$$f = \frac{A_{ie_i}}{(x - \alpha_i)^{e_i}} + \cdots + \frac{A_{i2}}{(x - \alpha_i)^2} + \frac{A_{i1}}{(x - \alpha_i)} + \cdots$$

where the A_{ij}'s are the same as those in (2.3). Thus, this approach can be seen as expanding the integrand into series around all its poles (including ∞), then integrating the series termwise, and then interpolating for the answer, by summing all the polar terms, obtaining (2.3).

Example 2.1.3. Consider $f = 1/(x^3 + x) \in \mathbb{Q}(x)$. The denominator of f factors over $\mathbb{Q}(\sqrt{-1})$ as $x^3 + x = x(x + \sqrt{-1})(x - \sqrt{-1})$, and the partial fraction decomposition of f is

$$\frac{1}{x^3 + x} = \frac{1}{x} - \frac{1/2}{x + \sqrt{-1}} - \frac{1/2}{x - \sqrt{-1}}.$$

So an integral of f is

$$\int \frac{dx}{x^3 + x} = \log(x) - \frac{1}{2} \log\left(x + \sqrt{-1}\right) - \frac{1}{2} \log\left(x - \sqrt{-1}\right).$$

Note that there exists an integral of f expressible without $\sqrt{-1}$, namely (2.2).

Since this algorithm can be based on power series expansions, we call it a *local* approach. Its major computational inconvenience is the requirement of computing with algebraic numbers over K that might not appear in the integral, namely the coefficients of the Laurent series. This is the case in the previous example, in which the algorithm computes in $\mathbb{Q}(\sqrt{-1})$, but there exists an integral that is expressible over $\mathbb{Q}(x)$ only. On the other hand, there are integrals that cannot be expressed without the introduction of new algebraic constants, like $\int (dx/(x^2 - 2))$, which cannot be expressed without using $\sqrt{2}$ ([60] Prop. 1.1), so in general we may need to introduce an algebraic extension of K at some point.

In order to have a complete and efficient algorithm, we have to answer the following questions:

Q1. How much of the integral can be computed with all calculations being done in $K(x)$?

Q2. Can we compute the minimal algebraic extension of K necessary to express the integral?

An algorithm that never makes an unnecessary algebraic extension and does not compute irreducible factorizations over K will be called "rational".

2.2 The Hermite Reduction

We can see from the variant of Bernoulli's algorithm discussed above, that any $f \in K(x)$ has an integral of the form

$$\int f = v + \sum_{i=1}^{m} c_i \log(u_i) \tag{2.4}$$

where $v, u_1, \ldots, u_m \in \overline{K}(x)$ and $c_1, \ldots, c_m \in \overline{K}$. v is called the *rational part* of the integral, and the sum of logarithms is called the *transcendental part* of the integral. Hermite [34] gave the following rational algorithm for computing v: write the integrand as $f = A/D$ where $A, D \in K[x]$ and $\gcd(A, D) = 1$. Let $D = D_1 D_2^2 \cdots D_n^n$ be a squarefree factorization of D. Using a partial fraction decomposition of f with respect to $D_1, D_2^2, \ldots, D_n^n$, write

$$f = P + \sum_{k=1}^{n} \frac{A_k}{D_k^k}$$

where P and the A_k's are in $K[x]$ and either $A_k = 0$ or $\deg(A_k) < \deg(D_k^k)$ for each k. Then,

$$\int f = \int P + \sum_{k=1}^{n} \int \frac{A_k}{D_k^k}$$

so the problem is now reduced to integrating a fraction of the form Q/V^k where $\deg(Q) < \deg(V^k)$ and V is squarefree, which implies that $\gcd(V, V') = 1$. Thus, if $k > 1$ we can use the extended Euclidean algorithm to find $B, C \in K[x]$ such that

$$\frac{Q}{1-k} = BV' + CV$$

and $\deg(B) < \deg(V)$. This implies that $\deg(BV') < \deg(V^2) \le \deg(V^k)$, hence that $\deg(C) < \deg(V^{k-1})$. Multiplying both sides by $(1-k)/V^k$ we get

$$\frac{Q}{V^k} = -\frac{(k-1)BV'}{V^k} + \frac{(1-k)C}{V^{k-1}}.$$

Adding and subtracting B'/V^{k-1} to the right hand side we get

$$\frac{Q}{V^k} = \left(\frac{B'}{V^{k-1}} - \frac{(k-1)BV'}{V^k} \right) + \frac{(1-k)C - B'}{V^{k-1}}.$$

And integrating both sides yields

$$\int \frac{Q}{V^k} = \frac{B}{V^{k-1}} + \int \frac{(1-k)C - B'}{V^{k-1}}.$$

Since $\deg((1-k)C - B') < \deg(V^{k-1})$, the integrand is thus reduced to a similar one with a smaller power of V in the denominator, so, repeating this until $k = 1$, we obtain $y \in K(x)$ and $E \in K[x]$ such that $\deg(E) < \deg(V)$ and $Q/V^k = y' + E/V$. Doing this to each term A_i/D_i^i, we get $g, h \in K(x)$ such that $f = g' + P + h$ and h has a squarefree denominator and no polynomial part, so $\int h$ is a linear combination of logarithms with constant coefficients. The v of (2.4) is then merely $g + \int P$. Hermite did not provide any new technique for integrating h, so question Q2 remained open at that point.

HermiteReduce(A, D) (* Hermite Reduction – original version *)

(* Given a field K and $A, D \in K[x]$ with D nonzero and coprime with A, return $g, h \in K(x)$ such that $\frac{A}{D} = \frac{dg}{dx} + h$ and h has a squarefree denominator. *)

$(D_1, \ldots, D_n) \leftarrow$ **SquareFree**(D)
$(P, A_1, A_2, \ldots, A_n) \leftarrow$ **PartialFraction**$(A, D_1, D_2^2, \ldots, D_n^n)$
$g \leftarrow 0$
$h \leftarrow P + A_1/D_1$
for $k \leftarrow 2$ **to** n such that $\deg(D_k) > 0$ **do**
    $V \leftarrow D_k$
    for $j \leftarrow k - 1$ **to** 1 **step** -1 **do**
        $(B, C) \leftarrow$ **ExtendedEuclidean**$(\frac{dV}{dx}, V, -A_k/j)$
        $g \leftarrow g + B/V^j$
        $A_k \leftarrow -jC - \frac{dB}{dx}$
    $h \leftarrow h + A_k/V$
return(g, h)

Example 2.2.1. Here is **HermiteReduce** on

$$f = \frac{x^7 - 24x^4 - 4x^2 + 8x - 8}{x^8 + 6x^6 + 12x^4 + 8x^2} \quad \in \mathbb{Q}(x).$$

A squarefree factorization of the denominator of f is

$$D = x^8 + 6x^6 + 12x^4 + 8x^2 = x^2(x^2 + 2)^3 = D_2^2 D_3^3$$

and the partial fraction decomposition of f is:

$$f = \frac{x - 1}{x^2} + \frac{x^4 - 6x^3 - 18x^2 - 12x + 8}{(x^2 + 2)^3}.$$

Here is the rest of the Hermite reduction for f:

i	V	j	A_i	B	C
2	x	1	$x - 1$	1	-1
3	$x^2 + 2$	2	$x^4 - 6x^3 - 18x^2 - 12x + 8$	$6x$	$-\frac{x^2}{2} + 3x - 2$
3	$x^2 + 2$	1	$x^2 - 6x - 2$	$-x + 3$	1

Thus,

$$\int \frac{x^7 - 24x^4 - 4x^2 + 8x - 8}{x^8 + 6x^6 + 12x^4 + 8x^2} \, dx = \frac{1}{x} + \frac{6x}{(x^2 + 2)^2} - \frac{x - 3}{x^2 + 2} + \int \frac{dx}{x}.$$

We also mention the following variant of Hermite's algorithm that does not require a partial fraction decomposition of f: let $D = D_1 D_2^2 \cdots D_m^m$ be a squarefree factorization of D and suppose that $m \geq 2$ (otherwise D is already squarefree). Let then $V = D_m$ and $U = D/V^m$. Since $\gcd(UV', V) = 1$, we can use the extended Euclidean algorithm to find $B, C \in K[x]$ such that

$$\frac{A}{1 - m} = BUV' + CV$$

and $\deg(B) < \deg(V)$. Multiplying both sides by $(1 - m)/(UV^m)$ gives

$$\frac{A}{UV^m} = \frac{(1 - m)BV'}{V^m} + \frac{(1 - m)C}{UV^{m-1}}$$

so, adding and subtracting B'/V^{m-1} to the right hand side, we get

$$\frac{A}{UV^m} = \left(\frac{B'}{V^{m-1}} - \frac{(m - 1)BV'}{V^m} \right) + \frac{(1 - m)C - UB'}{UV^{m-1}}$$

and integrating both sides yields

$$\int \frac{A}{UV^m} = \frac{B}{V^{m-1}} + \int \frac{(1 - m)C - UB'}{UV^{m-1}}$$

so the integrand is reduced to one with a smaller power of V in the denominator. This process is repeated until the denominator is squarefree. Since the exponent of one of the squarefree factors is reduced by 1 at every pass, the number of reduction steps in the worst case is $1 + 2 + \cdots + (m - 1)$, which is $\mathcal{O}(m^2)$ so we call this variant the *quadratic Hermite reduction*.

HermiteReduce(A, D) (* Hermite Reduction – quadratic version *)

(* Given a field K and $A, D \in K[x]$ with D nonzero and coprime with A, return $g, h \in K(x)$ such that $\frac{A}{D} = \frac{dg}{dx} + h$ and h has a squarefree denominator. *)

$g \leftarrow 0, (D_1, \ldots, D_m) \leftarrow$ **SquareFree**(D)
for $i \leftarrow 2$ **to** m such that $\deg(D_i) > 0$ **do**
    $V \leftarrow D_i, U \leftarrow D/V^i$
    for $j \leftarrow i - 1$ **to** 1 **step** -1 **do**
        $(B, C) \leftarrow$ **ExtendedEuclidean**$(U\frac{dV}{dx}, V, -A/j)$
        $g \leftarrow g + B/V^j, A \leftarrow -jC - U\frac{dB}{dx}$
    $D \leftarrow UV$
return$(g, A/D)$

Example 2.2.2. Given the same integrand as in example 2.2.1, the quadratic Hermite reduction makes the following steps, where $D_3 = x^2 + 2$:

i	V	U	j	B	C	A
2	x	D_3^3	1	1	$-x^6 - x^5 + 18x^3 - 8x - 8$	$x^6 + x^5 - 18x^3 + 8x + 8$
3	D_3	x	2	$6x$	$-\frac{x^4}{2} - \frac{x^3}{2} + x^2 - 2x - 2$	$x^4 + x^3 - 2x^2 - 2x + 4$
3	D_3	x	1	$-x + 3$	$-x^2 + x - 2$	$x^2 + 2$

Thus,

$$\int \frac{x^7 - 24x^4 - 4x^2 + 8x - 8}{x^8 + 6x^6 + 12x^4 + 8x^2} \, dx = \frac{1}{x} + \frac{6x}{(x^2 + 2)^2} + \frac{3 - x}{x^2 + 2} + \int \frac{dx}{x}$$

as in example 2.2.1, but no partial fraction decomposition was required.

Suppose that the denominator D of the integrand has a squarefree factorization of the form $D = D_1 D_2^2 \cdots D_m^m$ where each D_i has positive degree (this is the worst case for the Hermite reduction). In both of the above versions, the number of reduction steps needed is quadratic in m. There is however another variant, due to Mack [48], which requires only $m - 1$ reduction steps, so we call this variant the *linear Hermite reduction.* In addition, Mack's variant does not require either a partial fraction decomposition of f or a squarefree factorization of its denominator (which is computed during the reduction). Let $D = D_1 D_2^2 \cdots D_m^m$ be a squarefree factorization of the denominator of f (we do not need to actually compute it), and recall the notations defined in Definition 1.6.2, namely

$$P^* = \prod_{i=1}^n P_i \quad \text{and} \quad P^{-k} = \prod_{i=1}^n P_i^{\max(0, e_i - k)}$$

for any $P \in K[x] \setminus K$, where $\mathrm{pp}(P) = \prod_{i=1}^n P_i^{e_i}$ is a prime factorization of $\mathrm{pp}(P)$. Since we are working over a field K, we can assume that $D = \mathrm{pp}(D)$. As in the squarefree factorization algorithms, we first compute $D^- = \gcd(D, D')$ and $D^* = D/D^-$. If $\deg(D^-) = 0$, then D is squarefree, otherwise, since $D^- = D^{-*}D^{-2}$ by (1.11), $D^{-\prime} = D^{-2}Y_2$ by Lemma 1.7.2 where Y_2 is given by (1.16), and $D_1 = D^*/D^{-*}$ by Lemma 1.7.1, we get

$$\frac{D^* D^{-\prime}}{D^-} = \frac{D^* D^{-2} Y_2}{D^-} = \frac{D^* D^{-2} Y_2}{D^{-*}D^{-2}} = \frac{D^*}{D^{-*}} Y_2 = D_1 Y_2 \in K[x]. \qquad (2.5)$$

Furthermore, $\gcd(D_1, D^-) = 1$ as a consequence of Lemma 1.7.1, and $\gcd(Y_2, D^{-*}) = 1$ by Lemma 1.7.2, which implies that

$$\gcd\left(\frac{D^* D^{-\prime}}{D^-}, D^{-*}\right) = \gcd(D_1 Y_2, D^{-*}) = 1.$$

Therefore, we can use the extended Euclidean algorithm to find $B, C \in K[x]$ such that

$$A = B\left(-\frac{D^* D^{-'}}{D^-}\right) + CD^{-*}.$$

As previously, dividing both sides by $D = D^* D^- = D_1 D^{-*} D^-$ gives

$$\frac{A}{D} = -\frac{BD^{-'}}{D^{-2}} + \frac{C}{D_1 D^-}$$

so, adding and subtracting B'/D^- to the right hand side, we get

$$\frac{A}{D} = \left(\frac{B'}{D^-} - \frac{BD^{-'}}{D^{-2}}\right) + \frac{C - D_1 B'}{D_1 D^-}$$

and integrating both sides yields

$$\int \frac{A}{D} = \frac{B}{D^-} + \int \frac{C - D_1 B'}{D_1 D^-}.$$

Since $D_1 D^- = (D_1 D_2) D_3^2 \cdots D_m^{m-1}$, the integrand is reduced to one whose denominator has a squarefree factorization with at most $m-1$ different exponents, as opposed to m for the initial integrand. Thus, repeating this process at most $m-1$ times yields an integrand with a squarefree denominator. A further optimization is that the parameters of the next iteration can be computed from the current ones: the new integrand is

$$\frac{C - D_1 B'}{D_1 D^-} = \frac{\overline{A}}{\overline{D}}$$

where

$$\overline{A} = C - D_1 B' = C - \frac{D^*}{D^{-*}} B'$$

and

$$\overline{D} = D_1 D^- = D_1 D_2 D_3^2 \ldots D_m^{m-1}.$$

We then have

$$\overline{D}^* = D_1 D_2 \ldots D_m = D^*$$

which means that D^* remains unchanged throughout the reduction. In addition,

$$\overline{D}^- = D_3 D_4^2 \ldots D_m^{m-2} = D^{-2}$$

which means that D^- is replaced by its deflation at each step throughout the reduction.

We remark that it is possible to perform all the variants of Hermite's reduction over a UFD rather than a field, the result being expressed over its quotient field. In that case, it is necessary for Mack's variant that the denominator D be primitive (this is not necessary for the previous variants).

HermiteReduce(A, D) (* Hermite Reduction – Mack's linear version *)

(* Given a field K and $A, D \in K[x]$ with D nonzero and coprime with A, return $g, h \in K(x)$ such that $\frac{A}{D} = \frac{dg}{dx} + h$ and h has a squarefree denominator. *)

$g \leftarrow 0$

$D^- \leftarrow \gcd\left(D, \frac{dD}{dx}\right)$

$D^* \leftarrow D/D^-$

while $\deg(D^-) > 0$ **do**

$\quad D^{-2} \leftarrow \gcd\left(D^-, \frac{dD^-}{dx}\right)$

$\quad D^{-*} \leftarrow D^-/D^{-2}$

$\quad (B, C) \leftarrow$ **ExtendedEuclidean**$(-D^* \frac{dD^-}{dx}/D^-, D^{-*}, A)$

$\quad A \leftarrow C - \frac{dB}{dx} D^*/D^{-*}$ (* new numerator *)

$\quad g \leftarrow g + B/D^-$

$\quad D^- \leftarrow D^{-2}$ (* $\overline{D^-} = D^{-2}$ *)

return$(g, A/D^*)$

Example 2.2.3. Consider the same integrand as in example 2.2.1. Mack's algorithm proceeds as follows:

1. $g = 0$
2. $D^- = \gcd(D, dD/dx) = x^5 + 4x^3 + 4x$
3. $D^* = D/D^- = x^3 + 2x$
4. *First reduction step:*
   $D^{-2} = \gcd(x^5 + 4x^3 + 4x, 5x^4 + 12x^2 + 4) = x^2 + 2$
5. $D^{-*} = D^-/D^{-2} = x^3 + 2x$
6.

$$(B, C) = \textbf{ExtendedEuclidean}(-5x^2 - 2, x^3 + 2x, A)$$
$$= (8x^2 + 4, x^4 - 2x^2 + 16x + 4)$$

7. $A = x^4 - 2x^2 + 16x + 4 - 16x = x^4 - 2x^2 + 4$
8.

$$g = g + \frac{B}{D^-} = \frac{8x^2 + 4}{x^5 + 4x^3 + 4x}$$

9. $D^- = D^{-2} = x^2 + 2$
10. *Second reduction step:*
    $D^{-2} = \gcd(x^2 + 2, 2x) = 1$
11. $D^{-*} = D^-/S_3 = x^2 + 2$
12. $(B, C) = \textbf{ExtendedEuclidean}(-2x^2, x^2 + 2, x^4 - 2x^2 + 4) = (3, x^2 + 2)$
13. $A = x^2 + 2$
14.

$$g = g + \frac{B}{D^-} = \frac{8x^2 + 4}{x^5 + 4x^3 + 4x} + \frac{3}{x^2 + 2}$$

15. $D^- = D^{-2} = 1$

Thus,

$$\int \frac{x^7 - 24x^4 - 4x^2 + 8x - 8}{x^8 + 6x^6 + 12x^4 + 8x^2} \, dx = \frac{8x^2 + 4}{x^5 + 4x^3 + 4x} + \frac{3}{x^2 + 2} + \int \frac{dx}{x}$$

which is equivalent to the result obtained from the Hermite reduction, but only 2 reduction steps were needed instead of 3.

2.3 The Horowitz–Ostrogradsky Algorithm

Ostrogradsky's algorithm also computes rationally the rational part of the integral, but it reduces to solving systems of linear algebraic equations over K instead of solving polynomial diophantine equations of the form (1.5). Let the integrand f be of the form A/D and suppose in addition that $\deg(A) < \deg(D)$. As previously, let $D = D_1 D_2^2 \cdots D_m^m$ be a squarefree factorization of the denominator of f (the algorithm does not actually compute it). Using the notations P^* and P^- of Definition 1.6.2, we have $D^- = \gcd(D, D')$ and $D^* = D/D^-$. From looking at the steps in the Hermite reduction, it is clear that if $f = g' + h$ where h has a squarefree denominator, then the denominator of g divides D^- and the denominator of h divides D^*, so we can write $g = B/D^-$ and $h = C/D^*$ where $B, C \in K[x]$ are unknown. Furthermore, since $\deg(A) < \deg(D)$, we can assume that $\deg(B) < \deg(D^-)$ and $\deg(C) < \deg(D^*)$. Writing $f = g' + h$, we get

$$\frac{A}{D} = \frac{B'}{D^-} - \frac{BD^{-'}}{(D^-)^2} + \frac{C}{D^*}$$

and multiplying by $D = D^* D^-$,

$$A = B'D^* - B\left(\frac{D^* D^{-'}}{D^-}\right) + CD^- . \tag{2.6}$$

Since $D^- \mid D^* D^{-'}$ by (2.5), the above is a linear equation for B and C with polynomial coefficients. Furthermore, it must always have a solution in $K[x]$, since the Hermite reduction returns such a solution. Since we have bounds on the degrees of B and C, we can replace B and C in this equation by

$$\sum_{i=0}^{\deg(D^-)-1} b_i x^i \quad \text{and} \quad \sum_{j=0}^{\deg(D^*)-1} c_j x^j$$

where the b_i's and c_j's are undetermined constants in K. Equating both sides of (2.6) yields a system of linear algebraic equations for the b_i's and c_j's, and any solution of this system gives B and C, hence g and h.

HorowitzOstrogradsky(A, D) (* Horowitz–Ostrogradsky algorithm *)

(* Given a field K and $A, D \in K[x]$ with $\deg(A) < \deg(D)$, D nonzero and coprime with A, return $g, h \in K(x)$ such that $\frac{A}{D} = \frac{dg}{dx} + h$ and h has a squarefree denominator. *)

$D^- \leftarrow \gcd\left(D, \frac{dD}{dx}\right)$
$D^* \leftarrow D/D^-$
$n \leftarrow \deg(D^-) - 1$
$m \leftarrow \deg(D^*) - 1$
$B \leftarrow \sum_{i=0}^{n} b_i x^i$
$C \leftarrow \sum_{j=0}^{m} c_j x^j$
$H \leftarrow A - B'D^* + BD^*D^{-'}/D^- - CD^-$ (* exact division *)
$(b_0, \ldots, b_n, c_0, \ldots, c_m) \leftarrow$ **solve**(**coefficient**$(H, x^k) = 0, 0 \leq k \leq \deg(D))$
return$(\sum_{i=0}^{n} b_i x^i / D^-, \sum_{j=0}^{m} c_j x^j / D^*)$

Example 2.3.1. Given the same integrand as in example 2.2.1, the Horowitz–Ostrogradsky algorithm proceeds as follow:

1. $D^- = \gcd(D, dD/dx) = x^5 + 4x^3 + 4x$
2. $D^* = D/D^- = x^3 + 2x$
3. $n = \deg(D^-) - 1 = 4, m = \deg(D^*) - 1 = 2$
4.

$$H = A - D^* \left(\sum_{i=0}^{n} b_i x^i\right)' + \left(\sum_{i=0}^{n} b_i x^i\right) \frac{D^* D^{-'}}{D^-} - D^- \sum_{j=0}^{m} c_j x^j$$

$$= (1 - c_2) x^7 + (b_4 - c_1) x^6 + (2b_3 - c_0 - 4c_2) x^5$$
$$+ (3b_2 - 6b_4 - 4c_1 - 24) x^4 + 4(b_1 - b_3 - c_0 - c_2) x^3$$
$$+ (5b_0 - 2b_2 - 4c_1 - 4) x^2 + 4(2 - c_0) x + 2(b_0 - 4)$$

5. The system obtained from equating H to 0 has the unique solution

$$(b_0, b_1, b_2, b_3, b_4, c_0, c_1, c_2) = (4, 6, 8, 3, 0, 2, 0, 1).$$

Thus,

$$\int \frac{x^7 - 24x^4 - 4x^2 + 8x - 8}{x^8 + 6x^6 + 12x^4 + 8x^2} \, dx = \frac{3x^3 + 8x^2 + 6x + 4}{x^5 + 4x^3 + 4x} + \int \frac{x^2 + 2}{x^3 + x} \, dx$$

$$= \frac{3x^3 + 8x^2 + 6x + 4}{x^5 + 4x^3 + 4x} + \int \frac{dx}{x}$$

which corresponds to the result returned by the Hermite reduction.

While the complexity of this algorithm is very good for rational functions [36], it does not generalize as easily as the Hermite reduction to larger classes of functions, so we use the linear Hermite reduction in the general algorithm later.

2.4 The Rothstein–Trager Algorithm

Following the Hermite reduction, we consider now the integration of fractions of the form $f = A/D$ with $\deg(A) < \deg(D)$ and D squarefree. If $\alpha_1, \ldots, \alpha_n \in \overline{K}$ are the zeros of D in $\overline{K}$, the partial fraction decomposition of f must be of the form

$$f = \sum_{i=1}^{n} \frac{a_i}{x - \alpha_i}$$

where $a_1, \ldots, a_n \in \overline{K}$. By analogy with complex-valued functions, a_i is called the *residue of f at $x = \alpha_i$*. From the naive algorithm, we know that

$$\int f = \sum_{i=1}^{n} a_i \log(x - \alpha_i).$$

The problem is thus to compute the residues of f without factoring D. Trager [74] and Rothstein [68] independently proved the following theorem.

Theorem 2.4.1 ([68, 74]). *Let t be an indeterminate over $K(x)$ and A, D be in $K[x]$ with $\deg(D) > 0$, D squarefree and $\gcd(A, D) = 1$. Let*

$$R = \mathrm{resultant}_x(D, A - tD') \quad \in K[t]. \tag{2.7}$$

Then,

(i)   the zeros of R in $\overline{K}$ are exactly the residues of A/D at all the zeros of D in $\overline{K}$,

(ii) let $a \in \overline{K}$ be a zero of R, and $G_a = \gcd(D, A - aD') \in K(a)[x]$. Then, $\deg(G_a) > 0$, and the zeros of G_a in $\overline{K}$ are exactly the zeros of D at which the residue of A/D is equal to a.

(iii) Any field containing an integral of A/D in the form (2.4) also contains all the zeros of R in $\overline{K}$.

Since this theorem is a special case of results that will be proven in Chaps. 4 and 5, we delay its proof until then. A direct consequence of this theorem is that

$$\int \frac{A}{D} = \sum_{a \mid R(a) = 0} a \log(\gcd(D, A - aD')) \tag{2.8}$$

where the sum is taken over the *distinct* roots of R. The Rothstein–Trager algorithm for integrating rational functions with a squarefree denominator and no polynomial part is given by formulas (2.7) and (2.8). With appropriate modifications, the Rothstein–Trager algorithm can, like the Hermite reduction, be applied to rational functions over a UFD rather than a field. Part (iii) of Theorem 2.4.1 shows that the splitting field of R is the minimal algebraic extension of K necessary to express the integral of A/D using logarithms, thereby essentially answering question Q2. Of course it may be possible to express an integral over a smaller constant field using other functions than logarithms, for example $\int dx/(x^2 + 1) = \arctan(x)$, but since an antiderivative of a function can be formally adjoined to a field (Chap. 3), question Q2 is meaningful only when related to specific forms of the integral. If inverse trigonometric functions are allowed in the integral, then Rioboo's algorithm (Sect. 2.8) shows that the integral can be expressed in a field containing the real and imaginary parts of all the roots of R. This result, together with part (iii) of Theorem 2.4.1, provides a complete answer to question Q2 for elementary integrals of rational functions (elementary integrals will be defined formally in Chap. 5). Note that this algorithm requires a gcd computation in $K(a)[x]$ where a, a zero of R, is an algebraic constant over K. A prime factorization $R = u\, R_1^{e_1} \cdots R_m^{e_m}$ over K is thus required, and we must compute the corresponding gcd for a zero of each R_i. Since the answer can be presented as a formal sum over the zeros of each R_i, there is no need to actually compute the splitting field of R.

IntRationalLogPart(A, D) (* Rothstein–Trager algorithm *)

(* Given a field K of characteristic 0 and $A, D \in K[x]$ with $\deg(A) <$ $\deg(D)$, D nonzero, squarefree and coprime with A, return $\int A/D\, dx$. *)

$t \leftarrow$ a new indeterminate over K
$R \leftarrow \text{resultant}_x \left(D, A - t\, \frac{dD}{dx} \right)$
$u\, R_1^{e_1} \cdots R_m^{e_m} \leftarrow \mathbf{factor}(R)$ (* factorization into irreducibles *)
for $i \leftarrow 1$ **to** m **do**
  $a \leftarrow a \mid R_i(a) = 0$
  $G_i \leftarrow \gcd \left(D, A - a\, \frac{dD}{dx} \right)$ (* algebraic gcd computation *)
return$(\sum_{i=1}^{m} \sum_{a \mid R_i(a)=0} a \log(G_i))$

Example 2.4.1. Consider

$$f = \frac{x^4 - 3x^2 + 6}{x^6 - 5x^4 + 5x^2 + 4} \quad \in \mathbb{Q}(x).$$

The denominator of f, $D = x^6 - 5x^4 + 5x^2 + 4$ is squarefree (and in fact irreducible over $\mathbb{Q}$), and the Rothstein–Trager resultant is

$$\text{resultant}_x(x^6 - 5x^4 + 5x^2 + 4, x^4 - 3x^2 + 6 - t\,(6x^5 - 20x^3 + 10x)) = $$
$$45796(4t^2 + 1)^3.$$

Let a be an algebraic number such that $4a^2 + 1 = 0$, we find

$$\begin{aligned} G_a &= \gcd(x^6 - 5x^4 + 5x^2 + 4, x^4 - 3x^2 + 6 - a(6x^5 - 20x^3 + 10x)) \\ &= x^3 + 2ax^2 - 3x - 4a. \end{aligned}$$

Thus,

$$\int \frac{x^4 - 3x^2 + 6}{x^6 - 5x^4 + 5x^2 + 4} = \sum_{a|4a^2+1=0} a \log(x^3 + 2ax^2 - 3x - 4a)$$

$$= \frac{\sqrt{-1}}{2} \log\left(x^3 + x^2\sqrt{-1} - 3x - 2\sqrt{-1}\right)$$

$$- \frac{\sqrt{-1}}{2} \log\left(x^3 - x^2\sqrt{-1} - 3x + 2\sqrt{-1}\right) .$$

2.5 The Lazard–Rioboo–Trager Algorithm

While the Rothstein–Trager algorithm computes in the smallest algebraic extension required to express the integral, Trager[3] and Rioboo [42] independently discovered that the prime factorization of the Rothstein–Trager resultant, and the gcd computations over algebraic extensions of K can be avoided, if one uses the subresultant PRS algorithm to compute the resultant of (2.7). Their algorithm is justified by the following theorem.

Theorem 2.5.1 ([42, 52]). *Let $\overline{K}$ be the algebraic closure of K, t be an indeterminate over $K(x)$, and $A, B, C \in K[x] \setminus \{0\}$ be such that $\gcd(A, C) = \gcd(B, C) = 1$, $\deg(A) < \deg(C)$ and C is squarefree. Let*

$$R = \text{resultant}_x(C, A - tB) \in K[t]$$

and $(R_0, R_1, \ldots, R_k \neq 0, 0, \ldots)$ be the subresultant PRS with respect to x of C and $A - tB$ if $\deg(B) < \deg(C)$, or of $A - tB$ and C if $\deg(B) \geq \deg(C)$. Let $\alpha \in \overline{K}$ be a zero of multiplicity $n > 0$ of R. Then, either

(i) $n = \deg(C)$, in which case

$$\gcd(C, A - \alpha B) = C \in K(\alpha)[x] .$$

(ii) $n < \deg(C)$, in which case there exists a unique $m \geq 1$ such that $\deg_x(R_m) = n$, and

$$\gcd(C, A - \alpha B) = \text{pp}_x(R_m)(\alpha, x) \in K(\alpha)[x]$$

where $\text{pp}_x(R_m)$ is the primitive part of R_m with respect to x.

Proof. Let $R = \text{resultant}_x(C, A - tB)$ and $(R_0, R_1, \ldots, R_k \neq 0, 0, \ldots)$ be the subresultant PRS with respect to x (*i.e.* in $K[t][x]$) of C and $A - tB$ if $\deg(B) < \deg(C)$, or of $A - tB$ and C if $\deg(B) \geq \deg(C)$. Let $q = \deg(C)$, $c \in K^*$ be the leading coefficient of C, and $C = c \prod_{i=1}^{q}(x - \beta_i)$ be the linear

---

[3] Although he did not publish it, Trager programmed this algorithm in his axiom implementation of rational function integration.

factorization of C over $\overline{K}$, where the β_i's are distinct since C is squarefree. By Theorem 1.4.1 we have

$$R = c^p \prod_{i=1}^{q} (A(\beta_i) - t\,B(\beta_i))$$

where $p = \deg_x(A - tB)$. Hence, the leading coefficient of R is $\pm c^p \prod_{i=1}^{q} B(\beta_i)$, which is nonzero since $\gcd(B, C) = 1$. Thus $R \neq 0$, so let $\alpha \in \overline{K}$ be a zero of multiplicity $n > 0$ of R. We note that the trailing monomial of R is $c^p \prod_{i=1}^{q} A(\beta_i)$, which is nonzero since $\gcd(A, C) = 1$, so $\alpha \neq 0$. Since α has multiplicity n, then there is a subset $I_\alpha \subseteq \{1, \ldots, q\}$ of cardinality n such that $A(\beta_i) - \alpha B(\beta_i) = 0$ if and only if $i \in I_\alpha$. Hence $G_\alpha = \prod_{i \in I_\alpha} (x - \beta_i)$ divides $A - \alpha B$ in $\overline{K}[x]$. But $x - \beta_i \nmid A - \alpha B$ for $i \notin I_\alpha$, so G_α is a gcd of C and $A - \alpha B$ in $K(\alpha)[x]$. Hence, $\deg_x(\gcd(C, A - \alpha B)) = n$, which implies that $n \leq \deg(C)$.

(i) If $n = \deg(C)$, then $\gcd(C, A - \alpha B)$ is a divisor of C of degree $\deg(C)$, hence $\gcd(C, A - \alpha B) = C$.

(ii) Suppose that $n < \deg(C)$. Then, $A - \alpha B \neq 0$, otherwise we would have $\gcd(C, A - \alpha B) = \gcd(C, 0) = C$ which has degree greater than n. Let $S_n \in K[t][x]$ be the nth subresultant of C and $A - tB$ with respect to x, $\sigma : K[t] \to K(\alpha)$ be the ring-homomorphism that is the identity on K and maps t to α, and $\overline{\sigma} : K[t][x] \to K(\alpha)[x]$ be given by $\overline{\sigma}\left(\sum a_j x^j\right) = \sum \sigma(a_j) x^j$. Since A, B, C do not involve t, $\overline{\sigma}(C) = C$ and $\overline{\sigma}(A - tB) = A - \alpha B$, hence $\deg(\overline{\sigma}(C)) = q$, and Theorem 1.4.3 implies that $\overline{\sigma}(S_n) = c^r \overline{S_n}$ where r is a nonnegative integer and $\overline{S_n}$ is the nth subresultant of C and $A - \alpha B$. Let $(Q_0, Q_1, \ldots, Q_l \neq 0, 0, \ldots)$ be the subresultant PRS in $K(\alpha)[x]$ of C and $A - \alpha B$ if $\deg(B) < \deg(C)$, or of $A - \alpha B$ and C if $\deg(B) \geq \deg(C)$. By Theorem 1.5.1, Q_l is a gcd of C and $A - \alpha B$, so $\deg_x(Q_l) = n$. Hence, $\overline{\sigma}(S_n)$ is similar to Q_l by Theorem 1.5.2, which implies that $\overline{\sigma}(S_n) \neq 0$ and it is a gcd of C and $A - \alpha B$, and in particular $\deg(\overline{\sigma}(S_n)) = n$. Since $\deg_x(S_n) \leq n$ by definition, and $\deg(\overline{\sigma}(S_n)) \leq \deg_x(S_n)$, we have $\deg_x(S_n) = n$. By Theorem 1.5.2, S_n is similar to some R_m for $m \geq 0$, which implies that $\deg_x(R_m) = n$. Since $\deg(R_0) \geq \deg(C) > n$, we have $m \geq 1$, which implies that m is unique, since $\deg(R_i) > \deg(R_{i+1})$ for $i \geq 1$ in any PRS. Write $\rho_1 S_n = \rho_2 \, \mathrm{pp}_x(R_m)$ with $\rho_1, \rho_2 \in K[t]$ satisfying $\gcd(\rho_1, \rho_2) = 1$. Then, $\sigma(\rho_1)\overline{\sigma}(S_n) = \sigma(\rho_2)\overline{\sigma}(\mathrm{pp}_x(R_m))$. Note that $\overline{\sigma}(S_n) \neq 0$ and $\overline{\sigma}(\mathrm{pp}_x(R_m)) \neq 0$ since $\mathrm{pp}_x(R_m)$ is primitive. In addition we cannot have $\sigma(\rho_1) = \sigma(\rho_2) = 0$ since $\gcd(\rho_1, \rho_2) = 1$. Hence, $\sigma(\rho_1) \neq 0$ and $\sigma(\rho_2) \neq 0$, so $\overline{\sigma}(\mathrm{pp}_x(R_m)) = \mathrm{pp}_x(R_m)(\alpha, x)$ is a gcd of C and $A - \alpha B$.   □

Now let $A, D \in K[x] \setminus \{0\}$ with $\gcd(A, D) = 1$, D squarefree and $\deg(A) < \deg(D)$. Applying Theorem 2.5.1 with $A = A$, $B = D'$ and $C = D$, we get $R = \mathrm{resultant}_x(D, A - tD')$ and for any root α of R of multiplicity $i > 0$, we have $i \leq \deg(D)$ and:

(i) if $i = \deg(D)$, then $\gcd(D, A - \alpha D') = D$,

(ii) if $i < \deg(D)$, then $\gcd(D, A - \alpha D') = \mathrm{pp}_x(R_m)(\alpha, x)$ where $m \geq 1$ is the unique strictly positive integer such that $\deg_x(R_m) = i$.

Thus, it is not necessary to compute the gcd's appearing in the logarithms in (2.8), we can use the various remainders appearing in the subresultant PRS instead. Since $\deg(D') < \deg(D)$, we are in the case where $\deg(B) < \deg(C)$, so we use the subresultant PRS of D and $A - tD'$. As long as the result is returned as a formal sum over the roots of some polynomials, all the calculations are done over K, no algebraic extension is required, and the formal algebraic numbers introduced by the sum are in the smallest possible algebraic extension needed to express the integral.

In practice, we perform a squarefree factorization $R = \prod_{i=1}^n Q_i^i$, so the roots of multiplicity i of R are exactly the roots of Q_i. Evaluating $\mathrm{pp}_x(R_m)$ for $t = \alpha$ where α is a root of Q_i is equivalent to reducing each coefficient with respect to x of $\mathrm{pp}_x(R_m)$ modulo Q_i. We do not really need to compute $\mathrm{pp}_x(R_m)$, it is in fact sufficient to ensure that Q_i and the leading coefficient with respect to x of R_m do not have a nontrivial gcd, which implies then that the remainder by Q_i is nonzero[4].

Since multiplying the argument of any logarithm in (2.8) by an arbitrary constant does not change the derivative, we can make $\mathrm{pp}_x(R_m)(\alpha, x)$ monic in order to simplify the answer, although this requires computing an inverse in $K[\alpha]$, but not computing a gcd in $K[\alpha][x]$. Since the Q_i's are not necessarily irreducible over K, $K[\alpha]$ can have zero-divisors, but the leading coefficients of the $\mathrm{pp}_x(R_m)(\alpha, x)$'s are always invertible in $K[\alpha]$ (Exercise 2.7). This normalization step is optional.

IntRationalLogPart(A, D) (* Lazard–Rioboo–Trager algorithm *)

(* Given a field K of characteristic 0 and $A, D \in K[x]$ with $\deg(A) < \deg(D)$, D nonzero, squarefree and coprime with A, return $\int A/D\, dx$. *)

$t \leftarrow$ a new indeterminate over K
$(R, (R_0, R_1, \ldots, R_k, 0)) \leftarrow$ **SubResultant**$_x \left(D, A - t\frac{dD}{dx}\right)$
$(Q_1, \ldots, Q_n) \leftarrow$ **SquareFree**(R)
for $i \leftarrow 1$ **to** n such that $\deg_t(Q_i) > 0$ **do**
    if $i = \deg(D)$ **then** $S_i \leftarrow D$
    else
        $S_i \leftarrow R_m$ where $\deg_x(R_m) = i$, $\quad 1 \leq m \leq n$
        $(A_1, \ldots, A_q) \leftarrow$ **SquareFree**$(\mathrm{lc}_x(S_i))$
        for $j \leftarrow 1$ **to** q **do** $S_i \leftarrow S_i / \gcd(A_j, Q_i)^j$ (* exact quotient *)
return$(\sum_{i=1}^n \sum_{a|Q_i(a)=0} a \log(S_i(a, x)))$

---

[4] This requirement, which was overlooked in the original publication of Theorem 2.5.1, has been pointed out by Mulders [52] (see Exercise 2.5).

Example 2.5.1. Consider the same integrand as in example 2.4.1. D is square-free, and the subresultant PRS of D and $A - tD'$ is

i	R_i
0	$x^6 - 5x^4 + 5x^2 + 4$
1	$-6tx^5 + x^4 + 20tx^3 - 3x^2 - 10tx + 6$
2	$(-60t^2 + 1)\,x^4 + 2tx^3 + (120t^2 - 3)\,x^2 + 26tx + 144t^2 + 6$
3	$(800t^3 - 14t)\,x^3 - (400t^2 - 7)\,x^2 - (2440t^3 - 32t)\,x + 792t^2 - 16$
4	$(-11200t^4 - 2604t^2 + 49)\,x^2 + 25600t^4 + 5952t^2 - 112$
5	$(-119840t^5 - 59920t^3 - 7490t)\,x - 23968t^4 - 11984t^2 - 1498$
6	$2930944t^6 + 2198208t^4 + 549552t^2 + 45796$

The Rothstein–Trager resultant is $R = R_6$, and its squarefree factorization is

$$R = 2930944t^6 + 2198208t^4 + 549552t^2 + 45796 = 45796\,(4t^2 + 1)^3 = 45796\,Q_3^3$$

and the remainder of degree 3 in x in the PRS is

$$R_3 = (800t^3 - 14t)x^3 - (400t^2 - 7)x^2 - (2440t^3 - 32t)x + 792t^2 - 16.$$

Since

$$\gcd(\mathrm{lc}_x(R_3), Q_3) = \gcd(800t^3 - 14t, 4t^2 + 1) = 1\,,$$

$S_3 = R_3$. Evaluating for t at a root a of $Q_3(a) = 4a^2 + 1 = 0$ we get

$$S_3(a, x) = -214ax^3 + 107x^2 + 642ax - 214$$

so an integral is

$$\int \frac{x^4 - 3x^2 + 6}{x^6 - 5x^4 + 5x^2 + 4} = \sum_{a|4a^2+1=0} a \log(-214ax^3 + 107x^2 + 642ax - 214)\,.$$

Making $S_3(a, x)$ monic we get $S_3(a, x) = -214\,a\,(x^3 + 2ax^2 - 3x - 4a)$ so the integral is then as that in example 2.4.1.

IntegrateRationalFunction(f)   (* Rational function integration *)

(* Given a field K of characteristic 0 and $f \in K(x)$, return $\int f\,dx$. *)

$(g, h) \leftarrow$ **HermiteReduce**$(\mathrm{numerator}(f), \mathrm{denominator}(f))$
$(Q, R) \leftarrow$ **PolyDivide**$(\mathrm{numerator}(h), \mathrm{denominator}(h))$
if $R = 0$ then return$(g + \int Q\,dx)$
return$(g + \int Q\,dx +$ **IntRationalLogPart**$(R, \mathrm{denominator}(h)))$

Example 2.5.2. Let us compute the integral of

$$f = \frac{A}{D} = \frac{36}{x^5 - 2x^4 - 2x^3 + 4x^2 + x - 2} \in \mathbb{Q}(x).$$

1. **HermiteReduce**(A, D) returns $g = (12x + 6)/(x^2 - 1)$ and $h = 12/E$ where $E = x^2 - x - 2$,
2. **PolyDivide**$(12, E)$ returns $Q = 0$ and $R = 12$,
3. **IntRationalLogPart**$(12, E)$ returns $\sum_{\alpha|\alpha^2 - 16 = 0} \alpha \log(x - 1/2 - 3\alpha/8)$,

so we have

$$\int \frac{36}{x^5 - 2x^4 - 2x^3 + 4x^2 + x - 2} \, dx =$$
$$\frac{12x + 6}{x^2 - 1} + \sum_{\alpha|\alpha^2 - 16 = 0} \alpha \log\left(x - \frac{1}{2} - \frac{3\alpha}{8}\right). \quad (2.9)$$

A simpler form for the logarithmic part will be obtained by the algorithm of Sect. 2.7.

2.6 The Czichowski Algorithm

Czichowski has pointed out that the resultant and subresultant computations of the Rothstein–Trager and Lazard–Rioboo–Trager algorithms can be replaced by a Gröbner basis computation in $K[x, z]$. Since Gröbner bases are beyond the scope of this book, we do not present a proof of his theorem here, but give it without proof together with the corresponding algorithm. Interested readers can consult [7, 23] for an introduction to Gröbner bases, and [24] for a proof of the following theorem.

Theorem 2.6.1 ([24]). *Let t be an indeterminate over $K(x)$, A, D be in $K[x]$ with $\deg(D) > 0$, D squarefree and $\gcd(A, D) = 1$, and B be the reduced Gröbner basis with respect to pure lexicographic ordering with $x > t$ of the ideal generated by D and $A - tD'$ in $K[t, x]$. Write $B = \{P_1, \ldots, P_m\}$ where each P_i is in $K[t, x]$ and for each i the highest term of P_{i+1} is greater than the highest term of P_i in the pure lexicographic ordering with $x > t$. Then,*

(i) $\mathrm{lc}_x(\mathrm{pp}_x(P_i)) = 1$ *for* $1 \leq i \leq m$.
(ii) $\mathrm{content}_x(P_{i+1})$ *divides* $\mathrm{content}_x(P_i)$ *in* $K[t]$ *for* $1 \leq i < m$.
(iii)

$$\int \frac{A(x)}{D(x)} \, dx = \sum_{i=1}^{m-1} \sum_{a|Q_i(a) = 0} a \log\left(\mathrm{pp}_x(P_{i+1})(a, x)\right)$$

where $Q_i = \mathrm{content}_x(P_i)/\mathrm{content}_x(P_{i+1}) \in K[t]$.

Note that part (i) implies that this algorithm yields monic polynomials inside the logarithms.

IntRationalLogPart(A, D) (* Czichowski algorithm *)

(* Given a field K of characteristic 0 and $A, D \in K[x]$ with $\deg(A) <$ $\deg(D)$, D nonzero, squarefree and coprime with A, return $\int A/D\,dx$. *)

(* Compute the reduced Gröbner basis *)
$(P_1, \ldots, P_m) \leftarrow$ **ReducedGröbner** $\left(D, A - t\frac{dD}{dx}, \text{pure lex}, x > t\right)$
(* $(P_1, \ldots, P_m)$ must be sorted by increasing highest term *)
for $i \leftarrow 1$ **to** $m - 1$ **do**
$\quad Q_i \leftarrow \text{content}_x(P_i)/\text{content}_x(P_{i+1})$ (* exact quotient *)
$\quad S_i \leftarrow \text{pp}_x(P_{i+1})$
return$(\sum_{i=1}^{m-1} \sum_{a|Q_i(a)=0} a\log(S_i(a, x)))$

Example 2.6.1. Consider the same integrand as in example 2.4.1. The reduced Gröbner basis of

$$(D, A - tdD/dx) = (x^6 - 5x^4 + 5x^2 + 4, -6tx^5 + x^4 + 20tx^3 - 3x^2 - 10tx + 6)$$

w.r.t. pure lexicographic ordering with $x > t$ is

$$\mathcal{B} = \{P_1, P_2\} = \{4t^2 + 1, x^3 + 2tx^2 - 3x - 4t\}.$$

Thus, $Q_1 = 4t^2 + 1/1 = 4t^2 + 1$ and $S_1 = P_2$, which yields the integral

$$\int \frac{x^4 - 3x^2 + 6}{x^6 - 5x^4 + 5x^2 + 4} = \sum_{a|4a^2+1=0} a\log(x^3 + 2ax^2 - 3x - 4a)$$

which is the same integral that was obtained in example 2.4.1.

2.7 Newton–Leibniz–Bernoulli Revisited

We have seen that the difficulty with using formula (2.1), which dates back to Newton and Leibniz, was the computation of the Laurent series expansions at the poles of the integrand. However, Bronstein and Salvy [15] gave a rational algorithm for computing those series. That algorithm can be then used to make the full partial fraction approach rational, so we describe it here. The basic result is that the A_{ij}'s of

$$f = \frac{A_{ie_i}}{(x - \alpha_i)^{e_i}} + \cdots + \frac{A_{i2}}{(x - \alpha_i)^2} + \frac{A_{i1}}{(x - \alpha_i)} + \cdots$$

can be computed as functions of the α_i's without factoring.

Theorem 2.7.1. *Let $A, D \in K[x]$ with D monic and nonzero, $\gcd(A, D) = 1$, and let $D = D_1 D_2^2 \cdots D_n^n$ be a squarefree factorization of D. Then, using only rational operations over K, we can compute $H_{ij} \in K[x]$ for $1 \leq j \leq i \leq n$ such that the partial fraction decomposition of A/D is*

$$\frac{A}{D} = P + \sum_{i=1}^{n} \sum_{\alpha \mid D_i(\alpha)=0} \left(\frac{H_{ii}(\alpha)}{(x-\alpha)^i} + \cdots + \frac{H_{i1}(\alpha)}{x-\alpha} \right)$$

where P is the quotient of A by D.

Proof. We first describe the construction of the H_{ij}'s: let $i \in \{1, \ldots, n\}$, $E_i = D/D_i^i$, and

$$h_i = \frac{A}{u^i E_i} \in K(x)\langle u \rangle$$

where u is a differential variable over $K(x)$ (*i.e.* u and all its derivatives $u', u'', \ldots$ are independent indeterminates over $K(x)$). Each D_i is squarefree and coprime with the other D_k's by the definition of squarefree factorization, so $\gcd(E_i, D_i) = \gcd(D_i', D_i) = 1$. Thus, use the extended Euclidean algorithm to compute $B_i, C_i \in K[x]$ such that

$$B_i E_i \equiv 1 \pmod{D_i} \quad \text{and} \quad C_i D_i' \equiv 1 \pmod{D_i}. \tag{2.10}$$

For $j = 1, \ldots, i$, compute $h_i^{(i-j)}/(i-j)!$ and write it as

$$\frac{h_i^{(i-j)}}{(i-j)!} = \frac{P_{ij}(x, u, u', u'', \ldots, u^{(i-j)})}{u^{2i-j} E_i^{i-j+1}} \tag{2.11}$$

where P_{ij} is a polynomial with coefficients in K. Let then

$$Q_{ij} = P_{ij}\left(x, D_i', \frac{D_i''}{2}, \frac{D_i^{(3)}}{3}, \ldots, \frac{D_i^{(i-j+1)}}{i-j+1} \right) \in K[x]$$

and finally let

$$H_{ij} = Q_{ij} B_i^{i-j+1} C_i^{2i-j} \pmod{D_i} \tag{2.12}$$

where B_i and C_i are given by (2.10).

We prove now that the H_{ij}'s given by (2.12) satisfy the theorem. Let $\overline{K}$ be the algebraic closure of K, $\alpha \in \overline{K}$ be a root of D_i, $D_{i,\alpha} = D_i/(x-\alpha)$, and

$$h_{i,\alpha} = \frac{A}{D_{i,\alpha}^i E_i} = \frac{A}{D}(x-\alpha)^i.$$

Since $h_{i,\alpha}$ is just h_i evaluated at $u = D_{i,\alpha}$, we have

$$\frac{h_{i,\alpha}^{(i-j)}}{(i-j)!} = \frac{P_{ij}\left(x, D_{i,\alpha}, D_{i,\alpha}', D_{i,\alpha}'', \ldots, D_{i,\alpha}^{(i-j)} \right)}{D_{i,\alpha}^{2i-j} E_i^{i-j+1}}$$

where P_{ij} is the same polynomial as in (2.11). We have $D_i = (x - \alpha)D_{i,\alpha}$, so for $k > 0$,

$$D_i^{(k)} = \sum_{j=0}^{k} \binom{k}{j} (x - \alpha)^{(j)} D_{i,\alpha}^{(k-j)} = (x - \alpha) D_{i,\alpha}^{(k)} + k D_{i,\alpha}^{(k-1)}$$

since $(x - \alpha)^{(j)} = 0$ for $j > 1$. Hence,

$$\frac{D_i^{(k)}(\alpha)}{k} = D_{i,\alpha}^{(k-1)}(\alpha)$$

for $k > 0$, which implies that

$$
\begin{aligned}
Q_{ij}(\alpha) &= P_{ij}\left(\alpha, D_i'(\alpha), \frac{D_i''(\alpha)}{2}, \frac{D_i^{(3)}(\alpha)}{3}, \dots, \frac{D_i^{(i-j+1)}(\alpha)}{i-j+1}\right) \\
&= P_{ij}\left(\alpha, D_{i,\alpha}(\alpha), D_{i,\alpha}'(\alpha), D_{i,\alpha}''(\alpha), \dots, D_{i,\alpha}^{(i-j)}(\alpha)\right).
\end{aligned}
$$

In addition, (2.10) implies that $B_i(\alpha) = 1/E_i(\alpha)$ and $C_i(\alpha) = 1/D_i'(\alpha)$, so

$$H_{ij}(\alpha) = Q_{ij}(\alpha) B_i(\alpha)^{i-j+1} C_i(\alpha)^{2i-j} = \frac{h_{i,\alpha}^{(i-j)}(\alpha)}{(i-j)!}.$$

Since $x - \alpha$ does not divide the denominator of $h_{i,\alpha}$, $h_{i,\alpha}$ has a Taylor series at $x = \alpha$, and by Taylor's formula,

$$h_{i,\alpha} = \sum_{k \geq 0} \frac{h_{i,\alpha}^{(k)}(\alpha)}{k!} (x - \alpha)^k$$

so the Laurent series of A/D at $x = \alpha$ is

$$\frac{A}{D} = \frac{h_{i,\alpha}}{(x-\alpha)^i} = \sum_{j=1}^{i} \frac{h_{i,\alpha}^{(i-j)}(\alpha)}{(i-j)!} \frac{1}{(x-\alpha)^j} + \dots = \sum_{j=1}^{i} \frac{H_{ij}(\alpha)}{(x-\alpha)^j} + \dots$$

which proves the theorem.   □

This theorem yields an algorithm for computing Laurent series expansions of rational functions. We can make an additional improvement: it is possible that for given i and j's, $G_{ij} = \gcd(H_{ij}, D_i)$ is nontrivial. This means that for a root α of G_{ij}, the coefficient of $1/(x - \alpha)^j$ in the expansion of A/D is 0. When this happens, we replace D_i by $D_{ij} = D_i/G_{ij}$, and return the partial fraction decomposition of A/D in the form

$$\frac{A}{D} = P + \sum_{i=1}^{n} \sum_{j=1}^{i} \sum_{\alpha \mid D_{ij}(\alpha)=0} \frac{H_{ij}(\alpha)}{(x-\alpha)^j}$$

where all the summands are guaranteed to be nonzero.

LaurentSeries(A, D, F, n)
(* Contribution of F to the full partial fraction decomposition of A/D *)

(* Given a field K of characteristic 0 and $A, D, F \in K[x]$ with D monic, nonzero, coprime with A, and F the factor of multiplicity n in the squarefree factorization of D, return the principal parts of the Laurent series of A/D at all the zeros of F. *)

if $\deg(F) = 0$ **then return** 0
$u \leftarrow$ a differential indeterminate, $\sigma \leftarrow 0$
$E \leftarrow D/F^n$, $h \leftarrow A/(u^n E)$
$(B, G) \leftarrow$ **ExtendedEuclidean**$(E, F, 1)$ (* $BE + GF = 1$ *)
$(C, G) \leftarrow$ **ExtendedEuclidean**$(F', F, 1)$ (* $CF' + GF = 1$ *)
for $j \leftarrow 0$ **to** $n - 1$ **do**
    $P \leftarrow u^{n+j} E^{1+j} h$ (* $P \in K[x, u, u', u'', \ldots, u^{(j)}]$ *)
    $v_j \leftarrow F^{(j+1)}/(j+1)$
    $Q \leftarrow$ **eval**$(P, u \to v_0, \ldots, u^{(j)} \to v_j)$
    $h \leftarrow h'/(j+1)$
    $F^* \leftarrow F/\gcd(F, Q)$
    if $\deg(F^*) > 0$ **then**
      $H \leftarrow QB^{1+j}C^{n+j} \bmod F^*$
      $\sigma \leftarrow \sigma + \sum_{F^*(\alpha)=0} H(\alpha)/(x - \alpha)^{n-j}$
return σ

Example 2.7.1. Consider

$$f = \frac{36}{(x - 2)(x^2 - 1)^2} \in \mathbb{Q}(x).$$

Applying **LaurentSeries** to $A = 36$, $D = (x - 2)(x^2 - 1)^2$, $F = x^2 - 1$ and $n = 2$ we get:

1. $\sigma = 0$, $E = x - 2$, $h = 36/(u^2(x - 2))$,
2. $(x^2 - 1)/3 - (x/3 + 2/3)(x - 2) = 1$, so $B = -(x + 2)/3$,
3. $(x/2)2x - (x^2 - 1) = 1$, so $C = x/2$,
4. $P = u^2(x - 2)h = 36$,
5. $v_0 = F' = 2x$,
6. $Q = $ eval$(36, u \to v_0) = 36$, so $\gcd(x^2 - 1, Q) = 1$,
7. $F^* = x^2 - 1$,
8. $H = -36(x + 2)/3(x/2)^2 \bmod x^2 - 1 = -3x - 6$,
   so $\sigma = \sum_{\alpha^2 - 1 = 0}(-3\alpha - 6)/(x - \alpha)^2$,
9. $h = h' = ((-72x + 144)u' - 36u)/(u^3(x - 2)^2)$,
10. $P = u^3(x - 2)^2 h = (-72x + 144)u' - 36u$,
11. $v_1 = F''/2 = 1$,
12. $Q = $ eval$(P, u \to v_0, u' \to v_1) = -144x + 144$, so $\gcd(x^2 - 1, Q) = x - 1$,
13. $F^* = (x^2 - 1)/(x - 1) = x + 1$,
14. $H = (-144x + 144)((x + 2)/3)^2(x/2)^3 \bmod (x + 1) = -4$, so $\sigma = \sigma + \sum_{\alpha + 1 = 0} -4/(x - \alpha)$

Hence, the sum of the Laurent series of f at the roots of $x^2 - 1 = 0$ is

$$\frac{36}{x^5 - 2x^4 - 2x^3 + 4x^2 + x - 2} = \left(\sum_{\alpha^2 - 1 = 0} \frac{-3\alpha - 6}{(x - \alpha)^2} \right) - \frac{4}{x + 1} + \cdots$$

FullPartialFraction(f)
(* Full partial fraction decomposition of f *)

    (* Given a field K of characteristic 0 and $f \in K(x)$, return the full partial fraction decomposition of f. *)

    $D \leftarrow$ denominator(f)
    $(Q, R) \leftarrow$ **PolyDivide**(numerator(f), D)
    $(D_1, \ldots, D_m) \leftarrow$ **SquareFree**(D)
    return($Q + \sum_{i=1}^{m}$ **LaurentSeries**(R, D, D_i, i))

Example 2.7.2. Applying **FullPartialFraction** to

$$f = \frac{36}{x^5 - 2x^4 - 2x^3 + 4x^2 + x - 2} \in \mathbb{Q}(x)$$

we get:

1. $D = x^5 - 2x^4 - 2x^3 + 4x^2 + x - 2$,
2. $(Q, R) = $ **PolyDivide**($36, D$) $= (0, 36)$,
3. $D_1 D_2^2 = $ **SquareFree**(D) $= (x - 2)(x^2 - 1)^2$,
4. **LaurentSeries**($36, D, x - 2, 1$) returns $4/(x - 2)$,
5. **LaurentSeries**($36, D, x^2 - 1, 2$) returns

$$\left(\sum_{\alpha^2 - 1 = 0} \frac{-3\alpha - 6}{(x - \alpha)^2} \right) - \frac{4}{x + 1}$$

    as seen in example 2.7.1.

Hence, the full partial fraction decomposition of f is

$$\frac{36}{x^5 - 2x^4 - 2x^3 + 4x^2 + x - 2} = \left(\sum_{\alpha^2 - 1 = 0} \frac{-3\alpha - 6}{(x - \alpha)^2} \right) - \frac{4}{x + 1} + \frac{4}{x - 2}.$$

$$(2.13)$$

IntegrateRationalFunction(f)   (* Full partial fraction integration *)

(* Given a field K of characteristic 0 and $f \in K(x)$, return the full partial fraction decomposition of $\int f \, dx$. *)

$$P + \sum_{i=1}^{n} \sum_{j=1}^{i} \sum_{\alpha | D_{ij}(\alpha)=0} \frac{H_{ij}(\alpha)}{(x-\alpha)^j} \leftarrow \textbf{FullPartialFraction}(f)$$

$$\textbf{return} \int P \; + \; \sum_{i=1}^{n} \sum_{\alpha | D_{i1}(\alpha)=0} H_{i1}(\alpha) \log(x - \alpha)$$

$$+ \; \sum_{i=2}^{n} \sum_{j=2}^{i} \sum_{\alpha | D_{ij}(\alpha)=0} \frac{H_{ij}(\alpha)}{(1-j)(x-\alpha)^{j-1}}$$

Example 2.7.3. For the fraction f of example 2.7.2, **FullPartialFraction** returns (2.13), so the integral of f is

$$\int \frac{36}{x^5 - 2x^4 - 2x^3 + 4x^2 + x - 2} \, dx \; =$$

$$4 \log(x-2) - 4\log(x+1) + \sum_{\alpha^2 - 1 = 0} \frac{3\alpha + 6}{x - \alpha}.$$

Compare with the algorithm of the previous sections, which returns (2.9) for the same integrand.

Since the resulting integral is returned in the form (2.4) with the fraction v also expanded into partial fractions with linear denominators, this algorithm is not a better alternative than the other rational algorithms in this chapter, but it makes the partial fraction algorithm factor-free nonetheless. Thus, all the approaches to rational fraction integration can be implemented using only rational operations.

2.8 Rioboo's Algorithm for Real Rational Functions

The algorithms of this chapter give the integral of a rational function in the form (2.4), *i.e.* using logarithms whose arguments may involve algebraic quantities over the ground field. In the case where the ground field K is a subfield of the reals, those algebraic numbers can be complex, so complex arithmetic is necessary for computing a definite integral. This may cause branch problems in the numerical computation, since the arguments to the logarithms may have complex zeros, while the initial integrand has no pole in the path of integration. As a result, a direct application of the fundamental theorem of calculus can yield an incorrect value, since the antiderivative is not necessarily continuous on the interval of integration. For example, consider the definite integral

$$\int_1^2 \frac{x^4 - 3x^2 + 6}{x^6 - 5x^4 + 5x^2 + 4}\, dx\,. \tag{2.14}$$

It is easily checked that the integrand is continuous and positive on the real line, hence that the above integral must be a positive real number. The indefinite integral as computed by the algorithms of this chapter is

$$\int \frac{x^4 - 3x^2 + 6}{x^6 - 5x^4 + 5x^2 + 4}\, dx = \sum_{a|4a^2+1=0} a\, \log(x^3 + 2ax^2 - 3x - 4a)\,. \tag{2.15}$$

The zeros of $4a^2 + 1$ are $a = \pm i/2$ where $i^2 = -1$, so, applying (incorrectly) the fundamental theorem of calculus to the above integral with $x = 2$ and $x = 1$, we would get for the definite integral

$$\left(\frac{i}{2}\log(2 + 2i) - \frac{i}{2}\log(2 - 2i)\right) \quad - \quad \left(\frac{i}{2}\log(-2 - i) - \frac{i}{2}\log(-2 + i)\right)$$

$$= \quad -\frac{5\pi}{4} + \arctan\left(\frac{1}{2}\right) \approx -3.46\,.$$

As explained above, this result cannot be the correct area. Thus, it is preferable to return a real function given a real integrand. We describe in this section an algorithm of Rioboo [57] that expands a result of the form (2.4) into a real function without introducing new real poles, provided that the initial integrand is real. We use the following properties of fields which do not contain $\sqrt{-1}$: if $x^2 + 1$ is irreducible over K, then, for any $P, Q \in K[x]$,

$$P^2 + Q^2 = 0 \Longrightarrow P = Q = 0\,. \tag{2.16}$$

Indeed, if $P^2 + Q^2 = 0$ and $Q \neq 0$, then $(P/Q)^2 = -1$, so $Q \mid P$, which implies that $P/Q \in K$ is a square root of -1, in contradiction with $x^2 + 1$ irreducible over K. We first present the classical algorithm for rewriting complex logarithms as real arc-tangents.

Lemma 2.8.1. *Let $u \in K(x)$ be such that $u^2 \neq -1$. Then,*

$$\sqrt{-1}\,\frac{d}{dx}\log\left(\frac{u + \sqrt{-1}}{u - \sqrt{-1}}\right) = 2\,\frac{d}{dx}\arctan(u)\,. \tag{2.17}$$

Proof. Writing $i = \sqrt{-1}$, an immediate calculation yields

$$i\,\frac{d}{dx}\log\left(\frac{u + i}{u - i}\right) \quad = \quad i\left(\frac{u - i}{u + i}\right)\frac{d}{dx}\left(\frac{u + i}{u - i}\right)$$

$$= \quad i\left(\frac{u - i}{u + i}\right)\frac{du}{dx}\,\frac{(u - i) - (u + i)}{(u - i)^2}$$

$$= \quad 2\,\frac{du/dx}{u^2 + 1} = 2\,\frac{d}{dx}\arctan(u)\,.$$

$\square$

Directly using (2.17) for rewriting complex logarithms with real arc-tangents is possible, but does not eliminate the problem of obtaining discontinuous antiderivatives, since the resulting integral always has singularities at the poles of u, while its derivative does not. For example, applying it to the integral (2.15) gives (we write $f \sim g$ for $df/dx = dg/dx$):

$$\sum_{a|4a^2+1=0} a \log(x^3 + 2ax^2 - 3x - 4a)$$

$$\sim \frac{i}{2} \log\left(x^3 - 3x + i(x^2 - 2)\right) - \frac{i}{2} \log\left(x^3 - 3x - i(x^2 - 2)\right)$$

$$\sim \frac{i}{2} \log\left(\frac{x^3 - 3x + i(x^2 - 2)}{x^3 - 3x - i(x^2 - 2)}\right) \sim \arctan\left(\frac{x^3 - 3x}{x^2 - 2}\right). \quad (2.18)$$

Using this to compute the definite integral (2.14) via the fundamental theorem of calculus we get $\pi/4 - \arctan(2) \approx -0.32$, which is also incorrect. The reason is that (2.17) introduced discontinuities at $\pm\sqrt{2}$, as can be seen from the graph of $\arctan((x^3 - 3x)/(x^2 - 2))$ (Fig. 2.1).

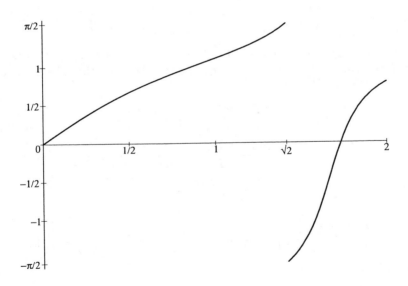

Fig. 2.1. A discontinuous formal integral of $\frac{x^4 - 3x^2 + 6}{x^6 - 5x^4 + 5x^2 + 4}$.

To avoid this problem, Rioboo gave an improvement to Lemma 2.8.1 where the argument of the arc-tangent is a polynomial in x instead of a fraction.

Theorem 2.8.1 ([57]). *Let $A, B \in K[x] \setminus \{0\}$ be such that $A^2 + B^2 \neq 0$. Then,*

$$\frac{d}{dx} \log\left(\frac{A + iB}{A - iB}\right) = \frac{d}{dx} \log\left(\frac{-B + iA}{-B - iA}\right)$$

and, for any $C, D \in K[x]$ such that $BD - AC = \gcd(A, B)$, $C \neq 0$ and $C^2 + D^2 \neq 0$,

$$i \frac{d}{dx} \log\left(\frac{A + iB}{A - iB}\right) = 2 \frac{d}{dx} \arctan\left(\frac{AD + BC}{\gcd(A, B)}\right) + i \frac{d}{dx} \log\left(\frac{D + iC}{D - iC}\right)$$

where $i^2 = -1$.

Proof. We have

$$\frac{A + iB}{A - iB} = \frac{(-i)(-B + iA)}{i(-B - iA)} = -\frac{-B + iA}{-B - iA}$$

so, taking logarithmic derivatives on both sides,

$$\frac{d}{dx} \log\left(\frac{A + iB}{A - iB}\right) = \frac{d}{dx} \log\left(\frac{-B + iA}{-B - iA}\right) .$$

Let now $G = \gcd(A, B)$ and $C, D \in K[x]$ be such that $C \neq 0$, $C^2 + D^2 \neq 0$ and $BD - AC = G$. Write $P = (AD + BC)/G$. We note that $P \in K[x]$ since $G \mid A$ and $G \mid B$. We have

$$\begin{aligned}
\frac{A + iB}{A - iB} &= \left(\frac{D - iC}{D + iC} \frac{A + iB}{A - iB}\right) \frac{D + iC}{D - iC} \\
&= \left(\frac{AD + BC + i(BD - AC)}{AD + BC - i(BD - AC)}\right) \frac{D + iC}{D - iC} \\
&= \left(\frac{P + i}{P - i}\right)\left(\frac{D + iC}{D - iC}\right),
\end{aligned}$$

so

$$i \frac{d}{dx} \log\left(\frac{A + iB}{A - iB}\right) = i \frac{d}{dx} \log\left(\frac{P + i}{P - i}\right) + i \frac{d}{dx} \log\left(\frac{D + iC}{D - iC}\right) .$$

Hence, by Lemma 2.8.1,

$$i \frac{d}{dx} \log\left(\frac{A + iB}{A - iB}\right) = 2 \frac{d}{dx} \arctan(P) + i \frac{d}{dx} \log\left(\frac{D + iC}{D - iC}\right) .$$

$\square$

Note that (2.16) implies that Theorem 2.8.1 is always applicable in fields not containing $\sqrt{-1}$. Furthermore, it provides an algorithm for rewriting

$$f = i \log\left(\frac{A + iB}{A - iB}\right) \tag{2.19}$$

as a sum of arc-tangents with polynomial arguments: since $G = \gcd(A, B)$, we have $\deg(G) \leq \deg(B)$. If $\deg(B) = \deg(G)$, then $B \mid A$, so $G = B$, which implies that $D = 1$ and $C = 0$, hence that $P = (AD + BC)/G = A/B \in K[x]$ and that

$$\frac{df}{dx} = 2\frac{d}{dx}\arctan(P)$$

by Lemma 2.8.1. If $\deg(A) < \deg(B)$, then

$$\frac{df}{dx} = i\frac{d}{dx}\log\left(\frac{-B + iA}{-B - iA}\right)$$

by Theorem 2.8.1, so we can assume that $\deg(A) \geq \deg(B) > \deg(G)$. By the extended Euclidean algorithm, we can find $C, D \in K[x]$ such that $BD - AC = G$ and $\deg(D) < \deg(A)$. In addition, $D \neq 0$ since $\deg(A) > \deg(G)$. This implies that $C \neq 0$, since $\deg(B) > \deg(G)$, hence that $C^2 + D^2 \neq 0$ as we have seen earlier. Hence, by Theorem 2.8.1, the derivatives of f and of

$$2\arctan\left(\frac{AD + BC}{G}\right) + i\log\left(\frac{D + iC}{D - iC}\right)$$

are equal. We can apply the algorithm recursively to the remaining logarithm, and $\max(\deg(C), \deg(D)) < \max(\deg(A), \deg(B))$ guarantees that this process terminates.

LogToAtan(A, B)
(* Rioboo's conversion of complex logarithms to real arc-tangents *)

(* Given a field K of characteristic 0 such that $\sqrt{-1} \notin K$, and $A, B \in K[x]$ with $B \neq 0$, return a sum f of arctangents of polynomials in $K[x]$ such that
$$\frac{df}{dx} = \frac{d}{dx} i \log\left(\frac{A + iB}{A - iB}\right).$$

*)

if $B \mid A$ then return$(2 \arctan(A/B))$
if $\deg(A) < \deg(B)$ then return **LogToAtan**$(-B, A)$
$(D, C, G) \leftarrow$ **ExtendedEuclidean**$(B, -A)$ (* $BD - AC = G$ *)
return$(2 \arctan((AD + BC)/G) +$ **LogToAtan**$(D, C))$

Example 2.8.1. Plugging in $a = \pm i/2$ in (2.15), we get

$$\sum_{a|4a^2+1=0} a \log(x^3 + 2ax^2 - 3x - 4a) = \frac{i}{2} \log\left(\frac{(x^3 - 3x) + i(x^2 - 2)}{(x^3 - 3x) - i(x^2 - 2)}\right).$$

Applying **LogToAtan** to $A = x^3 - 3x$ and $B = x^2 - 2$, we get

A	B	C	D	G	$(AD + BC)/G$
$x^3 - 3x$	$x^2 - 2$	$x/2$	$x^2/2 - 1/2$	1	$x^5/2 - 3x^3/2 + x/2$
$x^2/2 - 1/2$	$x/2$	2	$2x$	1	x^3
$2x$	2				

so the integral is

$$\int \frac{x^4 - 3x^2 + 6}{x^6 - 5x^4 + 5x^2 + 4}\, dx = \arctan\left(\frac{x^5 - 3x^3 + x}{2}\right) + \arctan(x^3) + \arctan(x) \tag{2.20}$$

which differ from (2.18) only by a step function (Fig. 2.2).

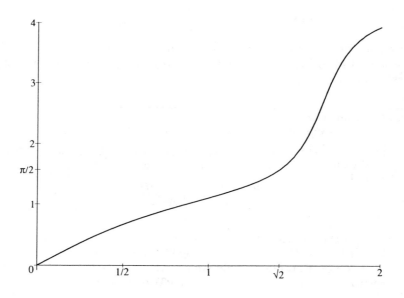

Fig. 2.2. A continuous formal integral of $\frac{x^4 - 3x^2 + 6}{x^6 - 5x^4 + 5x^2 + 4}$.

Using (2.20) to compute the definite integral (2.14), we get the correct answer:

$$\int_1^2 \frac{x^4 - 3x^2 + 6}{x^6 - 5x^4 + 5x^2 + 4}\, dx \;=\; \arctan(5) + \arctan(8) + \arctan(2)$$

$$- \arctan\left(-\frac{1}{2}\right) - \arctan(1) - \arctan(1)$$

$$= \; \arctan(5) + \arctan(8) + \frac{\pi}{2} - \frac{\pi}{4} - \frac{\pi}{4}$$

$$= \; \arctan(5) + \arctan(8) \approx 2.81\,.$$

The above algorithm returns a real primitive given an expression of the form (2.19). But the integration algorithms return a sum of terms of the form

$$\sum_{\alpha\,|\,R(\alpha)=0} \alpha \log(S(\alpha, x)) \tag{2.21}$$

where $R \in K[t]$ is squarefree, and $S \in K[t, x]$. In order to complete Rioboo's algorithm, we need to convert such a sum to one where all the complex logarithms are in the form (2.19). This conversion can be done whenever K is a real field, which is an algebraic generalization[5] of the subfields of the real numbers.

Definition 2.8.1 ([3]). *Let K be a field. K is a real field if -1 cannot be written as a sum of squares of elements of K. K is a* real closed field *if any real algebraic extension of K is isomorphic to K. E is a* real closure *of K if E is a real closed algebraic extension of K.*

Example 2.8.2. $\mathbb{R}$, $\mathbb{Q}$, $\mathbb{Q}(\sqrt{p})$ for any prime number $p \geq 2$, and $\mathbb{Q}(a)$ where a is an indeterminate over $\mathbb{Q}$, are all real fields. $\mathbb{Q}(\sqrt{-2})$ is not a real field since $-1 = 1^2 + \sqrt{-2}^2$ is a sum of squares. If K has characteristic $p > 0$, then $-1 = \sum_{i=1}^{p-1} 1^2$, so any real field must have characteristic 0.

Theorem 2.8.2 ([40], Chap. XI, §2). *Any real field has a real closure.*

This theorem is also proven in [77], §11.6 but for countable real fields only. Note that the real closure of K is not unique, even up to isomorphism, unless K is already ordered.

Theorem 2.8.3 ([40], Chap. XI, §2, [77], §11.5). *Let L be a real closed field. Then,*

(i) L has a unique ordering, given by: $x > 0 \iff x = y^2$ for some $y \in L$.
(ii) $L(\sqrt{-1})$ is the algebraic closure of L.

---

[5] The reader wishing to avoid this extra algebraic machinery can skip this definition and the following theorems, and think of K in the rest of this section as a given subfield of the real numbers, with real closure $\mathbb{K} = \mathbb{R}$.

Let K be a real field for the rest of this section, and let $\mathbb{K}$ be a real closure of K, and $\overline{K} = \mathbb{K}(i)$ where $i^2 = -1$. With a slight abuse of language, we say that $\alpha \in \overline{K}$ is "real" if $\alpha \in \mathbb{K}$. Let f be a sum of the form (2.21) where $R = \sum_j r_j x^j \in K[x]$, $S = \sum_{j,k} s_{jk} t^j x^k \in K[t,x]$, and let u, v be indeterminates over $K(x)$. We first separate the sum (2.21) into one over the real roots of R and one over the other roots:

$$f = g + \sum_{\alpha \notin \mathbb{K}, R(\alpha) = 0} \alpha \log(S(\alpha, x)). \qquad (2.22)$$

where

$$g = \sum_{\alpha \in \mathbb{K}, R(\alpha) = 0} \alpha \log(S(\alpha, x))$$

is a real function. We then compute $P, Q \in K[u, v]$ such that

$$R(u + iv) = \sum_j r_j (u + iv)^j = P(u, v) + i Q(u, v), \qquad (2.23)$$

and $A, B \in K[u, v, x]$ such that

$$S(u + iv, x) = \sum_{j,k} s_{jk} (u + iv)^j x^k = A(u, v, x) + i B(u, v, x). \qquad (2.24)$$

Since $\overline{K} = \mathbb{K}(i)$, it is a vector space of dimension 2 over $\mathbb{K}$ with basis $(1, i)$, so for $\alpha \in \overline{K}$, $R(\alpha) = 0$ if and only if $P(a, b) = Q(a, b) = 0$ where $\alpha = a + ib$. Furthermore, $\alpha \notin \mathbb{K}$ if and only if $b \neq 0$. Hence, we can rewrite (2.22) as

$$f = g + \sum_{\substack{a, b \in \mathbb{K}, b \neq 0 \\ P(a,b) = Q(a,b) = 0}} (a + ib) \log(S(a + ib, x)) \qquad (2.25)$$

Let σ be the field-automorphism of $\overline{K}$ such that $\sigma(i) = -i$ and $\sigma(z) = z$ for any $z \in \mathbb{K}$, and define $\overline{\sigma} : \overline{K}[x] \to \overline{K}[x]$ by $\overline{\sigma}(\sum a_j x^j) = \sum \sigma(a_j) x^j$. Let $a, b \in \mathbb{K}$. Applying $\overline{\sigma}$ to (2.24) we get

$$\begin{aligned} A(a, b, x) - i B(a, b, x) &= \overline{\sigma}(A(a, b, x) + i B(a, b, x)) = \overline{\sigma}(S(a + ib, x)) \\ &= S(\sigma(a + ib), x) = S(a - ib, x). \end{aligned}$$

Applying σ to (2.23) we get

$$\begin{aligned} P(a, b) - i Q(a, b) &= \sigma(P(a, b) + i Q(a, b)) \\ &= \sigma(R(a + ib)) = R(\sigma(a + ib)) = R(a - ib) \end{aligned}$$

which implies that $R(a + ib) = 0$ if and only if $R(a - ib) = 0$. Hence, for any pair (a, b) appearing in the sum (2.25) with $b \neq 0$, the pair $(a, -b)$ must appear also, and is a different pair, so we can rewrite (2.25) as

$$f \;=\; g + \sum_{\substack{a,b\in\mathbb{K},b>0 \\ P(a,b)=Q(a,b)=0}} \{(a+i\,b)\log(S(a+i\,b,x)) + (a-i\,b)\log(S(a-i\,b,x))\}$$

$$=\; g +$$

$$\sum_{\substack{a,b\in\mathbb{K},b>0 \\ P(a,b)=Q(a,b)=0}} \{a\;(\log(A(a,b,x)+i\,B(a,b,x)) + \log(A(a,b,x)-i\,B(a,b,x)))$$

$$+ i\,b\;(\log(A(a,b,x)+i\,B(a,b,x)) - \log(A(a,b,x)-i\,B(a,b,x)))\}\;.$$

Hence,

$$f = g + h + \sum_{\substack{a,b\in\mathbb{K},b>0 \\ P(a,b)=Q(a,b)=0}} i\,b\,\log\left(\frac{A(a,b,x)+i\,B(a,b,x)}{A(a,b,x)-i\,B(a,b,x)}\right) \qquad (2.26)$$

where

$$h = \sum_{\substack{a,b\in\mathbb{K},b>0 \\ P(a,b)=Q(a,b)=0}} a\,\log\left(A(a,b,x)^2 + B(a,b,x)^2\right)$$

is a real function. Since the remaining nonreal summands in (2.26) are all of the form (2.19), we can use Theorem 2.8.1 and its associated algorithm to convert them to real functions. Note that, since converting (2.19) to real functions requires computing the gcd of A and B, we have, in theory, to use algorithm **LogToAtan** over an algebraic extension $K(a,b)$ of K where $P(a,b) = Q(a,b) = 0$, which means that we have to solve this nonlinear algebraic system. However, the following theorem of Rioboo shows that, when the complex logarithms to expand arise from the integration of a real rational function, it is not necessary to solve this system.

Theorem 2.8.4 ([57]). *Let K be a real field, $\mathbb{K}$ be a real closure of K, $C, D \in K[x]$ with $\deg(D) > 0$, $\deg(D) > \deg(C)$, D squarefree and $\gcd(C, D) = 1$. Suppose that the R and S of (2.21) are produced by the Rothstein–Trager or Lazard–Rioboo–Trager algorithm applied to C/D, and let P, Q be given by (2.23) and A, B by (2.24). If $a, b \in \mathbb{K}$ satisfy $P(a,b) = Q(a,b) = 0$ and $b \neq 0$, then $\gcd(A(a,b,x), B(a,b,x)) = 1$ in $K(a,b)[x]$.*

Proof. Let $a, b \in \mathbb{K}$ be such that $P(a,b) = Q(a,b) = 0$ and $b \neq 0$ where P and Q are given by (2.23). Then, $R(a + ib) = 0$ where $i^2 = -1$ and $R = P + iQ$ is a squarefree factor of the Rothstein–Trager resultant of $C - tD'$ and D. Furthermore, since A and B are given by (2.24), $S(a + ib, x) = A(a,b,x) + iB(a,b,x)$ is a gcd in $K(a+ib)[x]$ of $C - (a + ib)D'$ and D, so there exist E and F in $K(a+ib)[x]$ such that

$$C(x) - (a+ib)D'(x) = E(a+ib,x)S(a+ib,x)$$

and
$$D(x) = F(a + ib, x)S(a + ib, x).$$

Writing $E = E_1 + iE_2$ and $F = F_1 + iF_2$ where $E_1, E_2, F_1, F_2 \in K(a, b)[x]$, we get

$$C(x) - (a + ib)D'(x) = (E_1(a, b, x) + iE_2(a, b, x))(A(a, b, x) + iB(a, b, x)) \tag{2.27}$$

and

$$D(x) = (F_1(a, b, x) + iF_2(a, b, x))(A(a, b, x) + iB(a, b, x)). \tag{2.28}$$

Taking the imaginary part of (2.27) and the real part of (2.28) we get

$$-bD'(x) = E_1(a, b, x)B(a, b, x) + E_2(a, b, x)A(a, b, x) \tag{2.29}$$

and

$$D(x) = F_1(a, b, x)A(a, b, x) - F_2(a, b, x)B(a, b, x). \tag{2.30}$$

Since D is squarefree, $\gcd(D, D') = 1$, so there exist $G_1, G_2 \in K[x]$ such that $G_1 D + G_2 D' = 1$. Multiplying (2.30) by bG_1, (2.29) by $-G_2$ and adding both yields

$$
\begin{aligned}
b &= (G_1 D + G_2 D')b \\
&= bG_1 F_1(a, b, x)A(a, b, x) - bG_1 F_2(a, b, x)B(a, b, x) \\
&\quad -G_2 E_1(a, b, x)B(a, b, x) - G_2 E_2(a, b, x)A(a, b, x) \\
&= (bG_1 F_1(a, b, x) - G_2 E_2(a, b, x))A(a, b, x) \\
&\quad -(G_2 E_1(a, b, x) + bG_1 F_2(a, b, x))B(a, b, x)
\end{aligned}
$$

which is a linear combination of $A(a, b, x)$ and $B(a, b, x)$ with coefficients in $K(a, b)[x]$. Since $b \neq 0$, this implies that $\gcd(A(a, b, x), B(a, b, x)) = 1$ in $K(a, b)[x]$. $\qquad\square$

As a consequence, we can perform Rioboo's conversion to arc-tangents generically, *i.e.* expand once

$$i \log \left(\frac{A(u, v, x) + i\, B(u, v, x)}{A(u, v, x) - i\, B(u, v, x)} \right)$$

where u and v are independent indeterminates, obtaining a real function $\phi(u, v, x)$. We can then rewrite (2.26) as

$$f = g + h + \sum_{\substack{a, b \in \mathbb{K}, b > 0 \\ P(a,b) = Q(a,b) = 0}} b\, \phi(a, b, x)$$

where Theorem 2.8.4 guarantees that $\phi(u, v, x)$ specializes well, *i.e.* that no division by 0 occurs when we replace u and v by the various solutions a and

$b > 0$ in $\mathbb{K}$ of $P(u,v) = Q(u,v) = 0$. By presenting the answer in terms of formal sums, we do not need to actually solve this system, or to introduce any algebraic number. In practice, whenever the real roots of $P(u,v) = Q(u,v) = 0$ can be computed efficiently (for example if they are all rational numbers), then it can be more efficient to first compute the roots, and then call **LogToAtan**, rather than perform the reduction with generic parameters.

LogToReal(R, S)
(* Rioboo's conversion of sums of complex logarithms to real functions *)

(* Given a real field K, $R \in K[t]$ and $S \in K[t, x]$, return a real function f such that

$$\frac{df}{dx} = \frac{d}{dx} \sum_{\alpha \mid R(\alpha) = 0} \alpha \log(S(\alpha, x)).$$

*)

write $R(u + iv)$ as $P(u, v) + i\,Q(u, v)$
write $S(u + iv, x)$ as $A(u, v, x) + i\,B(u, v, x)$
return

$$\sum_{\substack{a, b \in \mathbf{K}, b > 0 \\ P(a,b) = Q(a,b) = 0}} a \log\left(A(a, b, x)^2 + B(a, b, x)^2\right) + b\,\mathbf{LogToAtan}(A, B)(a, b, x)$$

$$+ \sum_{a \in \mathbf{K}, R(a) = 0} a \log(S(a, x)).$$

Example 2.8.3. Applying **LogToReal** to the integral (2.15), we have $R(t) = 4t^2 + 1 \in \mathbb{Q}[t]$, $S(t, x) = x^3 + 2tx^2 - 3x - 4t \in \mathbb{Q}[t, x]$, and

1. $R(u + iv) = 4(u + iv)^2 + 1 = 4u^2 - 4v^2 + 1 + 8i\,u\,v$, so $P = 4u^2 - 4v^2 + 1$ and $Q = 8uv$,
2. $S(u + iv, x) = x^3 + 2(u + iv)x^2 - 3x - 4(u + iv) = x^3 + 2ux^2 - 3x - 4u + i(2vx^2 - 4v)$, so $A = x^3 + 2ux^2 - 3x - 4u$ and $B = 2vx^2 - 4v$,
3. $H = \text{resultant}_v(p, q) = 256u^4 + 64u^2$ whose only real root is 0. $P(0, v) = 1 - 4v^2$, whose only real positive root is $1/2$,
4. $A(0, 1/2, x) = x^3 - 3x$, $B(0, 1/2, x) = x^2 - 2$, and **LogToAtan**$(x^3 - 3x, x^2 - 2)$ returns

$$2 \arctan\left(\frac{x^5 - 3x^3 + x}{2}\right) + 2 \arctan(x^3) + 2 \arctan(x)$$

as seen in example 2.8.1, so multiplying by $b = 1/2$ we get the same integral as in example 2.8.1.

Instead of solving the system $P(u,v) = Q(u,v) = 0$ in step 3, we can call **LogToAtan**$(x^3 + 2ux^2 - 3x - 4u, 2vx^2 - 4v)$, which returns

$$\phi(u,v,x) \;=\; 2\arctan\left(\frac{x}{2v} + \frac{u}{v}\right)$$

$$+ \; 2\arctan\left(\frac{x^3}{2v} + \frac{2u}{v}x^2 + \frac{4u^2 + 4v^2 - 1}{2v}x - \frac{u}{v}\right)$$

$$+ \; 2\arctan\left(\frac{x^5}{4v} + \frac{u}{v}x^4 + \frac{u^2 + v^2 - 1}{v}x^3 - \frac{3u}{v}x^2 - \frac{8u^2 + 8v^2 - 3}{4v}x + \frac{u}{v}\right)$$

and the integral would be returned formally as

$$\int \frac{x^4 - 3x^2 + 6}{x^6 - 5x^4 + 5x^2 + 4}\,dx = \sum_{\substack{a,b \in \mathbb{R}, b > 0 \\ 4a^2 - 4b^2 + 1 = 8ab = 0}} b\,\phi(a,b,x)$$

which is a real function. Plugging in $a = 0$ and $b = 1/2$ in this result, we get the same integral as previously.

IntegrateRealRationalFunction(f) (* Real rational function integration *)

(* Given a real field K and $f \in K(x)$, return a real function g such that $dg/dx = f$. *)

$$v + \sum_{i=1}^{m}\sum_{a\,|\,R_i(a)=0} a\,\log(S_i(a,x)) \leftarrow \textbf{IntegrateRationalFunction}(f)$$

return$(v + \sum_{i=1}^{m} \textbf{LogToReal}(R_i, S_i))$

2.9 In–Field Integration

We outline in this section minor variants of the integration algorithms that are used for deciding whether a rational function is either a

– derivative of a rational function,
– logarithmic derivative of a rational function.

Those problems are important because they arise from the integration of more general functions. Furthermore, deciding whether a rational function is a logarithmic derivative is useful when solving linear ordinary differential equations with rational function coefficients[6].

---

[6] The differential Galois group of $y^{(n)} + a_{n-1}(x)y^{(n-1)} + \ldots$ is unimodular if and only if a_{n-1} is the logarithmic derivative of a rational function.

Recognizing Derivatives

The first problem is, given $f \in K(x)$, to determine whether there exists $u \in K(x)$ such that $du/dx = f$. To compute such an u, we simply apply either the Horowitz–Ostrogradsky algorithm, or any variant of the Hermite reduction, to f, obtaining $g \in K(x)$ and $A, D \in K[x]$ such that D is squarefree and $f = dg/dx + A/D$. At that point, $f = du/dx$ for $u \in K(x)$ if and only if $D \mid A$, in which case $u = g + \int (A/D)dx$.

There are also a couple of criterions that can determine whether f is the derivative of a rational function without computing an integral of f:

— Compute the squarefree factorization $D_1 D_2^2 \ldots D_n^n$ of the denominator of f, and for each i the polynomial $H_{i1} \in K[x]$ of Theorem 2.7.1, using the **LaurentSeries** algorithm. Write $D_i = G_i E_i$ where $G_i = \gcd(H_{i1}, D_i)$ and $\gcd(E_i, H_{i1}) = 1$. Since the residues of f at the roots of G_i are all 0, and the residue of f at a root α of E_i is $H_{i1}(\alpha) \neq 0$, f is the derivative of a rational function if and only if $E_i = 1$ for each i, which is equivalent to $D_i \mid H_{i1}$ for each i.
— Compute the squarefree factorization $D_1 D_2^2 \ldots D_n^n$ of the denominator of f, and write f as a sum

$$f = \sum_{i=1}^{n} \frac{A_i}{D_i^i} .$$

If f is the derivative of a rational function, then $D_1 \mid A_1$, since the residues of f at the roots of D_1 would be nonzero otherwise. If $D_1 \mid A_1$, then f is the derivative of a rational function if and only if each A_i/D_i^i is the derivative of a rational function for $i > 1$, and we can use Mařík's criterion [49], which states that A/D^m is the derivative of a rational function for $m > 1$ if and only if D divides the Wronskian of $dD/dx, d(D^2)/dx, \ldots, d(D^{m-1})/dx$ and A.

While those criterions are not practical alternatives to either the Hermite reduction or the Horowitz–Ostrogradsky algorithm, they are of theoretical interest. No generalization of those criterions is known for more general functions, which makes the problem of recognizing derivatives more difficult in general (see Sect. 5.12).

Recognizing Logarithmic Derivatives

The second problem is, given $f \in K(x)$, to determine whether there exists $u \in K(x)^*$ such that $du/dx = uf$. It will be proven later (see Exercise 4.2) that f is the logarithmic derivative of a rational function if and only if f can be written as $f = A/D$ where D is squarefree, $\gcd(A, D) = 1$, and all the roots of the Rothstein–Trager resultant are integers. In that case, any of the Rothstein–Trager, Lazard–Rioboo–Trager or Czichowski algorithm produces $u \in K(x)$ such that $du/dx = uf$.

Exercises

Exercise 2.1. Compute

$$\int \frac{x^5 - x^4 + 4x^3 + x^2 - x + 5}{x^4 - 2x^3 + 5x^2 - 4x + 4} \, dx$$

using the Hermite reduction and the Rothstein-Trager algorithm.

Exercise 2.2. Compute

$$\int \frac{8x^9 + x^8 - 12x^7 - 4x^6 - 26x^5 - 6x^4 + 30x^3 + 23x^2 - 2x - 7}{x^{10} - 2x^8 - 2x^7 - 4x^6 + 7x^4 + 10x^3 + 3x^2 - 4x - 2} \, dx$$

using the Lazard-Rioboo-Trager algorithm.

Exercise 2.3.

a) Compute

$$\int \frac{72x^7 + 256x^6 - 192x^5 - 1280x^4 - 312x^3 + 1440x^2 + 576x - 96}{9x^8 + 36x^7 - 32x^6 - 252x^5 - 78x^4 + 468x^3 + 288x^2 - 108x + 9} \, dx$$

using the Rothstein-Trager or the Lazard-Rioboo-Trager algorithm. With that integral compute the symbolic definite integral for $-2 \le x \le -2/3$ and compare it with the result obtained by direct numerical integration.

b) Apply the Rioboo algorithm to the above result and compute again the definite integral for $-2 \le x \le -2/3$.

Exercise 2.4.

a) Compute

$$\int \frac{dx}{1 + x^4} .$$

b) Find a closed form for $\int dx/(1 + x^n)$ for $n \in \mathbb{N}$.

Exercise 2.5 ([52]). Compute

$$\int \frac{x^4 + x^3 + x^2 + x + 1}{x^5 + x^4 + 2x^3 + 2x^2 - 2 + 4\sqrt{-1 + \sqrt{3}}} \, dx$$

using the Lazard-Rioboo-Trager algorithm. What happens if the subresultants are not made primitive before evaluating them?

Exercise 2.6. Write procedures for the Hermite Reduction, the Lazard–Rioboo–Trager algorithm and the Rioboo algorithm using your favourite programming language or computer algebra system.

Exercise 2.7. Modify the Lazard–Rioboo–Trager algorithm so that in the result, the polynomials inside the logarithms are monic in x. Note that the polynomials $Q_i(t)$ indexing the sums of logarithms are not necessarily irreducible (show first that the leading coefficients of the polynomials inside the logarithms must be units in $K[t]/(Q_i(t))$).

3. Differential Fields

We develop in this chapter the algebraic machinery in which the integration algorithms can be presented and proved correct. The main idea, which originates from J. F. Ritt [63], is to define the notion of derivation in a pure algebraic setting (*i.e.* without using the notions of "function", "limit", and "tangent line" from analysis) and to study the properties of such formal derivations on arbitrary objects. This way, we can later translate an integration problem to solving an equation in some algebraic structure, which can be done using algebraic algorithms. Since an arbitrary transcendental function can be seen as a univariate rational function over a field with an arbitrary derivation, we first need to study the general properties of derivations over rings and fields. This will allow us to generalize the rational function integration algorithms to large classes of transcendental functions (Chap. 5).

3.1 Derivations

Although the integration algorithm we present in later chapters works only over differential fields of characteristic 0, the rings and fields in the first two sections of this chapter are of arbitrary characteristic. Given a map in any ring, we call it a derivation if it satisfies the usual rules for differentiating sums and products.

Definition 3.1.1. *Let R be a ring (resp. field). A* derivation *on R is a map $D : R \to R$ such that for any $a, b \in R$:*

(i) $D(a + b) = Da + Db$,
(ii) $D(ab) = aDb + bDa$.

The pair (R, D) is called a differential ring *(resp. field). The set*

$$\text{Const}_D(R) = \{a \in R \text{ such that } Da = 0\}$$

is called the constant subring *(resp. subfield) of R with respect to D. A subset $S \subseteq R$ is called a* differential subring *(resp. subfield) of R if S is a subring (resp. subfield) of R and $DS \subseteq S$.*

When there is no ambiguity about the derivation in use, we often say that R is a differential ring (field) rather than the pair (R, D). We first show that the usual algebraic properties of the derivations of analysis are consequences of the above definition.

Theorem 3.1.1. *Let (R, D) be a differential ring (resp. field). Then,*

(i) $D(ca) = cDa$ for any $a \in R$ and $c \in \mathrm{Const}_D(R)$.

(ii) If R is a field, then
$$D\frac{a}{b} = \frac{bDa - aDb}{b^2}$$
for any $a, b \in R$, $b \neq 0$.

(iii) $\mathrm{Const}_D(R)$ is a differential subring (resp. subfield) of R.

(iv) $Da^n = na^{n-1}Da$ for any $a \in R \setminus \{0\}$ and any integer $n > 0$ (resp. any integer n).

(v)   Logarithmic derivative identity:
$$\frac{D(u_1^{e_1} \ldots u_n^{e_n})}{u_1^{e_1} \ldots u_n^{e_n}} = e_1\frac{Du_1}{u_1} + \ldots + e_n\frac{Du_n}{u_n}$$
for any $u_1, \ldots, u_n \in R^*$ and any integers $e_1, \ldots, e_n$.

(vi)
$$DP(u_1, \ldots, u_n) = \sum_{i=1}^{n} \frac{\partial P}{\partial X_i}(u_1, \ldots, u_n)\, Du_i$$
for any $u_1, \ldots, u_n$ in R and polynomial P with coefficients in $\mathrm{Const}_D(R)$.

Proof. (i) Let $a \in R$ and $c \in \mathrm{Const}_D(R)$. Then, $D(ca) = cDa + aDc = cDa$ since $Dc = 0$.

(ii) Suppose that R is a field, and let $a, b \in R$ with $b \neq 0$, and $c = a/b$. Then, $a = bc$, so by property (ii) of Definition 3.1.1,
$$Da = D(bc) = bDc + cDb = bD\frac{a}{b} + \frac{a}{b}Db.$$

Hence,
$$D\frac{a}{b} = \frac{1}{b}\left(Da - \frac{a}{b}Db\right) = \frac{bDa - aDb}{b^2}.$$

(iii) Let $C = \mathrm{Const}_D(R)$. From property (i) of Definition 3.1.1, $D(0) = D(0 + 0) = D(0) + D(0)$, so $0 \in C$. From property (ii) of Definition 3.1.1, $D(1) = D(1 \times 1) = D(1) + D(1)$, so $1 \in C$. Since $DC = \{0\}$, this implies that $DC \subset C$. Let $a \in R$. Then, $Da + D(-a) = D(a + (-a)) = D(0) = 0$, so $D(-a) = -Da$. Let $c, d \in C$. Then, $D(c - d) = Dc + D(-d) = Dc - Dd = 0 - 0 = 0$, so $c - d \in C$. Also, $D(cd) = cDd + dDc = 0 + 0 = 0$, so $cd \in C$, hence C is a differential subring of R. Suppose that R is a field and that $d \neq 0$. Then, $D(1/d) = -Dd/d^2 = 0$, so $1/d \in C$, which implies that C is a differential subfield of R.

(iv) Let $a \in R \setminus \{0\}$. For $n = 1$, $Da^1 = Da = 1a^0 Da$. Suppose that $Da^n = na^{n-1} Da$ for some $n \geq 1$. Then,

$$Da^{n+1} = D(a^n a) = a^n Da + a Da^n = a^n Da + a(na^{n-1} Da) = (n+1)a^n Da$$

so (iv) holds for any integer $n \geq 1$. Suppose that R is a field. Then, $Da^0 = D(1) = 0 = 0a^{-1} Da$, so (iv) holds for $n = 0$. For $n < 0$ we have

$$Da^n = D\frac{1}{a^{-n}} = -\frac{Da^{-n}}{a^{-2n}} = -\frac{-na^{-n-1} Da}{a^{-2n}} = na^{n-1} Da.$$

(v) is left as Exercise 3.1 at the end of this chapter.
(vi) Let $X_1, \ldots, X_n$ be indeterminates, $P \in \mathrm{Const}_D(R)[X_1, \ldots, X_n]$ and write

$$P = \sum_{(e)=(e_1,\ldots,e_n)} c_{(e)} \prod_{i=1}^{n} X_i^{e_i}$$

where $a_{(e)} \in C = \mathrm{Const}_D(R)$. Using property (ii) of Definition 3.1.1 and the fact that D is C-linear, we get

$$
\begin{aligned}
DP(u_1, \ldots, u_n) &= D\left(\sum_{(e)=(e_1,\ldots,e_n)} c_{(e)} \prod_{i=1}^{n} u_i^{e_i} \right) \\
&= \sum_{(e)=(e_1,\ldots,e_n)} c_{(e)} \sum_{i=1}^{n} e_i u_i^{e_i-1} Du_i \prod_{\substack{j=1 \\ j \neq i}}^{n} u_j^{e_j} \\
&= \sum_{i=1}^{n} \frac{\partial P}{\partial X_i}(u_1, \ldots, u_n) Du_i .
\end{aligned}
$$

$\square$

In general, a ring can have more than one derivation defined on it. For example, $\mathbb{Q}[X, Y]$ has at least the derivations $0, d/dX$ and d/dY. But it has a lot more derivations, for instance $D = d/dX + d/dY$. In fact, any linear combination of derivations with coefficients in R is again a derivation on R.

Lemma 3.1.1. *The set $\Omega(R)$ of all the derivations on R is a left-module over R.*

Proof. Let $D_1, D_2 \in \Omega(R)$ and $c \in R$. Let $D = cD_1 + D_2$, i.e. $D : R \to R$ is defined by $Da = cD_1 a + D_2 a$ for any $a \in R$. Let $a, b \in R$. Then,

$$D(a + b) = cD_1(a + b) + D_2(a + b) = cD_1 a + cD_1 b + D_2 a + D_2 b = Da + Db,$$

and

$$\begin{aligned} D(ab) = cD_1(ab) + D_2(ab) \quad &= \quad caD_1b + cbD_1a + aD_2b + bD_2a \\ &= \quad a(cD_1b + D_2b) + b(cD_1a + D_2a) \\ &= \quad aDb + bDa \end{aligned}$$

so $D \in \Omega(R)$. Since the zero-map on R (which maps every element of R to 0) is a derivation on R, this implies that $\Omega(R)$ is a left-module over R. $\quad\square$

Definition 3.1.2. Let (R, D) be a differential ring. An ideal I of R is a differential ideal if $DI \subseteq I$.

Lemma 3.1.2. Let (R, D) be a differential ring, I be a differential ideal of R, and $\pi : R \to R/I$ be the canonical projection. Then, D induces a derivation D^* on R/I such that $D^* \circ \pi = \pi \circ D$.

Proof. Define D^* as follows: for $x \in R/I$, let $a \in R$ be such that $\pi(a) = x$, and set $D^*x = \pi(Da)$. Suppose that $\pi(a) = \pi(b) = x$ for $a, b \in R$. Then, $a - b \in I$, so $D(a - b) \in I$ since I is a differential ideal. This implies that $Da - Db \in I$, hence that $\pi(Da) = \pi(Db)$, so D^* is well-defined. We have $D^* \circ \pi = \pi \circ D$ by the definition of D^*. Let $x, y \in R/I$ and let $a, b \in R$ be such that $\pi(a) = x$ and $\pi(b) = y$. Then, $\pi(a + b) = x + y$ and $\pi(ab) = xy$, so

$$D^*(x + y) = \pi(D(a + b)) = \pi(Da + Db) = \pi(Da) + \pi(Db) = D^*a + D^*b$$

and

$$\begin{aligned} D^*(xy) = \pi(D(ab)) \quad &= \quad \pi(aDb + bDa) \\ &= \quad \pi(a)\pi(Db) + \pi(b)\pi(Da) = xD^*y + yD^*x \end{aligned}$$

so D^* is a derivation on R/I. $\quad\square$

Example 3.1.1. Let R be any ring and D be the zero-map on R. Then any ideal of R is a differential ideal, and the induced derivation D^* is the zero-map on R/I.

Example 3.1.2. Let X be an indeterminate and D be d/dX on $R = \mathbb{Q}[X]$. The only differential ideals of R are (0) and (1), and the induced derivations are D and the zero-map respectively.

Example 3.1.3. Let (R, D) be a differential ring, X be an indeterminate and $\Delta : R[X] \to R[X]$ be the map defined by

$$\Delta(\sum_n a_n X^n) = \sum_n (Da_n + na_n)X^n.$$

It can be checked that Δ is a derivation on $R[X]$ and that for any integer $m > 0$, the ideal $I_m = (X^m)$ is a differential ideal. For $m = 1$, the map $\pi : R[X] \to R[X]/(X) \simeq R$ is the substitution $X \to 0$, and the induced derivation Δ^* on R satisfies

$$\Delta^*\pi(p) = \pi(\Delta p) = D(p(0)) \quad \text{for any } p \in R[X]$$

so $\Delta^* = D$ on R.

3.2 Differential Extensions

We study in this section the problem of extending a given derivation to a larger ring or field. As in the previous section, the rings and fields in this section can have arbitrary characteristic. In classical algebra, roots of equations or new indeterminates are added to a given ring in order to create a larger ring. An obvious question is then, if the initial ring admits a derivation D, can it be extended to a new derivation on the larger ring? If this is the case, and the new derivation is compatible with D, we say that the larger differential ring is a differential extension of the initial one. The following definition formalizes the notion of "compatibility with D".

Definition 3.2.1. *Let (R, D) and (S, Δ) be differential rings. We say that (S, Δ) is a* differential extension *of (R, D) if R is a subring of S and $\Delta a = Da$ for any $a \in R$.*

We first show that any derivation on an integral domain has a unique extension to its quotient field, and this extension is given by the usual rule for differentiating quotients.

Theorem 3.2.1. *Let R be an integral domain, F the quotient field of R and D a derivation on R. Then there exists a unique derivation Δ on F such that (F, Δ) is a differential extension of (R, D).*

Proof. Define $\Delta : F \to F$ as follows: for any $x \in F$, write $x = a/b$ where $a, b \in R$, $b \neq 0$, and let $\Delta x = (bDa - aDb)/b^2$. Suppose that $x = a/b = c/d$ for $a, b, c, d \in R$. Then, $ad = bc$, therefore:

$$
\begin{aligned}
\frac{bDa - aDb}{b^2} &- \frac{dDc - cDd}{d^2} = \frac{d^2bDa - d^2aDb - b^2dDc + b^2cDd}{b^2d^2} \\
&= \frac{(bDd + dDb)(bc - ad) + abdDd - bcdDb + bd(dDa - bDc)}{b^2d^2} \\
&= \frac{D(bd)(bc - ad) + bd(dDa + aDd - bDc - cDb)}{b^2d^2} \\
&= \frac{D(bd)(bc - ad) + bdD(ad - bc)}{b^2d^2} = 0
\end{aligned}
$$

which implies that Δ is well defined. Let now $x, y \in F$ and write $x = a/b, y = c/d$ where $a, b, c, d \in R$. By a calculation similar to the one above, we get:

$$
\begin{aligned}
\Delta(x + y) = \Delta\frac{ad + bc}{bd} &= \frac{bdD(ad + bc) - (ad + bc)D(bd)}{b^2d^2} \\
&= \frac{bd^2Da + abdDd + bcdDb + b^2dDc - abdDd - ad^2Db - bcdDb - b^2cDd}{b^2d^2} \\
&= \frac{bDa - aDb}{b^2} + \frac{dDc - cDd}{d^2} = \Delta x + \Delta y
\end{aligned}
$$

and

$$\Delta(xy) = \Delta\frac{ac}{bd} \;=\; \frac{bdD(ac) - acD(bd)}{b^2 d^2}$$

$$= \frac{abdDc + bcdDa - abcDd - acdDb}{b^2 d^2}$$

$$= \frac{a(dDc - cDd)}{bd^2} + \frac{c(bDa - aDb)}{db^2} = x\Delta y + y\Delta x$$

so Δ is a derivation on F. Take $a \in R$, and write $a = a/1$. This implies that $\Delta a = (1Da - aD1)/1^2 = Da$, so (F, Δ) is a differential extension of (R, D).

Suppose that there are two derivations Δ_1 and Δ_2 on F such that (F, Δ_1) and (F, Δ_2) are both differential extensions of (R, D), and let $x \in F$. Write $x = a/b$ where $a, b \in R$ and $b \neq 0$. From part (ii) of Theorem 3.1.1 we have

$$\Delta_1 x = \Delta_1\frac{a}{b} = \frac{b\Delta_1 a - a\Delta_1 b}{b^2} = \frac{bDa - aDb}{b^2} = \frac{b\Delta_2 a - a\Delta_2 b}{b^2} = \Delta_2\frac{a}{b} = \Delta_2 x$$

so $\Delta_1 = \Delta_2$, which shows that Δ as defined above is the only derivation on F such that (F, Δ) is a differential extension of (R, D). $\qquad\square$

Definition 3.2.2. *Let R be a ring and X an indeterminate over R. For any derivation D on R, we define the* coefficient lifting *of D to be the map $\kappa_D : R[X] \to R[X]$ given by*

$$\kappa_D\left(\sum_{i=0}^{n} a_i X^i\right) = \sum_{i=0}^{n}(Da_i)X^i.$$

The map κ_D simply applies the derivation D to every coefficient of a polynomial over R. Note that the degree is not necessarily preserved under κ_D.

Lemma 3.2.1. *κ_D is a derivation on $R[X]$.*

Proof. Let $p, q \in R[X]$ and write $p = \sum_{i=0}^{n} a_i X^i$ and $q = \sum_{i=0}^{n} b_i X^i$. Then,

$$\kappa_D(p+q) = \sum_{i=0}^{n} D(a_i + b_i)X^i = \sum_{i=0}^{n}(Da_i)X^i + \sum_{i=0}^{n}(Db_i)X^i = \kappa_D(p) + \kappa_D(q)$$

and

$$\kappa_D(pq) \;=\; \sum_{k=0}^{2n} D\left(\sum_{\substack{i,j\geq 0 \\ i+j=k}} a_i b_j\right)X^k = \sum_{k=0}^{2n}\ \sum_{\substack{i,j\geq 0 \\ i+j=k}} D(a_i b_j)X^k$$

$$= \sum_{k=0}^{2n}\ \sum_{\substack{i,j\geq 0 \\ i+j=k}} a_i Db_j X^k + \sum_{k=0}^{2n}\ \sum_{\substack{i,j\geq 0 \\ i+j=k}} b_j Da_i X^k$$

$$= \; p\kappa_D(q) + q\kappa_D(p)$$

so κ_D is a derivation on $R[X]$. $\qquad\square$

If R is an integral domain, then $R[X]$ is an integral domain, so, by Theorem 3.2.1, κ_D can be extended uniquely to a derivation on its quotient field $R(X)$ (example 1.1.14), and we also write κ_D for this extension to $R(X)$.

Lemma 3.2.2. *Let (R, D) be a differential ring, (S, Δ) a differential extension of (R, D), and X an indeterminate over R. Then,*

$$\Delta(P(\alpha)) = \kappa_D(P)(\alpha) + (\Delta\alpha)\frac{dP}{dX}(\alpha)$$

for any $\alpha \in S$ and any $P \in R[X]$.

Proof. This follows directly from the sum and product derivation rules: write $P = \sum_{i=0}^{n} a_i X^i$ where the a_i's are in R. Then $\Delta a_i = Da_i$ for each i, so

$$\Delta(P(\alpha)) = \Delta\left(\sum_{i=0}^{n} a_i \alpha^i\right) = \sum_{i=0}^{n}(\Delta a_i)\alpha^i + \sum_{i=1}^{n} i a_i \alpha^{i-1}\Delta\alpha$$

$$= \kappa_D(P)(\alpha) + (\Delta\alpha)\frac{dP}{dX}(\alpha).$$

$\square$

We can now prove the main result about differential extensions: given a simple extension $F(t)$ of a differential field (F, D), if t is algebraic over F, then D can be extended in a unique way to $F(t)$, otherwise D can be extended in several ways to $F(t)$ but choosing a value for Dt makes the extension unique. We prove this in two theorems, one for the transcendental and one for the algebraic case.

Theorem 3.2.2. *Let (F, D) be a differential field, and t be transcendental over F. Then, for any $w \in F(t)$, there exists a unique derivation Δ on $F(t)$ such that $\Delta t = w$ and $(F(t), \Delta)$ is a differential extension of (F, D).*

Proof. By Lemma 3.2.1, κ_D is a derivation on $F[t]$, and by Theorem 3.2.1, it has a unique extension to a derivation on $F(t)$. Since d/dt is also a derivation on $F(t)$, the map $\Delta = \kappa_D + w \, d/dt$ is a derivation on $F(t)$ by Lemma 3.1.1. We have, $\Delta t = \kappa_D t + w \, dt/dt = D(1)t + w \cdot 1 = w$, and for $a \in F$, we get $\Delta a = \kappa_D a + w \, da/dt = Da + w \cdot 0 = Da$, so $(F(t), \Delta)$ is a differential extension of (F, D).

Suppose that there are two derivations Δ_1 and Δ_2 on $F(t)$ such that $(F(t), \Delta_1)$ and $(F(t), \Delta_2)$ are both differential extensions of (F, D), and that $\Delta_1 t = \Delta_2 t = w$. Let $x \in F(t)$ and write $x = a/b$ where $a, b \in F[t]$ and $b \neq 0$. Using part (ii) of Theorem 3.1.1 and Lemma 3.2.2 applied to both a and b with $\alpha = t$, we get

$$\Delta_1 x = \Delta_1\frac{a}{b} = \frac{b\Delta_1 a - a\Delta_1 b}{b^2} = \frac{b(\kappa_D a + w \, da/dt) - a(\kappa_D b + w \, db/dt)}{b^2}$$

$$= \frac{b\Delta_2 a - a\Delta_2 b}{b^2} = \Delta_2\frac{a}{b} = \Delta_2 x$$

so $\Delta_1 = \Delta_2$, which shows that Δ as defined above is the only derivation on $F(t)$ such that $\Delta t = w$ and $(F(t), \Delta)$ is a differential extension of (F, D).   □

Example 3.2.1. Let F be any field, 0_F be the map that sends every element of F to 0, and x be transcendental over F. Let D be an extension of 0_F to $F(x)$ satisfying $Dx = 1$. Since $(F(x), d/dx)$ is a differential extension of $(F, 0_F)$ and $dx/dx = 1$, Theorem 3.2.2 implies that $D = d/dx$, *i.e.* the only derivation on $F(x)$ that is 0 on F and maps x to 1 is d/dx.

Example 3.2.2. Let (F, D) be a differential field and t be transcendental over F. Let Δ be an extension of D to $F(t)$ satisfying $\Delta t = 0$. Since $(F(t), \kappa_D)$ is a differential extension of (F, D) and $\kappa_D t = 0$, Theorem 3.2.2 implies that $\Delta = \kappa_D$, *i.e.* the only extension of D to $F(t)$ for which t is constant is κ_D.

We now turn to algebraic extensions of differential fields. The assumption that E is separable over F in the next theorem is needed for the case where F has nonzero characteristic, and E separable over F means that the minimal irreducible polynomial over F for any element of E has no multiple roots. In characteristic 0, algebraic extensions are always separable, so the reader interested in this case only can ignore the separability hypothesis. In addition, we use Zorn's Lemma in the proof to allow for non-finitely generated extensions. That part of the proof can be skipped if one considers only finitely generated algebraic extensions.

Theorem 3.2.3. *Let (F, D) be a differential field, and E a separable algebraic extension of F. Then, there exists a unique derivation Δ on E such that (E, Δ) is a differential extension of (F, D).*

Proof. Suppose first that $E = F(\alpha)$ for some $\alpha \in E$. Let X be an indeterminate over F, and $P \in F[X]$ be the minimal irreducible polynomial for α over F. Then, since E is separable over F, $dP/dX(\alpha) \neq 0$, so let

$$w = -\frac{\kappa_D(P)(\alpha)}{dP/dX(\alpha)} \in E.$$

Since $E \simeq F[\alpha]$, there exists $Q \in F[X]$ such that $w = Q(\alpha)$. By Lemma 3.2.1, κ_D is a derivation on $F[X]$. Since d/dX is also a derivation on $F[X]$, the map $\Delta = \kappa_D + Q \cdot d/dX$ is a derivation on $F[X]$ by Lemma 3.1.1. Let $\pi : F[X] \to F[X]/(P) \simeq E$ be the canonical projection. We have

$$\pi(\Delta P) = \pi(\kappa_D P + Q \frac{dP}{dX}) = \kappa_D(P)(\alpha) + w \frac{dP}{dX}(\alpha)$$
$$= \kappa_D(P)(\alpha) - \kappa_D(P)(\alpha) = 0$$

hence $\Delta P \in \ker(\pi) = (P)$, so $\ker(\pi)$ is a differential ideal, which implies by Lemma 3.1.2 that Δ induces a derivation $\Delta^* : E \to E$ such that $\pi \circ \Delta = \Delta^* \circ \pi$. Finally, for $a \in F$, we get

$$\Delta^* a = \Delta^* \pi a = \pi \Delta a = \pi(\kappa_D a + Q \frac{da}{dX}) = \pi(Da) = Da$$

so (E, Δ^*) is a differential extension of (F, D).

Let now E be any algebraic extension of F, and let S be the set of all pairs (K, Δ) such that (K, Δ) is a differential extension of (F, D) and $K \subseteq E$. Define a partial ordering on S by $(K_1, \Delta_1) \le (K_2, \Delta_2)$ if (K_2, Δ_2) is a differential extension of (K_1, Δ_1). Since $(F, D) \in S$, S is not empty, so let $C = \{(K_i, \Delta_i)\}$ be a totally ordered subset of S, and let $K = \bigcup_i K_i$ and define Δ on K by $\Delta x = \Delta_i x$ if $x \in K_i$. Since C is totally ordered, (K, Δ) is a well-defined differential extension of (F, D). (K, Δ) is also a differential extension of (K_i, Δ_i) for each i, so $(K, \Delta) \in S$ is an upper bound for C with respect to $\le$. Hence every totally ordered subset of S has an upper bound in S, so there exists a maximal element $(K_{max}, \Delta_{max}) \in S$ by Zorn's Lemma ([40] App. 2, §2, [77] §9.2). By the definition of S, $K_{max} \subseteq E$ and (K_{max}, Δ_{max}) is a differential extension of (F, D). Let $x \in E$. By what we have just proven, there exists a derivation Δ on $K_{max}(x)$ such that $(K_{max}(x), \Delta)$ is a differential extension of (K_{max}, Δ_{max}), so $(K_{max}, \Delta_{max}) \le (K_{max}(x), \Delta)$ in S, which implies that $K_{max} = K_{max}(x)$ since (K_{max}, Δ_{max}) is a maximal element. Hence $x \in K_{max}$, so $E = K_{max}$, hence (E, Δ_{max}) is a differential extension of (F, D).

Suppose now that there are two derivations Δ_1 and Δ_2 on E such that (E, Δ_1) and (E, Δ_2) are both differential extensions of (F, D). Let $x \in E$ and $P \in F[X]$ be its minimal irreducible polynomial over F. Since $P(x) = 0$, we have by Lemma 3.2.2:

$$0 = \Delta_i(P(x)) = \kappa_D(P)(x) + (\Delta_i x)\frac{dP}{dX}(x).$$

Since E is separable over F, $dP/dX(x) \ne 0$, so

$$\Delta_1 x = -\frac{\kappa_D(P)(x)}{dP/dX(x)} = \Delta_2(x).$$

Since this holds for any $x \in E$, $\Delta_1 = \Delta_2$, so there exists a unique derivation Δ on E such that (E, Δ) is a differential extension of (F, D). $\qquad\square$

Example 3.2.3. Let (F, D) be a differential field of characteristic 0 and $C = \mathrm{Const}_D(F)$. Let α be algebraic over C and $P \in C[X]$ be its minimal irreducible polynomial over C. Then D has a unique extension to $F(\alpha)$ and we must have

$$0 = D(P(\alpha)) = \kappa_D(P)(\alpha) + (D\alpha)\frac{dP}{dX}(\alpha) = (D\alpha)\frac{dP}{dX}(\alpha)$$

so $D\alpha = 0$, which means that any algebraic element over the constants is itself a constant w.r.t D.

Example 3.2.4. Let $F = \mathbb{Q}(x)$ and α be a root of $Y^2 - x \in F[Y]$, *i.e.* α represents the function $\pm\sqrt{x}$. Then, d/dx has a unique extension to $\mathbb{Q}(x, \alpha)$ and we must have

$$0 = \frac{d}{dx}(\alpha^2 - x) = \kappa_{d/dx}(Y^2 - x)(\alpha) + \frac{d\alpha}{dx}\frac{d(Y^2 - x)}{dY}(\alpha) = -1 + 2\alpha\frac{d\alpha}{dx}$$

so

$$\frac{d\alpha}{dx} = \frac{1}{2\alpha}$$

which is the usual derivative w.r.t. x for $\alpha = \pm\sqrt{x}$.

As a consequence of Theorem 3.2.3, we can always replace any field in a tower of differential extensions by a separable algebraic extension, and we still have a valid tower of differential extensions.

Corollary 3.2.1. *Let (K, D) be a differential field, (F, Δ) be a differential extension of (K, D), $\overline{F}$ be the algebraic closure of F and $E \subseteq \overline{F}$ be a separable algebraic extension of K. Then, D can be extended uniquely to E, Δ can be extended uniquely to EF, and (EF, Δ) is a differential extension of (E, D).*

Proof. The picture is as follows:

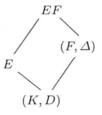

D can be extended uniquely to E by Theorem 3.2.3. Similarly, EF is a separable algebraic extension of F ([40] Chap. VII, §4), so Δ can be extended uniquely to EF. Let X be an indeterminate over F, and $Q \in K[X]$. Considering Q as an element of $F[X]$, we have $\kappa_\Delta(Q) = \kappa_D(Q)$ since Δ extends D. Let $x \in E$ and $P \in K[X]$ be its minimal irreducible polynomial over K. Since E is separable over K, $dP/dX(x) \neq 0$ and we have

$$0 = D(P(x)) = \kappa_D(P)(x) + \frac{dP}{dX}(x)\,Dx$$

so $Dx = -\kappa_D(P)(x)/(dP/dX)(x)$. Considering P as an element of $F[X]$, we get similarly $\Delta x = -\kappa_\Delta(P)(x)/(dP/dX)(x)$, so $Dx = \Delta x$ since $\kappa_D(P) = \kappa_\Delta(P)$. Hence, (EF, Δ) is a differential extension of (E, D). $\qquad\square$

It turns out that in an algebraic extension of a differential field, derivation commutes with conjugation. This technical point implies that the trace map commutes with the derivation, and gives a formula for the trace of a logarithmic derivative.

Theorem 3.2.4. *Let* (K, D) *be a differential field.*

(i) Let F be a separable algebraic extension of K. Then, any field automorphism of F over K commutes with D.

(ii) Let E be a finitely generated separable algebraic extension of K, and $Tr : E \to K$ and $N : E \to K$ be the trace and norm maps from E to K. Then, Tr commutes with D and

$$Tr\left(\frac{Da}{a}\right) = \frac{DN(a)}{N(a)} \quad \text{for any } a \in E^*.$$

Proof. (i) Let F be a separable algebraic extension of K. By Theorem 3.2.3, D extends uniquely to a derivation of F. Let σ be a field automorphism of F over K and $D_\sigma = \sigma^{-1} \circ D \circ \sigma$. Since σ is an automorphism, it follows that D_σ is a derivation of F. In addition, σ is the identity on K, so $Dx = D_\sigma x$ for any $x \in K$. By the unicity clause of Theorem 3.2.3, $D = D_\sigma$, which implies that $\sigma \circ D = \sigma \circ D_\sigma = D \circ \sigma$.

(ii) Let E be a finitely generated separable algebraic extension of K, and $Tr : E \to K$ and $N : E \to K$ be the trace and norm maps from E to K. Let $\overline{K}$ be an algebraic closure of K containing E, $\sigma_1, \ldots, \sigma_n$ be the distinct embeddings of E in $\overline{K}$ over K, and $F = (\sigma_1 E) \cdots (\sigma_n E)$ be the normal closure of E in $\overline{K}$. F is also separable over K ([40] Chap. VII, §4), and for each i, σ_i can be extended to a field automorphism of F over K. By Theorem 3.2.3, D extends uniquely to a derivation on F such that (F, D) is a differential extension of (K, D). Let $a \in E$. By part (i) applied to F we have $D(a^{\sigma_i}) = (Da)^{\sigma_i}$, so

$$D(Tr(a)) = D(\sum_{i=1}^{n} a^{\sigma_i}) = \sum_{i=1}^{n} D(a^{\sigma_i}) = \sum_{i=1}^{n} (Da)^{\sigma_i} = Tr(Da)$$

so $D \circ Tr = Tr \circ D$. Furthermore,

$$
\begin{aligned}
Tr\left(\frac{Da}{a}\right) = \sum_{i=1}^{n} \left(\frac{Da}{a}\right)^{\sigma_i} &= \sum_{i=1}^{n} \frac{(Da)^{\sigma_i}}{a^{\sigma_i}} \\
&= \sum_{i=1}^{n} \frac{D(a^{\sigma_i})}{a^{\sigma_i}} = \frac{D\left(\prod_{i=1}^{n} a^{\sigma_i}\right)}{\prod_{i=1}^{n} a^{\sigma_i}} = \frac{D(N(a))}{N(a)}.
\end{aligned}
$$

$\square$

3.3 Constants and Extensions

We study in this section the effect of extending differential fields on their constant subfields. We first mention the obvious fact that constants remain constants in an extension.

Lemma 3.3.1. *Let (F, D) be a differential field and (E, Δ) a differential extension of (F, D). Then, $\mathrm{Const}_D(F) \subseteq \mathrm{Const}_\Delta(E)$.*

Proof. Let $c \in \mathrm{Const}_D(F)$. Then $c \in E$ and $\Delta c = Dc = 0$ since Δ extends D, so $c \in \mathrm{Const}_\Delta(E)$. $\square$

In the next few lemmas, we consider the new algebraic constants that can appear in a differential extension. We first show that an algebraic constant must in fact be algebraic over the initial constant field, and conversely that any separable algebraic element over the initial constant field must also be a constant.

Lemma 3.3.2. *Let (F, D) be a differential field and (E, Δ) a differential extension of (F, D). Then,*

(i) $c \in \mathrm{Const}_\Delta(E)$ is algebraic over $F \Longrightarrow c$ is algebraic over $\mathrm{Const}_D(F)$.
(ii) $c \in E$ is algebraic and separable over $\mathrm{Const}_D(F) \Longrightarrow c \in \mathrm{Const}_\Delta(E)$.

Proof. (i) Suppose that $c \in \mathrm{Const}_\Delta(E)$ is algebraic over F, and let

$$P = X^n + b_{n-1}X^{n-1} + \ldots + b_0$$

be the minimal polynomial for c over F. We have $P(c) = 0$, so

$$
\begin{aligned}
0 = \Delta(P(c)) \;&=\; \kappa_D(P)(c) + (\Delta c)\frac{dP}{dX}(c) \\
&=\; \kappa_D(P)(c) = (Db_{n-1})c^{n-1} + \ldots + Db_0
\end{aligned}
$$

so $Db_i = 0$ for $i = 0 \ldots n - 1$ by the minimality of P. Hence P is in $\mathrm{Const}_D(F)[X]$, so c is algebraic over $\mathrm{Const}_D(F)$.
(ii) Let $c \in E$ be algebraic and separable over $\mathrm{Const}_D(F)$, and let $P \in \mathrm{Const}_D(F)[X]$ be its minimal polynomial over $\mathrm{Const}_D(F)$. Then,

$$0 = D(P(c)) = \kappa_D(P)(c) + (Dc)\frac{dP}{dX}(c) = (Dc)\frac{dP}{dX}(c).$$

Since c is separable over $\mathrm{Const}_D(F)$, $dP/dX(c) \neq 0$, so $Dc = 0$, which means that $c \in \mathrm{Const}_\Delta(E)$. $\square$

As a consequence, when making a separable algebraic extension of a differential field, the new constants are exactly the elements of the extension that are algebraic over the initial constant subfield. In particular, the constants of the algebraic closure of a differential field of characterisric 0 form exactly an algebraic closure of the initial constant subfield.

Corollary 3.3.1. *Let (F, D) be a differential field and E a separable algebraic extension of F. Let also $C = \mathrm{Const}_D(F)$ and $\overline{C}^E$ be the algebraic closure of C in E, i.e. the subfield of all the elements of E that are algebraic over C. Then D can be extended uniquely to E and $\mathrm{Const}_D(E) = \overline{C}^E$. In addition, if E is algebraically closed, then $\mathrm{Const}_D(E)$ is an algebraic closure of C.*

Proof. The picture is as follows:

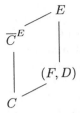

D can be extended uniquely to E by Theorem 3.2.3. Let $c \in \text{Const}_D(E)$. Since E is algebraic over F, c is algebraic over F, so $c \in \overline{C}$ by Lemma 3.3.2. Hence, $\text{Const}_D(E) \subseteq \overline{C} \cap E$. Conversely, let $c \in \overline{C} \cap E$. Then, since E is separable over F, $c \in \text{Const}_D(E)$ by Lemma 3.3.2, so $\text{Const}_D(E) = \overline{C} \cap E$. Suppose that E is algebraically closed. Then, $\overline{C} \subseteq E$ since $C \subseteq F \subseteq E$. Hence, $\text{Const}_D(E) = \overline{C} \cap E = \overline{C}$. □

As a consequence of Corollary 3.2.1, we can always replace any field in a tower of differential extensions by its algebraic closure if the latter is separable. We show now that if the constant fields were equal in the initial tower, then they remain equal after such a replacement. The hypothesis that F is perfect in the next lemma just ensures that its algebraic closure is separable over it. All fields of characteristic 0 are perfect, so the reader interested in the characteristic 0 case only can ignore that hypothesis.

Lemma 3.3.3. *Let (F, D) be a perfect differential field, (E, Δ) be a differential extension of (F, D), $\overline{E}$ be an algebraic closure of E, and $\overline{F} \subseteq \overline{E}$ be an algebraic closure of F. Then, $(E\overline{F}, \Delta)$ is a differential extension of $(\overline{F}, D)$, and*
$$\text{Const}_D(F) = \text{Const}_\Delta(E) \implies \text{Const}_D(\overline{F}) = \text{Const}_\Delta(E\overline{F}).$$

Proof. Since F is perfect, $\overline{F}$ is separable over F, so $(E\overline{F}, \Delta)$ is a differential extension of $(\overline{F}, D)$ by Corollary 3.2.1. Suppose that $\text{Const}_D(F) = \text{Const}_\Delta(E) = C$, and let $\overline{C}$ be the algebraic closure of C. Since $\overline{F}$ is algebraically closed, $\text{Const}_D(\overline{F}) = \overline{C}$ by Corollary 3.3.1. Since $E\overline{F}$ is algebraic over E, $\text{Const}_\Delta(E\overline{F}) = \overline{C} \cap E\overline{F}$ by Corollary 3.3.1. But $\overline{C} \subseteq \overline{F}$, so $\text{Const}_\Delta(E\overline{F}) = \overline{C} = \text{Const}_D(\overline{F})$. □

As expected, adjoining a constant to a differential field extends the constant field by that constant only.

Lemma 3.3.4. *Let (F, D) be a differential field and (E, Δ) be a differential extension of (F, D). Then, $\text{Const}_\Delta(F(t)) = \text{Const}_D(F)(t)$ for any $t \in \text{Const}_\Delta(E)$.*

Proof. Let $C = \text{Const}_D(F)$. Since $\text{Const}_\Delta(F(t))$ is a field containing C and t, it contains $C(t)$. Since $\Delta t = 0$, we have $\Delta = \kappa_D$ on $F[t]$ by Lemma 3.2.2.

Suppose first that t is algebraic over F and let $m = [F(t) : F]$. Then, any $u \in F(t)$ can be written as $u = \sum_{i=0}^{m-1} a_i t^i$ with $a_i \in F$, so $\Delta u = \kappa_D u = \sum_{i=0}^{m-1} (Da_i) t^i$. Since $1, t, \ldots, t^{m-1}$ are linearly independent over F, $\Delta u = 0$ if and only if $a_i \in C$ for each i, so $\mathrm{Const}_\Delta(F(t)) = C(t)$.

Suppose now that t is transcendental over F, and let $u = \sum_{i=0}^{n} a_i t^i \in F[t]$. Then, $\Delta u = \kappa_D u = \sum_{i=0}^{n} (Da_i) t^i$, so $\Delta u = 0$ if and only if $a_i \in C$ for each i. This implies that $\mathrm{Const}_\Delta(F[t]) = C[t]$. Let now $c \in \mathrm{Const}_\Delta(F(t))$ and write $c = u/v$ where $u, v \in F[t]$, $\gcd(u, v) = 1$ and $v \neq 0$. Dividing u and v by the leading coefficient of v if necessary, we can assume that the leading coefficient of v is 1, which implies that either $\Delta v = 0$ or $\deg(\Delta v) < \deg(v)$. Suppose that $\Delta v \neq 0$. Since u/v is a constant, we have

$$0 = \Delta \frac{u}{v} = \frac{v \Delta u - u \Delta v}{v^2}$$

so $z = v \Delta u = u \Delta v$ is a common multiple of u and v in $F[t]$. Since

$$\deg(z) = \deg(u) + \deg(\Delta v) < \deg(u) + \deg(v) = \deg(uv)$$

and $\mathrm{lcm}(u, v) \mid z$, we have $\deg(\mathrm{lcm}(u, v)) < \deg(uv)$. But $\mathrm{lcm}(u, v) \gcd(u, v) = uv$, so $\deg(\gcd(u, v)) > 0$ in contradiction with $\gcd(u, v) = 1$. Hence $\Delta v = 0$, which implies that $v \Delta u = 0$, hence that $\Delta u = 0$. Therefore $u, v \in \mathrm{Const}_\Delta(F[t])$. But $\mathrm{Const}_\Delta(F[t]) = C[t]$, so $c = u/v \in C(t)$.   □

Finally, we need a few results from differential algebra about properties involving constants which are preserved under differential extensions.

Definition 3.3.1. *Let (F, D) be a differential field and $y_1, \ldots, y_n \in F$. The Wronskian of $y_1, \ldots, y_n$ is $W(y_1, \ldots, y_n) = \det(\mathcal{M}(y_1, \ldots, y_n))$ where*

$$\mathcal{M}(y_1, \ldots, y_n) = \begin{pmatrix} y_1 & y_2 & \cdots & y_n \\ Dy_1 & Dy_2 & \cdots & Dy_n \\ \vdots & \vdots & & \vdots \\ D^{n-1}y_1 & D^{n-1}y_2 & \cdots & D^{n-1}y_n \end{pmatrix} \tag{3.1}$$

The vanishing of the Wronskian is a well-known test in analysis for linear dependence of functions over the constants. It turns out to have the same property in arbitrary differential fields.

Lemma 3.3.5 ([37]). *Let (F, D) be a differential field. Then, $y_1, \ldots, y_n \in F$ are linearly dependent over $\mathrm{Const}_D(F)$ if and only if $W(y_1, \ldots, y_n) = 0$.*

Proof. We have $W(y_1, \ldots, y_n) = 0$ if and only if $\ker(\mathcal{M}) \neq \{0\}$ where $\mathcal{M}$ is the matrix given by (3.1). Write $C = \mathrm{Const}_D(F)$ and suppose first that $y_1, \ldots, y_n$ are linearly dependent over C. Then there are $c_1, \ldots, c_n \in C$, not all 0, such that $\sum_{i=1}^{n} c_i y_i = 0$. Differentiating this an arbitrary number of times, we get that

$$\sum_{i=1}^{n} c_i D^m y_i = 0$$

for any $m \geq 0$. This implies that $(c_1, \ldots, c_n) \in \ker(\mathcal{M})$, hence that $\ker(\mathcal{M}) \neq \{0\}$ and $W(y_1, \ldots, y_n) = 0$.

We proceed by induction on n for the converse. For $n = 1$, we have $W(y_1) = y_1$ so if $W(y_1) = 0$, then $y_1 = 0$ is linearly dependent over C. Suppose now that $n > 1$, that the lemma holds for any $n - 1$ elements in F, and that $W(y_1, \ldots, y_n) = 0$. Then, $\ker(\mathcal{M}) \neq \{0\}$, so let $(x_1, \ldots, x_n)$ be in $\ker(\mathcal{M})$ where $x_i \neq 0$ for some i. Renumbering the y_i's if necessary, we can assume that $x_1 \neq 0$, hence that $x_1 = 1$ since $\ker(\mathcal{M})$ is a vector space over F. Since $(x_1, \ldots, x_n) \in \ker(\mathcal{M})$, we have $\sum_{i=1}^{n} x_i D^j y_i = 0$ for $0 \leq j < n$. Differentiating those equations for $0 \leq j < n - 1$ and using them together with $Dx_1 = D1 = 0$, we get

$$0 = D\left(\sum_{i=1}^{n} x_i D^j y_i\right) = \sum_{i=1}^{n} Dx_i D^j y_i + \sum_{i=1}^{n} x_i D^{j+1} y_i = \sum_{i=2}^{n} Dx_i D^j y_i$$

so $(Dx_2, \ldots, Dx_n) \in \ker(\mathcal{M}(y_2, \ldots, y_n))$. If $Dx_2 = \ldots = Dx_n = 0$, then $x_1, \ldots, x_n \in C$, so $y_1, \ldots, y_n$ are linearly dependent over C. If $Dx_i \neq 0$ for some $i > 1$, then $\ker(\mathcal{M}(y_2, \ldots, y_n)) \neq \{0\}$, so $y_2, \ldots, y_n$ are linearly dependent over C by induction, which implies that $y_1, \ldots, y_n$ are linearly dependent over C. $\square$

As a consequence, linear independence over the constants is independent of the constant field, hence preserved under differential extensions.

Corollary 3.3.2. *Let (F, D) be a differential field and (E, Δ) be a differential extension of (F, D). If $S \subset F$ is linearly independent over $\mathrm{Const}_D(F)$, then S is linearly independent over $\mathrm{Const}_\Delta(E)$.*

Proof. Let $S \subset F$ be linearly independent over $\mathrm{Const}_D(F)$ and $\{s_1, \ldots, s_n\}$ be any finite subset of S. Then, $W(s_1, \ldots, s_n) \neq 0$ by Lemma 3.3.5. But $s_1, \ldots, s_n \in E$, so by Lemma 3.3.5 applied to (E, Δ), $\{s_1, \ldots, s_n\}$ is linearly independent over $\mathrm{Const}_\Delta(E)$. Since this holds for any finite subset of S, S is linearly independent over $\mathrm{Const}_\Delta(E)$. $\square$

The following lemma states that if an algebraic system of equations and inequations is satisfied by constants, then it is also satisfied by algebraic constants.

Lemma 3.3.6 ([37]). *Let (F, D) be a differential field with algebraically closed constant field, (E, Δ) be a differential extension of (F, D), $X_1, \ldots, X_m$ be independent indeterminates over F, $g \in F[X_1, \ldots, X_m]$ and S be any subset of $F[X_1, \ldots, X_m]$. If there are $c_1, \ldots, c_m \in \mathrm{Const}_\Delta(E)$ such that $g(c_1, \ldots, c_m) \neq 0$ and $f(c_1, \ldots, c_m) = 0$ for any f in S, then there are also such $c_1, \ldots, c_m$ in $\mathrm{Const}_D(F)$.*

Proof. Let $C = \mathrm{Const}_D(F)$ and

$$V(S) = \{(a_1, \ldots, a_m) \in C^m \text{ such that } f(a_1, \ldots, a_m) = 0 \text{ for all } f \in S\}.$$

Since F is a field containing C, it is a vector space over C, so let $\mathcal{B}$ be a vector space basis for F over C. Then, $\mathcal{B}$ generates $F[X_1, \ldots, X_m]$ as a free module over $C[X_1, \ldots, X_m]$ so write each f in S as $f = \sum_{b \in \mathcal{B}} h_{f,b} b$ where $h_{f,b} \in C[X_1, \ldots, X_m]$ and all but finitely many of the $h_{f,b}$ are identically 0. Let $I \subseteq C[X_1, \ldots, X_m]$ be the ideal generated by all the $h_{f,b}$ and

$$V(I) = \{(a_1, \ldots, a_m) \in C^m \text{ such that } h(a_1, \ldots, a_m) = 0 \text{ for all } h \in I\}.$$

By construction, we have $V(I) \subseteq V(S)$. Let $c_1, \ldots, c_m \in \mathrm{Const}_\Delta(E)$ be such that $g(c_1, \ldots, c_m) \neq 0$, which implies that $g \neq 0$, and $f(c_1, \ldots, c_m) = 0$ for all $f \in S$. Then, for each $f \in S$,

$$\sum_{b \in \mathcal{B}} h_{f,b}(c_1, \ldots, c_m) b = 0$$

which implies that $h_{f,b}(c_1, \ldots, c_m) = 0$ for each $b \in \mathcal{B}$, since $\mathcal{B}$ is linearly independent over $\mathrm{Const}_\Delta(E)$ by Corollary 3.3.2. Suppose that $1 \in I$. Then, there are polynomials $a_{f,b} \in C[X_1, \ldots, X_m]$, all but finitely many of which identically 0, such that

$$1 = \sum_{b \in \mathcal{B}} a_{f,b} h_{f,b}.$$

Evaluating that equality at $(X_1, \ldots, X_m) = (c_1, \ldots, c_m)$ yields $1 = 0$. Therefore $1 \notin I$, so by Hilbert's Nullstellensatz (Theorem 1.1.10) $V(I) \neq \emptyset$.

Write now $g = \sum_{b \in \mathcal{B}} g_b b$ where $g_b \in C[X_1, \ldots, X_m]$ and all but finitely many of the g_b are identically 0. Suppose that $g(a_1, \ldots, a_m) = 0$ for every $(a_1, \ldots, a_m) \in V(I)$. As previously, this implies that $g_b(a_1, \ldots, a_m) = 0$ for every $b \in \mathcal{B}$ and every $(a_1, \ldots, a_m) \in V(I)$, hence, by Hilbert's Nullstellensatz (Theorem 1.1.11), that there exist positive integers n_b such that $g_b^{n_b} \in I$ for each $b \in \mathcal{B}$. Since $h(c_1, \ldots, c_m) = 0$ for every $h \in I$, we get $g_b^{n_b}(c_1, \ldots, c_m) = 0$, hence $g_b(c_1, \ldots, c_m) = 0$ for every $b \in \mathcal{B}$, in contradiction with $g(c_1, \ldots, c_m) \neq 0$. Hence there exist $(a_1, \ldots, a_m) \in V(I)$ such that $g(a_1, \ldots, a_m) \neq 0$. Since $V(I) \subseteq V(S)$, this proves the lemma. $\qquad \square$

3.4 Monomial Extensions

We want to study simple transcendental differential extensions of the form $k(t)$ where there is some amount of similarity between the derivations D and d/dt, which will allow us to apply the algorithms for integrating rational functions to such extensions. Recall that if k is a differential field, K a differential extension of k, and t an element of K, then $k(t)$ is a differential field itself if it is closed under the derivation D of K. A condition for some similarity with d/dt is that D transforms polynomials in t into polynomials in t, *i.e.* that $k[t]$ is closed under D[1]. Therefore, we study here differential extensions where the derivatives of polynomials are polynomials. In addition, we now restrict our study to fields of characteristic 0, so for the rest of this chapter, k is a differential field of characteristic 0, K is a differential extension of k, and D denotes the derivation on K. We first show that the requirement that $Dt \in k[t]$ is equivalent to $k[t]$ being a differential subring of $k(t)$.

Lemma 3.4.1. *Let* $t \in K$. *Then,* $Dt \in k[t] \iff k[t]$ *is closed under* D.

Proof. Suppose that $Dt \in k[t]$, and let $p \in k[t]$. By Lemma 3.2.2,

$$Dp = \kappa_D(p) + (Dt)\frac{dp}{dt} \in k[t]$$

so $k[t]$ is closed under D. Conversely, if $k[t]$ is closed under D, then $Dt \in k[t]$ since $t \in k[t]$. □

Note that we did not require that t be transcendental over k in the above lemma. We can now define the class of differential extensions for which the integration algorithm will be presented later. This class is general enough to model the usual elementary transcendental functions of calculus. It consists of simple transcendental extensions for which $k[t]$ is closed under D.

Definition 3.4.1. *We say that* $t \in K$ *is a* monomial over k *(w.r.t.* D*), if*

(i) t is transcendental over k,
(ii) $Dt \in k[t]$.

In addition, we define then the D-degree *of t to be* $\delta(t) = \deg_t(Dt)$, *and the* D-leading coefficient *of t to be* $\lambda(t) = \mathrm{lc}_t(Dt)$. *We call t* linear *if* $\delta(t) \le 1$, nonlinear *otherwise. Furthermore we let* $\mathcal{H}_t \in k[X]$ *be the polynomial such that* $Dt = \mathcal{H}_t(t)$.

Since the derivative of polynomials are polynomials in monomial extensions, we often need to know the degree and leading coefficient of a derivative.

Lemma 3.4.2. *Let t be a monomial over k, and $p \in k[t]$.*

(i) $\deg(Dp) \le \deg(p) + \max(0, \delta(t) - 1)$.
(ii) If t is nonlinear and $\deg(p) > 0$, then equality holds in (i), and the leading coefficient of Dp is $\deg(p)\,\mathrm{lc}(p)\,\lambda(t)$.

Proof. If $p = 0$, then $Dp = 0$ and (i) is satisfied under the convention that $\deg(0) = -\infty$, so suppose that $p \ne 0$ and let $n = \deg(p)$.
(i) We have $Dp = \kappa_D(p) + (Dt)(dp/dt)$ by Lemma 3.2.2. If $n = 0$, then $dp/dt = 0$, so $\deg(Dp) = \deg(\kappa_D(p)) \le n \le n + \max(0, \delta(t) - 1)$. Otherwise

---

[1] This condition is probably not even necessary (Exercises 3.6 to 3.10) but the integration algorithms have not been generalized to such extensions as described in [56].

$n > 0$, so $\deg(dp/dt) = n-1$, which implies that $\deg((Dt)dp/dt) = n+\delta(t)-1$, hence

$$\deg(Dp) \leq \max\left(\deg(\kappa_D(p)), \deg((Dt)\frac{dp}{dt})\right) \leq \max(n, n + \delta(t) - 1)$$
$$= n + \max(0, \delta(t) - 1).$$

(ii) Suppose that t is nonlinear and $n > 0$. Then, $\delta(t) > 1$ and

$$\deg((Dt)\, dp/dt) = n + \delta(t) - 1 > n \geq \deg(\kappa_D(p))$$

so $\deg(Dp) = n + \delta(t) - 1$. Furthermore, the leading coefficient of dp/dt is $n\,a$, where a is the leading coefficient of p, so the leading coefficient of Dp is $n\,a\,\lambda(t)$.   □

Let $t \in K$ be a monomial over k for the rest of this section. It is well-known that for $D = d/dt$, every squarefree polynomial has no common factor with its derivative, and this fact forms the basis of the various squarefree factorization algorithms. This fact is not always true for more arbitrary derivations, so we introduce a name for the polynomials for which it still holds.

Definition 3.4.2. *We say that $p \in k[t]$ is* normal *with respect to D if $\gcd(p, Dp) = 1$. We say that p is* special *with respect to D if $\gcd(p, Dp) = p$ i.e. $p \mid Dp$.*

In addition, we introduce the following notations for the sets of special and special monic irreducible polynomials:

$$\mathcal{S}_{k[t]:k} = \{p \in k[t] \text{ such that } p \text{ is special}\},$$

$$\mathcal{S}^{\text{irr}}_{k[t]:k} = \{p \in \mathcal{S}_{k[t]:k} \text{ such that } p \text{ is monic and irreducible}\}.$$

When the monomial extension is clear from the context, we omit the subscripts and simply write $\mathcal{S}$ and $\mathcal{S}^{\text{irr}}$. A polynomial is not necessarily normal or special, but an irreducible polynomial $p \in k[t]$ must be either normal or special, since $\gcd(p, Dp)$ is a factor of p. Note that $k \subseteq \mathcal{S}$, and that $p \in k[t]$ is both normal and special if and only if $(p) = (1)$, which is equivalent to say that $p \in k^*$. Special polynomials generate differential ideals, so there is an induced derivation on the quotient rings (Lemma 3.1.2). More importantly, this induced derivation turns out to be an extension of D.

Lemma 3.4.3. *Let $p \in \mathcal{S} \setminus k$. Then, (p) is a differential ideal of $k[t]$ and $(k[t]/(p), D^*)$ is a differential extension of (k, D) where D^* is the induced derivation.*

Proof. Let $p \in k[t] \setminus k$ be special. Then, $p \mid Dp$ by definition, so (p) is a differential ideal of $k[t]$. By Lemma 3.1.2, $D^* \circ \pi = \pi \circ D$, where D^* is the induced derivation on $k[t]/(p)$ and $\pi : k[t] \to k[t]/(p)$ is the canonical projection. Hence, $D^*a = D^*\pi(a) = \pi(Da) = Da$ for any $a \in k$, which implies that $(k[t]/(p), D^*)$ is a differential extension of (k, D).   □

Lemma 3.4.4. *Let $p_1, \ldots, p_m \in k[t]$ be such that $\gcd(p_i, p_j) = 1$ for $i \neq j$, and let $p = \prod_{i=1}^{m} p_i^{e_i}$ where the e_i's are positive integers. Then,*

$$\gcd(p, Dp) = \left(\prod_{i=1}^{m} p_i^{e_i - 1}\right) \prod_{i=1}^{m} \gcd(p_i, Dp_i).$$

Proof. Let $a, b \in k[t]$ and suppose that $\gcd(a, b) = 1$. Then,

$$
\begin{aligned}
\gcd(ab, D(ab)) &= \gcd(a, D(ab)) \gcd(b, D(ab)) \\
&= \gcd(a, aDb + bDa) \gcd(b, aDb + bDa) \\
&= \gcd(a, bDa) \gcd(b, aDb) = \gcd(a, Da) \gcd(b, Db).
\end{aligned}
$$

So by induction, $\gcd(p, Dp) = \prod_{i=1}^{m} \gcd(p_i^{e_i}, D(p_i^{e_i}))$. In addition,

$$
\begin{aligned}
\gcd\left(p_i^{e_i}, D(p_i^{e_i})\right) &= \gcd(p_i^{e_i}, e_i p_i^{e_i - 1} Dp_i) \\
&= p_i^{e_i - 1} \gcd(p_i, e_i Dp_i) = p_i^{e_i - 1} \gcd(p_i, Dp_i)
\end{aligned}
$$

which proves the lemma. $\qquad\square$

As a consequence, any normal polynomial must be squarefree. In addition, we get the multiplicative properties of special and normal polynomials, in particular that S is a multiplicative semigroup generated by k and S^{irr}.

Theorem 3.4.1.

(i) *Any finite product of normal and two by two relatively prime polynomials is normal. Any factor of a normal polynomial is normal.*

(ii) $p_1, \ldots, p_n \in S \implies \prod_{i=1}^{n} p_i \in S$.

(iii) $p \in S \setminus \{0\} \implies q \in S$ *for any $q \in k[t]$ which divides p.*

Proof. (i) Let $p_1, \ldots, p_m \in k[t]$ be normal and such that $\gcd(p_i, p_j) = 1$ for $i \neq j$, and let $p = \prod_{i=1}^{m} p_i$. By Lemma 3.4.4 we have

$$\gcd(p, Dp) = \left(\prod_{i=1}^{m} p_i^0\right) \prod_{i=1}^{m} \gcd(p_i, Dp_i) = 1$$

since each p_i is normal. Hence, p is normal.

Let $p \in k[t]$ be normal and write $p = qh$ where $q, h \in k[t]$. Since p is squarefree, we have $\gcd(q, h) = 1$, hence by Lemma 3.4.4, $1 = \gcd(p, Dp) = \gcd(q, Dq) \gcd(h, Dh)$, so $\gcd(q, Dq) = 1$, which implies that q is normal.

(ii) Let $a, b \in S$, then $Da = ap$ and $Db = bq$ for some $p, q \in k[t]$. Hence,

$$D(ab) = aDb + bDa = abq + bap = ab(p + q)$$

so $ab \in S$. Part (ii) follows by induction.

(iii) Let $p \in S \setminus \{0\}$, $r \in k[t]$ be an irreducible factor of p, and n be the maximal exponent such that $r^n \mid p$. Then, $n \geq 1$, since $r \mid p$, and $p = r^n h$ for some $h \in k[t]$ with $\gcd(r, h) = 1$, so by Lemma 3.4.4,

$$r^n h = p = \gcd(p, Dp) = \gcd(r^n h, D(r^n h)) = r^{n-1} \gcd(r, Dr) \gcd(h, Dh).$$

Hence $rh = \gcd(r, Dr) \gcd(h, Dh)$, which implies that $\gcd(h, Dh) = h$ and $\gcd(r, Dr) = r$, hence that $r \in \mathcal{S}$. Therefore, every irreducible factor of p must be special. Let now $q \in k[t]$ be any factor of p. If $q \in k$, then $q \in \mathcal{S}$ by definition. Otherwise, q is a nonempty finite product of irreducible factors of p, so it is special by part (ii). □

As mentioned above, every normal polynomial must be squarefree. The converse is not always true however, and there is an important connection between the normality of a squarefree polynomial and the differential properties of its roots. This relationship is described by the following two theorems.

Theorem 3.4.2. *Let $\overline{k}$ be the algebraic closure of k, and $p \in k[t]$ be square-free. Then,*

$$p \text{ normal} \iff D\alpha \neq \mathcal{H}_t(\alpha) \text{ for all roots } \alpha \in \overline{k} \text{ of } p.$$

Proof. Let $p \in k[t]$ be squarefree, and let $\alpha_1, \dots, \alpha_n \in \overline{k}$ be the distinct roots of p, where $n = \deg(p) \geq 0$. The factorization of p over $\overline{k}$ is then

$$p = c \prod_{i=1}^{n} (t - \alpha_i)$$

where $c \in k^*$ is the leading coefficient of p. By Lemma 3.4.4 we have

$$\gcd(p, Dp) = c \prod_{i=1}^{n} \gcd(t - \alpha_i, D(t - \alpha_i)) = c \prod_{i=1}^{n} \gcd(t - \alpha_i, \mathcal{H}_t(t) - D\alpha_i).$$

Hence p is normal if and only if $\gcd(t - \alpha_i, \mathcal{H}_t(t) - D\alpha_i) = 1$ for each i. This is equivalent to $t - \alpha_i$ does not divide $\mathcal{H}_t(t) - D\alpha_i$ in $\overline{k}[t]$, hence to $D\alpha_i \neq \mathcal{H}_t(\alpha_i)$ for all i. □

Theorem 3.4.3. *Let $\overline{k}$ be the algebraic closure of k, and $p \in k[t] \setminus \{0\}$. Then,*

$$p \in \mathcal{S} \iff D\alpha = \mathcal{H}_t(\alpha) \text{ for all roots } \alpha \in \overline{k} \text{ of } p.$$

Proof. Let $p \in k[t] \setminus \{0\}$, and let

$$p = c \prod_{i=1}^{n} (t - \alpha_i)^{e_i}$$

be the irreducible factorization of p over $\overline{k}$, where $c \in k^*$ is the leading coefficient of p and $e_i > 0$ for each i. By Lemma 3.4.4 we have

$$
\begin{aligned}
\gcd(p, Dp) &= c \prod_{i=1}^{n} (t - \alpha_i)^{e_i - 1} \prod_{i=1}^{n} \gcd(t - \alpha_i, D(t - \alpha_i)) \\
&= c \prod_{i=1}^{n} (t - \alpha_i)^{e_i - 1} \prod_{i=1}^{n} \gcd(t - \alpha_i, \mathcal{H}_t(t) - D\alpha_i).
\end{aligned}
$$

Hence p is special if and only if $\gcd(t - \alpha_i, \mathcal{H}_t(t) - D\alpha_i) = t - \alpha_i$ for each i. This is equivalent to $t - \alpha_i \mid \mathcal{H}_t(t) - D\alpha_i$ in $\overline{k}[t]$, hence to $D\alpha_i = \mathcal{H}_t(\alpha_i)$ for all i. $\qquad\square$

We make frequent use in the future of the trivial remark that special and normal polynomials remain such when we make an algebraic extension of k.

Corollary 3.4.1. *Let E be an algebraic extension of k. Then, t is a monomial over E. Furthermore, $\mathcal{S}_{k[t]:k} \subseteq \mathcal{S}_{E[t]:E}$ and if $p \in k[t]$ is normal, then it remains normal when viewed as an element of $E[t]$.*

Proof. t is transcendental over k and E is algebraic over k, so t is transcendental over E. Also, $Dt \in k[t] \subseteq E[t]$ so t is a monomial over E. Let $p \in k[t]$, $\overline{k}$ be the algebraic closure of k containing E, and $\alpha_1, \ldots, \alpha_m$ be the distinct roots of p in $\overline{k}$. If p is normal (resp. special), then $D\alpha_i \neq \mathcal{H}_t(\alpha_i)$ (resp. $D\alpha_i = \mathcal{H}_t(\alpha_i)$) for each i by Theorem 3.4.2 (resp. Theorem 3.4.3), so p is normal (resp. special) when viewed as an element of $E[t]$ again by Theorem 3.4.2 (resp. Theorem 3.4.3). $\qquad\square$

As a consequence, in the case where all the elements of k are constants, the special irreducible polynomials are exactly the factors of $\mathcal{H}_t$, and the normal polynomials are exactly the squarefree polynomials that are coprime with $\mathcal{H}_t$.

Corollary 3.4.2. *Suppose that $Da = 0$ for any $a \in k$.*

(i) Let $p \in k[t]$ be monic and irreducible. Then, $p \in \mathcal{S}^{\mathrm{irr}} \iff p \mid \mathcal{H}_t$.
(ii) Let $p \in k[t]$ be squarefree. Then, p normal $\iff \gcd(p, \mathcal{H}_t) = 1$.

Proof. (i) Let $p \in k[t]$ be monic and irreducible, and suppose first that $p \in \mathcal{S}^{\mathrm{irr}}$. Let $\alpha \in \overline{k}$ be any root of p. Then, $D\alpha = \mathcal{H}_t(\alpha)$ by Theorem 3.4.3. But α is algebraic over k, so $D\alpha = 0$ by Lemma 3.3.2, hence $\mathcal{H}_t(\alpha) = 0$. Since this holds for any root of p and p is irreducible, $p \mid \mathcal{H}_t$. Conversely, let $p \in k[t]$ be a monic irreducible factor of $\mathcal{H}_t$, and let $\alpha \in \overline{k}$ be any root of p. Then, $\mathcal{H}_t(\alpha) = 0$. But α is algebraic over $\overline{k}$ so $D\alpha = 0$ as before, hence $D\alpha = \mathcal{H}_t(\alpha)$ so $p \in \mathcal{S}^{\mathrm{irr}}$ by Theorem 3.4.3.
(ii) Let $p \in k[t]$ be squarefree. Suppose first that p is normal and let $\alpha \in \overline{k}$ be any root of p. Then, $D\alpha \neq \mathcal{H}_t(\alpha)$ by Theorem 3.4.2. But $D\alpha = 0$ as before since α is algebraic over k, so $\mathcal{H}_t(\alpha) \neq 0$. Since this holds for any root of p, $\gcd(p, \mathcal{H}_t) = 1$. Conversely, suppose that $\gcd(p, \mathcal{H}_t) = 1$ and let $\alpha \in \overline{k}$ be any root of p. Then $\mathcal{H}_t(\alpha) \neq 0$. But $D\alpha = 0$, so p is normal by Theorem 3.4.2. $\qquad\square$

In particular, applying the above corollary to the case $D = d/dt$, we have $Dt = 1 = \mathcal{H}_t$, so every squarefree polynomial in $k[t]$ is normal with respect to d/dt.

We have made no assumptions on the possible extensions of the constant field in a monomial extension, so we now look at the possible new constants of $k(t)$. It turns out that new constants and special polynomials are closely related. Recall that a monomial t is called *nonlinear* when $\deg_t(Dt) \geq 2$.

Lemma 3.4.5. *If $c \in \mathrm{Const}_D(k(t))$ then both the numerator and denominator of c are in S. Furthermore, if $c \neq 0$ and t is nonlinear, then both the numerator and denominator of c have the same degree.*

Proof. Let $c \in \mathrm{Const}_D(k(t))$ and write $c = a/b$ where $a, b \in k[t]$, $b \neq 0$ and $\gcd(a, b) = 1$. Then,

$$0 = Dc = \frac{bDa - aDb}{b^2}$$

so $bDa = aDb$, which implies that $a \mid Da$ and $b \mid Db$, hence that $a, b \in S$. Suppose now that $c \neq 0$, t is nonlinear, and that $\deg(a) \neq \deg(b)$. Since $1/c \in \mathrm{Const}_D(k(t))$, we can assume that $\deg(a) > \deg(b)$. Write then $c = p + e/b$ where $p, e \in k[t]$, $\deg(p) = \deg(a) - \deg(b) > 0$, and $e = 0$ or $\deg(e) < \deg(b)$. Then,

$$0 = Dc = Dp + \frac{bDe - eDb}{b^2} = Dp + q + \frac{r}{b^2} \tag{3.2}$$

where $q, r \in k[t]$, $(bDe - eDb) = qb^2 + r$ and either $r = 0$ or $\deg(r) < 2\deg(b)$. Since t is nonlinear, we have $\delta(t) > 1$ and $\deg(Dp) = \deg(p) + \delta(t) - 1$ by Lemma 3.4.2, so $\deg(Dp) > \delta(t) - 1$, which means in particular that $Dp \neq 0$. Hence, $e \neq 0$, so $\deg(b) > 0$, which implies that $\deg(eDb) = \deg(e) + \deg(b) + \delta(t) - 1$ by Lemma 3.4.2, so $\deg(eDb) < 2\deg(b) + \delta(t) - 1$. Either $e \in k$, in which case $\deg(bDe) \leq \deg(b)$, or $e \notin k$, in which case $\deg(bDe) = \deg(b) + \deg(e) + \delta(t) - 1$ by Lemma 3.4.2, so $\deg(bDe) < 2\deg(b) + \delta(t) - 1$ in both cases. Hence $\deg(bDe - eDb) < 2\deg(b) + \delta(t) - 1$, which implies that

$$\deg(q) = \deg(bDe - eDb) - 2\deg b < \delta(t) - 1 < \deg(Dp)$$

in contradiction with (3.2), so $\deg(a) = \deg(b)$.   □

Thus, the existence of new constants in $k(t) \setminus k$ implies that S^{irr} is nonempty. The converse, whether the existence of nontrivial special polynomials imply the existence of a new constant, is a more difficult problem. A theorem of Darboux [25, 70, 78] essentially states that if S^{irr} is large enough, then there exists a new constant in $k(t) \setminus k$. The situation is easier for the key monomial extensions of the integration problem, where any element of S^{irr} produces a new constant, as the next lemmas show.

Lemma 3.4.6. *Suppose that $Dt \in k$, and let $p \in k[t]$ be nonzero. Then,*

$$p \in S \iff D\left(\frac{p}{\mathrm{lc}(p)}\right) = 0.$$

Proof. Let $p \in k[t]$ be nonzero, and write $q = p/\mathrm{lc}(p)$. If $Dq = 0$, then $q \mid Dq$, so $q \in S$, which implies that $p \in S$ by Theorem 3.4.1. Conversely, suppose that $p \in S$. Then, $q \in S$ by Theorem 3.4.1, and write $q = \prod_{i=1}^{n}(t - \alpha_i)^{e_i}$ where the α_i's are in the algebraic closure of k and the e_i's are positive integers. Then, $D\alpha_i = \mathcal{H}_t(\alpha_i) = Dt$ for each i by Theorem 3.4.3, so

$$Dq = \sum_{i=1}^{n} e_i(Dt - D\alpha_i)(t - \alpha_i)^{e_i-1} \prod_{j \neq i}(t - \alpha_j)^{e_j} = 0.$$

$\square$

Lemma 3.4.7. *Suppose that $Dt/t \in k$, and let $p \in k[t]$ be nonzero. Then,*

$$p \in \mathcal{S} \iff D\left(\frac{p}{\operatorname{lc}(p)\, t^{\deg(p)}}\right) = 0.$$

Proof. Let $p \in k[t]$ be nonzero, and write $q = p/\operatorname{lc}(p)$. We have $\deg(q) = \deg(p)$ and

$$D\left(\frac{q}{t^n}\right) = \frac{Dq - nqDt/t}{t^n}$$

for any integer n. Suppose that $D(q/t^{\deg(q)}) = 0$. Then, $Dq = \deg(q)qDt/t$, so $q \mid Dq$, which implies that $q \in \mathcal{S}$, hence that $p \in \mathcal{S}$ by Theorem 3.4.1. Conversely, suppose that $p \in \mathcal{S}$, and write $q = \prod_{i=1}^{n}(t - \alpha_i)^{e_i}$ where the α_i's are in the algebraic closure of k and the e_i's are positive integers. Then, $D\alpha_i = (Dt/t)\alpha_i$ for each i by Theorem 3.4.3, so

$$
\begin{aligned}
Dq &= \sum_{i=1}^{n} e_i(Dt - D\alpha_i)(t - \alpha_i)^{e_i-1} \prod_{j \neq i}(t - \alpha_j)^{e_j} \\
&= \sum_{i=1}^{n} e_i \frac{Dt}{t}(t - \alpha_i)^{e_i} \prod_{j \neq i}(t - \alpha_j)^{e_j} \\
&= \left(\sum_{i=1}^{n} e_i\right) q\frac{Dt}{t} = \deg(q)q\frac{Dt}{t}
\end{aligned}
$$

which implies that $D(q/t^{\deg(q)}) = 0$. $\qquad\square$

We need for later use to define one particularly interesting class of special polynomials. We first define some useful terminology.

Definition 3.4.3. *We say that $u \in k$ is a* logarithmic derivative of a k-radical *if there exist $v \in k^*$ and an integer $n \neq 0$ such that $nu = Dv/v$.*

Note that if $n < 0$, then we can write $(-n)u = Dw/w$ where $w = v^{-1}$, so we can always assume that the coefficient n is positive in the above definition.

Example 3.4.1. Let $k = \mathbb{Q}(x)$ with derivation $D = d/dx$, and $u = 1/(2x) \in k$. Since $2u = Dx/x$, u is a logarithmic derivative of a $\mathbb{Q}(x)$-radical. In fact, u is the logarithmic derivative of $\sqrt{x}$, which is a radical over $\mathbb{Q}(x)$. On the other hand, $Dv/v \notin \mathbb{Z}$ for any $v \in k^*$, so 1 is not a logarithmic derivative of a $\mathbb{Q}(x)$-radical.

It is clear from the definition that if we extend k to some extension field E, then the logarithmic derivatives of k-radicals become logarithmic derivatives of E-radicals. However, when E is algebraic over k, then an element of k that is not a logarithmic derivative of a k-radical cannot become a logarithmic derivative of an E-radical.

Lemma 3.4.8. *Let E be algebraic over k, and $a \in k$. If a is not a logarithmic derivative of a k-radical, then it is not a logarithmic derivative of an E-radical.*

Proof. Suppose that $a \in k$ is not a logarithmic derivative of a k-radical, and that there exist $\alpha \in E^*$ and an integer $n \neq 0$ such that $na = D\alpha/\alpha$. Since E is algebraic over k, let $p \in k[X]$ be the minimal polynomial of α over k, and write $p = X^m + \sum_{i=0}^{m-1} a_i X^i$ where the a_i's are in k and $m \geq 1$. Then,

$$0 = D(p(\alpha)) = mn\,a\,\alpha^m + \sum_{i=0}^{m-1}(Da_i + i\,n\,a\,a_i)\alpha^i = q(\alpha)$$

where $q = m\,n\,a\,X^m + \sum_{i=0}^{m-1}(Da_i + i\,n\,a\,a_i)X^i \in k[X]$. Since p is the minimal polynomial for α over k, $p \mid q$, so $q = m\,n\,a\,p$, which implies that $Da_i + i\,n\,a\,a_i = m\,n\,a\,a_i$ for $i = 0, \ldots, m-1$. Since p is irreducible and $\alpha \neq 0$, $a_j \neq 0$ for some j in $\{0, \ldots, m-1\}$. We then have,

$$\frac{Da_j}{a_j} = n(m-j)a$$

in contradiction with a not a logarithmic derivative of a k-radical since $n \neq 0$ and $m \neq j$.   □

Definition 3.4.4. *We say that $q \in k[t]$ is special of the first kind (with respect to D) if $q \in S$ and for any root α of q in the algebraic closure of k, $p_\alpha(\alpha)$ is not a logarithmic derivative of a $k(\alpha)$-radical, where*

$$p_\alpha = \frac{Dt - D\alpha}{t - \alpha} \in k(\alpha)[t] \,.$$

In addition, we introduce the following notations:

$$S_{1,k[t]:k} = \{p \in S_{k[t]:k} \text{ such that } p \text{ is special of the first kind}\} \,,$$

$$S^{\text{irr}}_{1,k[t]:k} = \{p \in S_{1,k[t]:k} \text{ such that } p \text{ is monic and irreducible}\} \,.$$

When the monomial extension is clear from the context, we omit the extension subscripts and simply write S_1 and S_1^{irr}. Note that since α is a root of the polynomial $Dt - D\alpha$, $t - \alpha \mid Dt - D\alpha$ in $k(\alpha)[t]$, so $p_\alpha(\alpha)$ is always defined. In addition, we remark that $k^* \subseteq S_1$ by definition, and that we could have replaced "$k(\alpha)$-radical" by "k-radical" in the above definition in view of Lemma 3.4.8. Theorem 3.4.1 and Corollary 3.4.1 are easily generalized to special polynomials of the first kind, showing that S_1 is a multiplicative semigroup generated by k^* and S_1^{irr}.

Theorem 3.4.4.

(i) $p_1, \ldots, p_n \in \mathcal{S}_1 \Longrightarrow \prod_{i=1}^n p_i \in \mathcal{S}_1$.
(ii) $p \in \mathcal{S}_1 \Longrightarrow q \in \mathcal{S}_1$ *for any* $q \in k[t]$ *which divides* p.
(iii) If E *is algebraic over* k, *then* $\mathcal{S}_{1,k[t]:k} \subseteq \mathcal{S}_{1,E[t]:E}$

Proof. Let $\overline{k}$ be the algebraic closure of k.
(i) Let $p_1, \ldots, p_n \in \mathcal{S}_1$. Then, $q = p_1 \cdots p_n \in \mathcal{S}$ by Theorem 3.4.1. Let $\alpha \in \overline{k}$ be a root of q. Then, α is a root of p_i for some i and $p_i \in \mathcal{S}_1$, so $p_\alpha(\alpha)$ is not a logarithmic derivative of a $k(\alpha)$-radical, which implies that $q \in \mathcal{S}_1$.
(ii) Let $p \in \mathcal{S}_1$ and $q \in k[t]$ be any factor of p. Then, $q \in \mathcal{S}$ by Theorem 3.4.1. If $q \in k$, then $q \neq 0$ (since $p \neq 0$), so $q \in \mathcal{S}_1$. Otherwise, $q \notin k$, so let $\alpha \in \overline{k}$ be a root of q. Then, α is a root of p, so $p_\alpha(\alpha)$ is not a logarithmic derivative of a $k(\alpha)$-radical, which implies that $q \in \mathcal{S}_1$.
(iii) Let $p \in \mathcal{S}_1$ and E be an algebraic extension of k. Then, p is special in $E[t]$ by Corollary 3.4.1. Let $\alpha \in \overline{k}$ be a root of p. Then, $p_\alpha(\alpha)$ is not a logarithmic derivative of a $k(\alpha)$-radical, so it is not a logarithmic derivative of an $E(\alpha)$-radical by Lemma 3.4.8. Hence, p is special of the first kind when viewed as an element of $E[t]$. $\square$

3.5 The Canonical Representation

Given $p \in k[t]$, we want to separate the special and normal components of p. The following definition formalizes that separation.

Definition 3.5.1. *Let* $p \in k[t]$. *We say that* $p = p_s p_n$ *is a splitting factorization of* p *if* $p_n, p_s \in k[t]$, $p_s \in \mathcal{S}$, *and every squarefree factor of* p_n *is normal.*

A consequence of Theorems 3.4.2 and 3.4.3 is that a splitting factorization of p over k is also a splitting factorization of p over any algebraic extension of k, since $D\alpha = \mathcal{H}_t(\alpha)$ for all the roots of p_s and $D\alpha \neq \mathcal{H}_t(\alpha)$ for all the roots of p_n in $\overline{k}$. For the same reason, we always have $\gcd(p_n, p_s) = 1$ in a splitting factorization of p, and such a factorization is unique up to multiplication by units in k, like a prime factorization. It is clear that a prime factorization of p yields a splitting factorization of p, but it turns out that a splitting factorization can always be computed by gcd's only, like a squarefree factorization.

Theorem 3.5.1. *Let* $p \in k[t]$. *Then,*

(i)

$$\frac{\gcd(p, Dp)}{\gcd(p, dp/dt)}$$

is the product of all the coprime special irreducible factors of p.

(ii) If p is squarefree, then $p = p_s p_n$ is a splitting factorization of p, where $p_s = \gcd(p, Dp)$ and $p_n = p/p_s$.

Proof. (i) Let $p \in k[t]$, $N_1, \ldots, N_m$ be all its coprime normal irreducible factors, and $S_1, \ldots, S_n$ be all its coprime special irreducible factors in $k[t]$. The prime factorization of p has then the form $p = u \prod_{j=1}^{n} S_j^{d_j} \prod_{i=1}^{m} N_i^{e_i}$, so by Lemma 3.4.4 applied to both D and d/dt, we have

$$
\begin{aligned}
S &= \frac{\gcd(p, Dp)}{\gcd(p, dp/dt)} \\
&= \frac{\prod_{j=1}^{n} S_j^{d_j-1} \prod_{i=1}^{m} N_i^{e_i-1} \prod_{j=1}^{n} \gcd(S_j, DS_j) \prod_{i=1}^{m} \gcd(N_i, DN_i)}{\prod_{j=1}^{n} S_j^{d_j-1} \prod_{i=1}^{m} N_i^{e_i-1} \prod_{j=1}^{n} \gcd(S_j, dS_j/dt) \prod_{i=1}^{m} \gcd(N_i, dN_i/dt)} \\
&= \frac{\prod_{j=1}^{n} \gcd(S_j, DS_j) \prod_{i=1}^{m} \gcd(N_i, DN_i)}{\prod_{j=1}^{n} \gcd(S_j, dS_j/dt) \prod_{i=1}^{m} \gcd(N_i, dN_i/dt)}.
\end{aligned}
$$

Each N_i and S_j is irreducible, so $\gcd(N_i, dN_i/dt) = \gcd(S_j, dS_j/dt) = 1$. Each N_i is normal with respect to D, so $\gcd(N_i, DN_i) = 1$. Each S_j is special, so $\gcd(S_j, DS_j) = S_j$. Therefore, $S = \prod_{j=1}^{n} S_j$, which is the product of all the coprime special irreducible factors of p.

(ii) Suppose that $p \in k[t]$ is squarefree. Then $\gcd(p, dp/dt) = 1$, so, by (i), $p_s = \gcd(p, Dp)$ is the product of all the coprime special irreducible factors of p. But p is squarefree, so $p_n = p/p_s$ has no special irreducible factor, which implies by Theorem 3.4.1 that p_n is normal. $\square$

This theorem gives us two algorithms for computing splitting factorizations: the first is to compute $S = \gcd(p, Dp)/\gcd(p, dp/dt)$ and $q = p/S$. If $S \in k$, then p has no special irreducible factor, so return $p_n = p, p_s = 1$. Otherwise $\deg(q) < \deg(p)$, so recursively compute a splitting factorization $q = q_n q_s$ of q and return $p_n = q_n, p_s = Sq_s$.

SplitFactor(p, D) (* Splitting Factorization *)

(* Given a derivation D on $k[t]$ and $p \in k[t]$, return $(p_n, p_s) \in k[t]^2$ such that $p = p_n p_s$, p_s is special, and each squarefree factor of p_n is normal. *)

$S \leftarrow \gcd(p, Dp)/\gcd(p, dp/dt)$ (* exact division *)
if $\deg(S) = 0$ **then return**$(p, 1)$
$(q_n, q_s) \leftarrow$ **SplitFactor**$(p/S, D)$ (* exact division *)
return(q_n, Sq_s)

Example 3.5.1. Let $k = \mathbb{Q}(x)$ with $D = d/dx$, and let t be a monomial over k satisfying

$$
Dt = -t^2 - \frac{3}{2x}t + \frac{1}{2x} \tag{3.3}
$$

i.e. t represents a transcendental function solution of the above differential equation. Applying **SplitFactor** to

$$p = 4x^4t^5 - 4x^3(x+1)t^4 + x^2(2x-3)t^3 + x(2x^2+7x+2)t^2 - (4x^2+4x-1)t + 2x - 1$$

we get:

1.

$$\begin{aligned} Dp \;=\;\; & -20x^4t^6 + 2x^3(8x+1)t^5 + 3x^2(4x+7)t^4 \\ & -(12x^3 + 25x^2 - \frac{7}{2}x)t^3 + (7x^2 - \frac{15}{2}x - 5)t^2 \\ & +(2x^2 + 5x + 4 - \frac{3}{2x})t - 2x + \frac{1}{2x} \end{aligned}$$

2. $\gcd(p, Dp) = t^3 - (2x+3)t/(4x^2) + (2x-1)/(4x^3)$
3. $dp/dt = 20x^4t^4 - 16x^3(x+1)t^3 + 3x^2(2x-3)t^2 + 2x(2x^2+7x+2)t - 4x^2 - 4x + 1$
4. $\gcd(p, dp/dt) = t - 1/x$
5. $S = t^2 + t/x - (2x-1)/(4x^2)$
6. $p_1 = p/S = 4x^4t^3 - 4x^3(x+2)t^2 + 4x^2(2x+1)t - 4x^2$
7. recursive call, **SplitFactor**(p_1, D):
   a) $Dp_1 = -12x^4t^4 + 2x^3(4x+7)t^3 - 2x^2(3x+2)t^2 - 2x(2x^2-2x-1)t + 4x^2 - 6x$
   b) $\gcd(p_1, Dp_1) = t - 1/x$
   c) $dp_1/dt = 12x^4t^2 - 8x^3(x+2)t + 8x^3 + 4x^2$
   d) $\gcd(p_1, dp_1/dt) = t - 1/x$
   e) $S_1 = 1$
8. $q_n = p_1,\; q_s = S \cdot 1 = S$

So a splitting factorization of p is

$$\begin{aligned} p \;=\;\; & p_n p_s \\ =\;\; & \left(4x^4t^3 - 4x^3(x+2)t^2 + 4x^2(2x+1)t - 4x^2\right)\left(t^2 + \frac{1}{x}t - \frac{2x-1}{4x^2}\right). \end{aligned}$$

In addition, the roots of p_s are

$$\alpha = -\frac{1}{2x} \pm \frac{1}{2}\sqrt{\frac{2}{x}}$$

which are indeed algebraic functions solutions of (3.3), as expected from Theorem 3.4.3.

The second algorithm is to compute first a squarefree factorization $p = p_1 p_2^2 \cdots p_m^m$ of p, and then compute $S_i = \gcd(p_i, Dp_i)$ and $N_i = p_i/S_i$. By Theorem 3.5.1, $p = p_s p_n$ is a splitting factorization of p, where $p_s = S_1 S_2^2 \cdots S_m^m$ and $p_n = N_1 N_2^2 \cdots N_m^m$. This approach has the advantage of also giving us squarefree factorizations for p_s and p_n. Furthermore, Yun's algorithm can be used for the initial squarefree factorization.

SplitSquarefreeFactor(p, D) (* Splitting Squarefree Factorization *)

(* Given a derivation D on $k[t]$ and $p \in k[t]$, return $(N_1, \ldots, N_m)$ and $(S_1, \ldots, S_m)$ in $k[t]^m$ such that $p = (N_1 N_2^2 \cdots N_m^m)(S_1 S_2^2 \cdots S_m^m)$ is a splitting factorization of p and the N_i and S_i are squarefree and coprime. *)

$(p_1, \ldots, p_m) \leftarrow$ **Squarefree**(p)
for $i \leftarrow 1$ **to** m **do**
   $S_i \leftarrow \gcd(p_i, Dp_i)$
   $N_i \leftarrow p_i / S_i$ (* exact division *)
return$((N_1, \ldots, N_m), (S_1, \ldots, S_m))$

Example 3.5.2. Applying **SplitSquarefreeFactor** to the polynomial p of the previous example with the same monomial extension, we get:

1. $p = p_1 p_2^2 = \left(4x^2 t^3 - 4x(x-1)t^2 - (6x-1)t + 2x - 1\right)(xt - 1)^2$
2.

$$Dp_1 = -12x^2 t^4 + 2x(4x-9)t^3 + (16x-9)t^2 - \left(4x - 7 + \frac{3}{2x}\right)t - 1 + \frac{1}{2x}$$

3. $S_1 = \gcd(p_1, Dp_1) = t^2 + t/x - (2x-1)/(4x^2)$
4. $N_1 = p_1/S_1 = 4x^2 t - 4x^2$
5. $Dp_2 = -xt^2 - t/2 + 1/2$
6. $S_2 = \gcd(p_2, Dp_2) = 1$
7. $N_2 = p_2/S_2 = xt - 1$

So we get the splitting factorization $p = p_n p_s$ with squarefree factorizations of p_n and p_s:

$$p_n = N_1 N_2^2 = 4x^2(t-1)(xt-1)^2$$

and

$$p_s = S_1 = t^2 + \frac{1}{x}t - \frac{2x-1}{4x^2}.$$

We can now define a decomposition of the elements of $k(t)$ that generalizes the canonical representations $f = p + a/d$ of rational functions. Let $f \in k(t) \setminus \{0\}$ and write $f = a/d$ where $a, d \in k[t]$, $\gcd(a, d) = 1$ and d is monic (such a representation is unique). Let $d = d_s d_n$ be a splitting factorization of d with respect to D with d_s and d_n monic, which makes this factorization unique. Then, there are unique $p, b, c \in k[t]$ such that $\deg(b) < \deg(d_s)$, $\deg(c) < \deg(d_n)$, and

$$f = \frac{a}{d} = p + \frac{b}{d_s} + \frac{c}{d_n}.$$

We call this decomposition, which is unique, the *canonical representation of* f with respect to D. We also introduce the notations $f_p = p$ (the *polynomial part* of f), $f_s = b/d_s$ (the *special part* of f), and $f_n = c/d_n$ (the *normal part* of f).

CanonicalRepresentation(f, D) (* Canonical Representation *)

(* Given a derivation D on $k[t]$ and $f \in k(t)$, return $(f_p, f_s, f_n) \in k[t] \times k(t)^2$ such that $f = f_p + f_s + f_n$ is the canonical representation of f. *)

$(a, d) \leftarrow (\text{numerator}(f), \text{denominator}(f))$ (* d is monic *)
$(q, r) \leftarrow$ **PolyDivide**(a, d)
$(d_n, d_s) \leftarrow$ **SplitFactor**(d, D)
$(b, c) \leftarrow$ **ExtendedEuclidean**(d_n, d_s, r) (* $\deg(b) < \deg(d_s)$ *)
return$(q, b/d_s, c/d_n)$

We need to define a few more terms that are often used later. A rational function over $\mathbb{C}$ is called simple if it has only simple affine poles, *i.e.* poles of order one only. This is equivalent to having a squarefree denominator. Since normal polynomials are the analogue of squarefree polynomials in monomial extensions, it is natural to call an element of $k(t)$ simple if it has a normal denominator. Similarly, a usual polynomial can be seen as a rational function with no affine poles, or a rational function with no denominator. The useful analogue in monomial extensions is a function with no normal affine poles, *i.e.* with poles at most at infinity and at special polynomials. This means a function whose denominator is special.

Definition 3.5.2. *Let $f \in k(t)$. We say that f is simple with respect to D if the denominator of f is normal w.r.t. D. We say that f is reduced with respect to D if the denominator of f is special w.r.t. D. In addition we write $k\langle t \rangle$ for the set of all the reduced elements of $k(t)$.*

Obviously, $k[t] \subseteq k\langle t \rangle$. It will be shown in the next chapter that, like $k[t]$, $k\langle t \rangle$ is a differential subring of $k(t)$.

There is an application of splitting factorization that will be useful in the sequel: its use in separating the constant from the nonconstant roots of a polynomial over a differential field. Let K be a differential field of characteristic 0, X an indeterminate over K, $p \in K[X]$ and suppose that we want to separate the constant roots of p from the others. It turns out that this is just a splitting factorization with respect to the coefficient lifting κ_D of D on $K[X]$.

Theorem 3.5.2. *Let (K, D) be a differential field of characteristic 0, $\overline{K}$ the algebraic closure of K and X an indeterminate over K. For any $p \in K[X] \setminus \{0\}$, let $p = p_s p_n$ be a splitting factorization of p w.r.t. κ_D. Then, for any root α of p in $\overline{K}$,*

$$\begin{cases} D\alpha = 0 & \Longleftrightarrow & p_s(\alpha) = 0, \\ D\alpha \neq 0 & \Longleftrightarrow & p_n(\alpha) = 0. \end{cases}$$

Proof. Let $\alpha \in \overline{K}$ be a root of p. Then, by Theorems 3.4.2 and 3.4.3,

$$\begin{cases} D\alpha = \mathcal{H}_t(\alpha) & \Longleftrightarrow \quad p_s(\alpha) = 0, \\ D\alpha \neq \mathcal{H}_t(\alpha) & \Longleftrightarrow \quad p_n(\alpha) = 0. \end{cases}$$

But $\kappa_D X = 0$, so $\mathcal{H}_t = 0$, which proves the theorem. $\qquad\qquad\square$

Exercises

Exercise 3.1 (Logarithmic derivative identity). Let (R, D) be a differential ring, $u_1, \ldots, u_n \in R^*$ and $e_1, \ldots, e_n \in \mathbb{Z}$ be integers. Show that

$$\frac{D(u_1^{e_1} \cdots u_n^{e_n})}{u_1^{e_1} \cdots u_n^{e_n}} = e_1 \frac{Du_1}{u_1} + \cdots + e_n \frac{Du_n}{u_n}.$$

Exercise 3.2. Let $\mathbb{Q}$ be the field of rational numbers, x be an indeterminate over $\mathbb{Q}$, and D be the derivation d/dx on $\mathbb{Q}(x)$. Let $P = Y^2 - 2x^2 \in \mathbb{Q}(x)[Y]$, and y be a root of P, which is irreducible over $\mathbb{Q}(x)$. Show that

$$\mathrm{Const}_D(\mathbb{Q}(x, y)) = \mathbb{Q}(\alpha) \quad \text{where } \alpha = \frac{y}{x}.$$

Exercise 3.3. Let (F, D) be a differential field and (E, Δ) a differential extension of (F, D). Show that if $S \subset \mathrm{Const}_\Delta(E)$ is algebraically independent over $\mathrm{Const}_D(F)$, then S is algebraically independent over F.

Exercise 3.4. Let (k, D) be a differential field of characteristic 0, and $u \in k$ be such that $u^2 \neq -1$. Let E be a differential extension of k such that $\sqrt{-1} \in E$, and let $t_1, t_2 \in E$ be solutions of the following differential equations:

$$Dt_1 = \frac{Dv}{v} \quad \text{where} \quad v = \frac{u + \sqrt{-1}}{u - \sqrt{-1}}$$

i.e. t_1 is a logarithm of v, and

$$Dt_2 = \frac{Du}{1 + u^2}$$

i.e. t_2 is an arc-tangent of u. Show that $t_1\sqrt{-1} - 2t_2$ is a constant with respect to D.

Exercise 3.5. Let (k, D) be a differential field of characteristic 0, t be a monomial over k, and E be an algebraic extension of k. Show that

$$S_{E[t]:E}^{\mathrm{irr}} \neq \emptyset \Longrightarrow S_{k[t]:k}^{\mathrm{irr}} \neq \emptyset.$$

(Compare with Corollary 3.4.1).

Although we have defined normal and special polynomials in monomial extensions only, Rao [56] has defined them in any simple transcendental differential extension as follows: let (k, D) be a differential field of characteristic 0 and $(k(t), \Delta)$ be a differential extension of (k, D) where t is transcendental over k. Then, $\Delta t \in k(t)$, so let $a, b \in k[t]$ be such that $b \neq 0$, $\gcd(a, b) = 1$ and $\Delta t = a/b$. Define then $p \in k[t]$ to be *normal with respect to* Δ if $\gcd(p, b\Delta p) = 1$, and *special with respect to* Δ if $p \mid b\Delta p$. The following exercices all relate to this definition.

Exercise 3.6. Prove that if $\Delta t = a/b$ for $a, b \in k[t]$, then $b\Delta p \in k[t]$ for any $p \in k[t]$.

Exercise 3.7. Prove that all the parts of Theorem 3.4.1 remain true with the above definition.

Exercise 3.8. Prove the following analogue of Theorem 3.4.2: let $\overline{k}$ be the algebraic closure of k, and $p \in k[t]$ be squarefree. Then,

$$p \text{ normal} \iff b(\alpha)\Delta\alpha \neq a(\alpha) \text{ for all roots } \alpha \in \overline{k} \text{ of } p.$$

Exercise 3.9. Prove the following analogue of Theorem 3.4.3: let $\overline{k}$ be the algebraic closure of k, and $p \in k[t] \setminus \{0\}$. Then,

$$p \text{ special} \iff b(\alpha)\Delta\alpha = a(\alpha) \text{ for all roots } \alpha \in \overline{k} \text{ of } p.$$

Exercise 3.10. Prove that if $p \in k[t]$ is special, then $\gcd(p, b) = 1$.

4. The Order Function

We introduce in this chapter the *order function* at an element, which will be our main tool later when we prove the correctness of the integration algorithm. The usefulness of this function is that it maps elements of arbitrary unique factorization domains into integers, so applying it on both sides of an equation produces equations and inequalities involving integers, making it possible to either prove that the original equation cannot have a solution, or to compute estimates for the orders of its solutions. Therefore it is used in many contexts besides integration, for example in algorithms for solving differential equations. While we use only the order function at a polynomial in the integration algorithm, we introduce it here in unique factorization domains of arbitrary characteristic, and study its properties in the general case when the order is taken at an element that is not necessarily irreducible.

4.1 Basic Properties

Throughout this section and the next one, let D be a unique factorization domain of arbitrary characteristic, D^* be its group of units (Definition 1.1.4), F be its quotient field (example 1.1.14), and $a \in D$ be such that $a \neq 0$ and $a \notin D^*$.

Definition 4.1.1. *The* order at a *is the map* $\nu_a : D \to \mathbb{Z} \cup \{+\infty\}$ *given by:*

(i) $\nu_a(0) = +\infty$,
(ii) for $x \in D \setminus \{0\}, \nu_a(x) = \max\{n \in \mathbb{N}$ *such that* $a^n \mid x\}$.

Even though the map ν_a takes only nonnegative values, we define it as a map into $\mathbb{Z} \cup \{+\infty\}$ in order to extend it eventually to the quotient field of D. We first show that the set $S_a(x) = \{n \in \mathbb{N}$ such that $a^n \mid x\}$ is finite and nonempty for $x \in D \setminus \{0\}$. Since $a \neq 0$ and a is not a unit in D, let $p \in D$ be an irreducible factor of a. Then there is an irreducible factorization of x in which p appears with some exponent $e \geq 0$. Let $n > e$ and suppose that $p^n \mid x$. Then $x = p^n y$ for some $y \in D$. Let $y = u \prod_{i=1}^m p_i^{e_i}$ be the irreducible factorization of y, where the p_i's are coprime and u is a unit. We have then a factorization $x = up^n \prod_{i=1}^m p_i^{e_i}$ of x where p appears with exponent at least $n > e$, in contradiction with D being a unique factorization domain. Thus any q in $S_a(x)$ satisfies $q \leq e$, so $S_a(x)$ is finite. In addition $0 \in S_a(x)$, so ν_a is well-defined on D.

Lemma 4.1.1. *Let $x, y \in D$. Then,*

(i) $\nu_a(xy) \geq \nu_a(x) + \nu_a(y)$ *and equality holds if a is irreducible.*
(ii) $\nu_a(x + y) \geq \min(\nu_a(x), \nu_a(y))$ *and equality holds if $\nu_a(x) \neq \nu_a(y)$.*
(iii) *If $x \mid y$, then $\nu_a(x) \leq \nu_a(y)$.*
(iv) $\nu_a(\gcd(x, y)) = \min(\nu_a(x), \nu_a(y))$.

Proof. All the statements are trivial if either x or y is 0, so suppose that $x \neq 0 \neq y$. Let $n = \nu_a(x)$ and $m = \nu_a(y)$. Then $x = ca^n$ and $y = da^m$ for some $c, d \in D$, and a divides neither c nor d.
(i) we have $xy = cda^{n+m}$ so $\nu_a(xy) \geq n + m$. Suppose that a is irreducible. Then $a \nmid cd$ since it does not divide c or d, so $a^{n+m+1} \nmid xy$, which implies that $\nu_a(xy) = n + m$.
(ii) we can assume without loss of generality that $n \leq m$. We have then $x + y = a^n(c + da^{m-n})$ so $\nu_a(x + y) \geq n = \min(n, m)$. Suppose that $n \neq m$, then $m - n > 0$, so $a \mid da^{m-n}$, which implies that $a \nmid (c + da^{m-n})$ since $a \nmid c$. Hence, $\nu_a(x + y) = n$.
(iii) Suppose that $x \mid y$. Then $y = xz$ for some $z \in D$. Hence $\nu_a(y) = \nu_a(xz) \geq \nu_a(x) + \nu_a(z)$ by part (i), so $\nu_a(y) \geq \nu_a(x)$.
(iv) Let $g = \gcd(x, y)$. Then $g \mid x$ and $g \mid y$, so $\nu_a(g) \leq \nu_a(x)$ and $\nu_a(g) \leq \nu_a(y)$ by part (iii). Hence $\nu_a(g) \leq \min(\nu_a(x), \nu_a(y))$. Let $z = a^{\min(\nu_a(x), \nu_a(y))} \in D$. Then, $z \mid x$ and $z \mid y$, so $z \mid g$. Hence, $\nu_a(g) \geq \nu_a(z) = \min(\nu_a(x), \nu_a(y))$ by part (iii), so $\nu_a(g) = \min(\nu_a(x), \nu_a(y))$. $\square$

Example 4.1.1. In $\mathbb{Z}$ we have $\nu_6(12) = \nu_6(18) = 1$ and $\nu_6(12 \times 18) = \nu_6(216) = 3$, which shows that the equality in (i) above does not always hold if a is not irreducible. On the other hand, $\nu_3(12) = 1, \nu_3(18) = 2$ and $\nu_3(216) = 3 = 1 + 2$, as well as $\nu_2(12) = 2, \nu_2(18) = 1$ and $\nu_2(216) = 3 = 1 + 2$.

The following lemma shows that multiplying a or the argument of ν_a by a unit does not change the order function. This property is necessary in order to extend the definition of ν_a to F.

Lemma 4.1.2. *Let $u \in D^*$ and $x \in D$. Then:*

(i) $\nu_a(ux) = \nu_a(x) = \nu_{ua}(x)$.
(ii) $\nu_a(u) = 0$.

Proof. (i) If $x = 0$, then $\nu_a(ux) = \nu_a(x) = \nu_{ua}(x) = +\infty$, so suppose that $x \neq 0$. Then $a^{\nu_a(x)} \mid x$, so $a^{\nu_a(x)} \mid ux$ so $\nu_a(x) \leq \nu_a(ux)$. Since this inequality holds for any unit, and u^{-1} is also a unit in D, we have $\nu_a(ux) \leq \nu_a(u^{-1}ux) = \nu_a(x)$, so $\nu_a(x) = \nu_a(ux)$. Similarly, $a^{\nu_a(x)} \mid x$ implies that $(ua)^{\nu_a(x)} \mid x$ since $u^{\nu_a(x)}$ is a unit, so $\nu_a(x) \leq \nu_{ua}(x)$. As previously, this inequality applied to ua and u^{-1} implies that $\nu_{ua}(x) \leq \nu_{u^{-1}ua}(x) = \nu_a(x)$, so $\nu_a(x) = \nu_{ua}(x)$.
(ii) By (i), $\nu_a(u) = \nu_a(u^2)$. But $\nu_a(u^2) \geq 2\nu_a(u)$ by Lemma 4.1.1, so $\nu_a(u) \in \{0, +\infty\}$. Since $u \neq 0$, we must have $\nu_a(u) = 0$. $\square$

In the following definition and the rest of this chapter, we say that any two elements y and z in D have *no common factor* when $\gcd(y, z)$ is a unit in D.

Definition 4.1.2. *Let $x \in F^*$ and write $x = y/z$ where $y, z \in D$ have no common factor and $z \neq 0$. We then define $\nu_a(x) = \nu_a(y) - \nu_a(z)$.*

Let $x \in F^*$ and $y, z, t, w \in D$ be such that y and z have no common factor, t and w have no common factor, and $x = y/z = t/w$. Then $y/t = z/w = u \in D^*$, so $\nu_a(y) = \nu_a(ut) = \nu_a(t)$ and $\nu_a(z) = \nu_a(uw) = \nu_a(w)$ by Lemma 4.1.2, so $\nu_a(y) - \nu_a(z) = \nu_a(t) - \nu_a(w)$, which shows that ν_a is well-defined on F. In addition, $\nu_a(1) = 0$ by Lemma 4.1.2, so choosing $y = x$ and $z = 1$ when $x \in D$, we see that the above definition is compatible with the definition of ν_a on D. We note that parts (i) and (ii) of Lemma 4.1.1 do not remain valid over F: $\nu_6(5/3) = \nu_6(1/2) = 0$, but $\nu_6(5/3 \times 1/2) = \nu_6(5/3 + 1/2) = -1 < 0$. Those statements remain however true when a is irreducible.

Theorem 4.1.1. *Let $x, y \in F$ and suppose that a is irreducible in D. Then,*

(i) $\nu_a(xy) = \nu_a(x) + \nu_a(y)$.
(ii) if $x \neq 0$ then $\nu_a(x^m) = m\nu_a(x)$ for any $m \in \mathbb{Z}$.
(iii) $\nu_a(x + y) \geq \min(\nu_a(x), \nu_a(y))$ *and equality holds if $\nu_a(x) \neq \nu_a(y)$.*

Proof. Let $x, y \in F$ and write $x = b/c, y = d/e$ where $b, c, d, e \in D$, b and c have no common factor, d and e have no common factor and $c \neq 0 \neq e$. Since a is irreducible, we have $\nu_a(fg) = \nu_a(f) + \nu_a(g)$ for any $f, g \in D$ by Lemma 4.1.1.
(i) Let $h = \gcd(bd, ce)$, $f = bd/h$ and $g = ce/h$. We have $f, g, h \in D$, f and g have no common factors, and $xy = bd/ce = f/g$, so

$$
\begin{aligned}
\nu_a(xy) &= \nu_a(f) - \nu_a(g) = \nu_a(f) + \nu_a(h) - (\nu_a(g) + \nu_a(h)) \\
&= \nu_a(fh) - \nu_a(gh) = \nu_a(bd) - \nu_a(ce) \\
&= (\nu_a(b) - \nu_a(c)) + (\nu_a(d) - \nu_a(e)) = \nu_a(x) + \nu_a(y).
\end{aligned}
$$

(ii) $x^0 = 1$ is a unit in D, so $\nu_a(1) = 0$ by Lemma 4.1.2. Suppose that the statement holds for $m \geq 0$. Then,

$$\nu_a(x^{m+1}) = \nu_a(x^m x) = \nu_a(x^m) + \nu_a(x) = m\nu_a(x) + \nu_a(x) = (m + 1)\nu_a(x)$$

so it holds for $m + 1$. Thus (ii) holds for all $m \geq 0$. For $m < 0$ we have $0 = \nu_a(1) = \nu_a(x^m x^{-m}) = \nu_a(x^m) - m\nu_a(x)$, so $\nu_a(x^m) = m\nu_a(x)$. Thus (ii) holds for all $m \in \mathbb{Z}$.
(iii) $x + y = (be + cd)(ce)^{-1}$. Although $be + cd$ and ce may have common factors, we have

$$\nu_a(x + y) = \nu_a(be + cd) + \nu_a\left((ce)^{-1}\right) = \nu_a(be + cd) - \nu_a(ce)$$

by parts (i) and (ii). We can suppose without loss of generality that $\nu_a(x) \leq \nu_a(y)$, which implies that $\nu_a(b) - \nu_a(c) \leq \nu_a(d) - \nu_a(e)$, hence that $\nu_a(b) +$

$\nu_a(e) \leq \nu_a(d) + \nu_a(c)$. Thus, $\nu_a(be) \leq \nu_a(dc)$ so $\nu_a(be + cd) \geq \nu_a(be)$ by Lemma 4.1.1, so $\nu_a(x+y) \geq \nu_a(be) - \nu_a(ce) = \nu_a(b) - \nu_a(c) = \nu_a(x)$. Suppose that $\nu_a(x) < \nu_a(y)$, then $\nu_a(be) < \nu_a(dc)$ as above, so $\nu_a(be + cd) = \nu_a(be)$ by Lemma 4.1.1, so $\nu_a(x + y) = \nu_a(be) - \nu_a(ce) = \nu_a(x)$.   □

Parts (i) and (ii) of the above theorem show that the restriction that y and z have no common factor in Definition 4.1.2 can be removed if a is irreducible: for any $y, z \in F$ such that $x = y/z$, we have $\nu_a(x) = \nu_a(yz^{-1}) = \nu_a(y) - \nu_a(z)$.

In the case of polynomial rings, we need to study the effect of enlarging the constant field on the order function. It turns out that when an irreducible polynomial splits in an algebraic extension, then the order at the new irreducible factors remains the same as before for arguments that are defined over the ground field.

Theorem 4.1.2. *Let F be a field, E be a separable algebraic extension of F and x be an indeterminate over E. If $p \in F[x]$ is irreducible over F, then $\nu_p(f) = \nu_q(f)$ for any irreducible factor $q \in E[x]$ of p in $E[x]$, and any $f \in F(x)$.*

Proof. Let $q \in E[x]$ be any irreducible factor of p in $E[x]$ and write $p = qr$ with $r \in E[x]$. Let $h \in F[x]$ and $n = \nu_p(h) \geq 0$. Then $p^n \mid h$, so $h = p^n s = q^n r^n s$ with $s \in F[x]$, which implies that $q^n \mid h$. In addition, $p^{n+1} \nmid h$, so p does not divide s, which implies that $\gcd(p, s) = 1$ since p is irreducible in $F[x]$. Thus, $1 = ap + bs = arq + bs$ for some $a, b \in F[x]$, so $\gcd(q, s) = 1$. Suppose now that $q^m \mid h$ for some $m > n$. Then, $h = p^n s = q^n r^n s = q^m t$ for some $t \in E[x]$, so $r^n s = q^{m-n} t$, which implies that $q \mid r^n s$ in $E[x]$. Since q is irreducible in $E[x]$ and $\gcd(q, s) = 1$, $q \mid r^n$, which implies that $n > 0$ (otherwise q would be a unit) and that $q \mid r$, hence that $q^2 \mid p$, in contradiction with p squarefree in $E[x]$ (since E separable over F). Hence $q^m \nmid h$ for $m > n$, so $\nu_q(h) = n$.

Let now $f \in F(x)$ and write $f = a/b$ for $a, b \in F[x]$ and $b \neq 0$. Then, by Theorem 4.1.1 and the above proof,

$$\nu_p(f) = \nu_p(a) - \nu_p(b) = \nu_q(a) - \nu_q(b) = \nu_q(f).$$

□

4.2 Localizations

Definition 4.2.1. *We define the localization at a to be*

$$\mathcal{O}_a = \bigcap_{p \mid a} \{x \in F \text{ such that } \nu_p(x) \geq 0\}$$

where the intersection is taken over all the irreducible factors of a in D.

Intuitively, the localization at a is the set of all the fractions in F that can be written with a denominator having no common factor with a. If a is irreducible, the localization, which is then a local ring, can also be seen as the set of all the fractions in F with nonnegative order at a.

Lemma 4.2.1.

(i) $\mathcal{O}_a$ is a subring of F containing D.

(ii) $x \in \mathcal{O}_a \Longrightarrow \nu_a(x) \geq 0$.

(iii) $x \in a\mathcal{O}_a \Longleftrightarrow \nu_a(x) \geq 1$, where $a\mathcal{O}_a$ is the ideal generated by a in $\mathcal{O}_a$.

(iv)   If a is irreducible, then $x \in \mathcal{O}_a \Longleftrightarrow \nu_a(x) \geq 0$.

(v)   If a is irreducible, then $xa^{-\nu_a(x)} \in \mathcal{O}_a$ for any $x \in F^*$.

(vi)   If Δ is any derivation on D, then $\Delta\mathcal{O}_a \subseteq \mathcal{O}_a$.

Proof. (i) Let $p \in D$ be any irreducible factor of a, and $x, y \in \mathcal{O}_a$. Then $\nu_p(x) \geq 0$ and $\nu_p(y) \geq 0$. By Lemma 4.1.2, $\nu_p(-y) = \nu_p(y)$, so $\nu_p(x - y) \geq 0$ and $\nu_p(xy) \geq 0$ by Theorem 4.1.1. Since this holds for any irreducible factor p of a, $x - y \in \mathcal{O}_a$ and $xy \in \mathcal{O}_a$. Let $c \in D$. Then, $\nu_p(c) \geq 0$ for any irreducible factor p of a, hence $D \subseteq \mathcal{O}_a$, so in particular $0, 1 \in \mathcal{O}_a$, and $\mathcal{O}_a$ is a subring of F containing D.

(ii) Let $x \in \mathcal{O}_a$ and write $x = b/c$ where $b, c \in D$ have no common factor. Let $p \in D$ be any irreducible factor of a. Then $\nu_p(x) \geq 0$, so $\nu_p(b) - \nu_p(c) \geq 0$. Since $\nu_p(b)$ and $\nu_p(c)$ cannot be both nonzero (otherwise p would divide both b and c) and since they are both nonnegative, this implies that $\nu_p(c) = 0$, hence $p \nmid c$, so $a \nmid c$, so $\nu_a(c) = 0$, which implies that $\nu_a(x) = \nu_a(b) \geq 0$.

(iii) Let $x \in a\mathcal{O}_a$, then $x = ay$ for some $y \in \mathcal{O}_a$. Write $y = b/c$ where $b, c \in D$ have no common factor. From the proof of part (ii) we have $p \nmid c$ for any irreducible factor p of a, so c and ab have no common factor. Hence, $\nu_a(x) = \nu_a(ab/c) = \nu_a(ab) - \nu_a(c)$. But $\nu_a(c) = 0$ from the proof of part (ii), and $\nu_a(ab) \geq \nu_a(a) + \nu_a(b) \geq 1$ by Lemma 4.1.1. Hence, $\nu_a(x) \geq 1$. Conversely, let $x \in F^*$ be such that $\nu_a(x) \geq 1$, and write $x = b/c$ where $b, c \in D$ have no common factor. At most one of $\nu_a(b)$ and $\nu_a(c)$ can be nonzero, and $\nu_a(x) = \nu_a(b) - \nu_a(c) \geq 1$, so $\nu_a(c) = 0$ and $\nu_a(b) \geq 1$, which implies that $a \mid b$, so $b = ad$ for $d \in D$. Let $p \in D$ be any irreducible factor of a. Then, $p \mid b$, so $p \nmid c$, hence $\nu_p(d/c) = \nu_p(d) - \nu_p(c) = \nu_p(d) \geq 0$. Since this holds for any irreducible factor of a, we get $d/c \in \mathcal{O}_a$, hence $x = b/c = ad/c \in a\mathcal{O}_a$.

(iv) We have $x \in \mathcal{O}_a \Longrightarrow \nu_a(x) \geq 0$ by part (ii). Conversely, suppose that a is irreducible and let $x \in F$ be such that $\nu_a(x) \geq 0$. Let p be any irreducible factor of a in D. Then $p = ua$ for $u \in D^*$, so $\nu_p(x) = \nu_{ua}(x) = \nu_a(x)$ by Lemma 4.1.2. Thus $\nu_p(x) \geq 0$, so $x \in \mathcal{O}_a$.

(v) Suppose that a is irreducible, and let $x \in F^*$. Then, $\nu_a(xa^{-\nu_a(x)}) = \nu_a(x) - \nu_a(x) = 0$, so $xa^{-\nu_a(x)} \in \mathcal{O}_a$ by part (iv).

(vi) Let Δ be any derivation on D. Then, Δ can be extended uniquely to a derivation on F by Theorem 3.2.1. Let $x \in \mathcal{O}_a$ and write $x = b/c$ where $b, c \in D$ have no common factor and $c \neq 0$. Let $p \in D$ be any irreducible factor of a. Then, $\nu_p(x) = \nu_p(b) - \nu_p(c) \geq 0$ since $x \in \mathcal{O}_a$, so $\nu_p(c) = 0$, which implies that

$$\nu_p(\Delta x) = \nu_p\left(\frac{c\Delta b - b\Delta c}{c^2}\right) = \nu_p(c\Delta b - b\Delta c) - 2\nu_p(c) = \nu_p(c\Delta b - b\Delta c) \geq 0.$$

Since this holds for any irreducible factor of a, we get $\Delta x \in \mathcal{O}_a$, hence $\Delta \mathcal{O}_a \subset \mathcal{O}_a$.   □

Example 4.2.1. In $D = \mathbb{Z}$,

$$\mathcal{O}_6 = \mathcal{O}_2 \cap \mathcal{O}_3 = \{x \in \mathbb{Q} \text{ such that } x = b/c \text{ where } b, c \in \mathbb{Z}, 2 \nmid c \text{ and } 3 \nmid c\}$$

so $1/3 \notin \mathcal{O}_6$ although $\nu_6(1/3) = 0$, which shows that parts (iv) and (v) of the above lemma do not always hold if a is not irreducible. This makes it worth noticing that both directions of part (iii) of the lemma hold for non-irreducible a's.

When D is a principal ideal domain, for any proper nonzero ideal I of D, the canonical projection $\pi_I : D \to D/I$ can be extended naturally to the localization $\mathcal{O}_a$ for any generator a of I. The next definition constructs this extension.

Definition 4.2.2. *Let D be a principal ideal domain, and I be a proper nonzero ideal of D, i.e. $I \neq D$ and $I \neq (0)$, and $a \in D$ be a generator of I, i.e. $I = (a)$. We define the* value at a *to be the map $\pi_a : \mathcal{O}_a \to D/I$ given by: let $x \in \mathcal{O}_a$ and write $x = b/c$ where $b, c \in D$ have no common factor. We define $\pi_a(x)$ to be $\pi_I(bd)$ where $d, e \in D$ are such that $cd + ae = 1$ and π_I is the canonical projection from D onto D/I.*

In order to show that π_a is well-defined, we need to show that such d and e always exist, and that the value of $\pi_a(x)$ is independent of the choice of b, c, d and e. First, $a \neq 0$ since $I \neq (0)$, and $a \notin D^*$ since $I \neq D$, so $\mathcal{O}_a$ is defined. Let $x \in \mathcal{O}_a$ and write $x = b/c$ where $b, c \in D$ have no common factor. Let p be any irreducible factor of a. Since $x \in \mathcal{O}_a$ we have $\nu_p(x) = \nu_p(b) - \nu_p(c) \geq 0$. But at least one of $\nu_p(b)$ and $\nu_p(c)$ must be 0 since b and c have no common factor, so $\nu_p(c) = 0$, which implies that $p \nmid c$. Since this holds for any irreducible factor p of a, we have $\gcd(a, c) = 1$, so there are $d, e \in D$ such that $cd + ae = 1$. Suppose now that $cd + ae = cf + ag = 1$ for some $d, e, f, g \in D$. Then, $a(g - e) = c(d - f)$. Let p be any irreducible factor of a. We then have

$$\nu_p(c) + \nu_p(d - f) = \nu_p(c(d - f)) = \nu_p(a(g - e)) = \nu_p(a) + \nu_p(g - e) \geq \nu_p(a).$$

But $\nu_p(c) = 0$ as previously, so $\nu_p(d - f) \geq \nu_p(a)$, which implies that any irreducible $p \in D$ that appears in the factorization of a with a positive exponent n must appear with an exponent $m \geq n$ in the factorization of $d - f$, hence that $a \mid d - f$, i.e. $d - f \in I$, so $\pi_I(d - f) = 0$. Since π_I is a ring-homomorphism, we get $\pi_I(bd) = \pi_I(bf)$ so the value of $\pi_a(x)$ does not depend on the choice of d and e. Suppose finally that $x = b/c = b'/c'$ where $b, c, b', c' \in D$ and $\gcd(b, c) = \gcd(b', c') = 1$. As previously, this implies that $b' = ub$ and $c' = uc$ for some $u \in D^*$. Let $d, e \in D$ be such that $cd + ae = 1$. Then, $c'd' + ae = 1$ for $d' = u^{-1}d \in D$, and we have $b'd' = ubu^{-1}d = bd$, so the value of $\pi_a(x)$ does not depend on the choice of b and c, so π_a is well-defined on $\mathcal{O}_a$.

We next show that π_a is an extension of π_I to $\mathcal{O}_a$ which induces an isomorphism between $\mathcal{O}_a/a\mathcal{O}_a$ and D/I, i.e. that we have the following diagram:

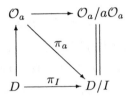

Theorem 4.2.1. *Let D be a principal ideal domain, I be a proper nonzero ideal of D and $a \in D$ be a generator of I. Then,*

(i) $\pi_a(b) = \pi_I(b)$ *for any $b \in D$ (i.e. π_a extends π_I).*

(ii) $\ker(\pi_a) = a\mathcal{O}_a$.

(iii) π_a *is a surjective ring-homomorphism from $\mathcal{O}_a$ onto D/I, hence a ring-isomorphism between $\mathcal{O}_a/a\mathcal{O}_a$ and D/I (a field-isomorphism if I is maximal).*

(iv) *If Δ is a derivation of D and $\Delta I \subseteq I$, then $\Delta^* \circ \pi_a = \pi_a \circ \Delta$ where Δ^* is the induced derivation on D/I (Lemma 3.1.2).*

Proof. (i) Let $b \in D$ and write $b = b/c$ with $c = 1$. Then $cd + ae = 1$ for $d = 1$ and $e = 0$, so $\pi_a(b) = \pi_I(bd) = \pi_I(b)$.

(ii) Let $x \in \mathcal{O}_a$ and write $x = b/c$ where $b, c \in D$ have no common factor. Let $d, e \in D$ be such that $cd + ae = 1$. Suppose first that $x \in a\mathcal{O}_a$. Then, $\nu_a(x) = \nu_a(b) - \nu_a(c) > 0$ by Lemma 4.2.1, so $\nu_a(b) > \nu_a(c) \geq 0$, so $a \mid b$, hence $b \in I$, which implies that $bd \in I$, therefore that $\pi_a(x) = \pi_I(bd) = 0$. Conversely, suppose that $\pi_a(x) = 0$. Then $\pi_I(bd) = 0$, so $bd \in I$, which implies that $a \mid bd$. But $\gcd(a, d) = 1$ since $cd + ae = 1$, hence $a \mid b$, so $\nu_a(b) > 0$. Also, $\gcd(a, c) = 1$ since $cd + ae = 1$, so $a \nmid c$, so $\nu_a(c) = 0$, hence $\nu_a(x) = \nu_a(b) - \nu_a(c) > 0$, so $x \in a\mathcal{O}_a$ by Lemma 4.2.1.

(iii) Since π_I is surjective and π_a is an extension of π_I by (i), it follows that π_a is surjective. Another consequence of (i) is $\pi_a(1) = \pi_I(1) = 1$. Let $x, x' \in \mathcal{O}_a$ and write $x = b/c, x' = b'/c'$ where $b, c, b', c' \in D$, b and c have no common factor, and b' and c' have no common factor. Write also $xx' = b''/c''$ where $b'', c'' \in D$ have no common factor. Then, $bb' = gb''$ and $cc' = gc''$ for some $g \in D$. Let $d, e, d', e' \in D$ be such that $cd + ae = 1$ and $c'd' + ae' = 1$. Multiplying those two equalities together, we get $cc'dd' + ah = 1$ where

$h = c'd'e + cde' + aee' \in D$. Hence, $c''(gdd') + ah = 1$, so, using the fact that π_I is a ring-homomorphism:

$$\pi_a(xx') = \pi_a(\frac{b''}{c''}) = \pi_I(b''gdd') = \pi_I(bb'dd') = \pi_I(bd)\pi_I(b'd') = \pi_a(x)\pi_a(x').$$

Write now $x + x' = b''/c''$ where $b'', c'' \in D$ have no common factor. Then, $bc' + b'c = gb''$ and $cc' = gc''$ for some $g \in D$. Let $d, e, d', e' \in D$ be such that $cd + ae = c'd' + ae' = 1$. As above, this implies that $c''(gdd') + ah = 1$ for some $h \in D$, so

$$
\begin{aligned}
\pi_a(x + x') &= \pi_a(\frac{b''}{c''}) = \pi_I(b''gdd') = \pi_I((bc' + b'c)dd') \\
&= \pi_I(bd)\pi_I(c'd') + \pi_I(b'd')\pi_I(cd) \\
&= \pi_a(x)\pi_I(c'd') + \pi_a(x')\pi_I(cd).
\end{aligned}
$$

From $1 = cd + ae$, we get $1 = \pi_I(1) = \pi_I(cd) + \pi_I(ae) = \pi_I(cd)$ since $a \in I$. Similarly, $\pi_I(c'd') = 1$, hence $\pi_a(x + x') = \pi_a(x) + \pi_a(x')$, so π_a is a ring-homomorphism. Since $\ker(\pi_a) = a\mathcal{O}_a$ by part (ii), this implies that π_a is a ring-isomorphism between $\mathcal{O}_a/a\mathcal{O}_a$ and D/I. If I is maximal, then D/I is a field, so π_a must be a field-isomorphism.

(iv) Let Δ be a derivation on D and suppose that $\Delta I \subseteq I$. Then, the induced derivation Δ^* on D/I satisfies $\Delta^* \circ \pi_I = \pi_I \circ \Delta$ by Lemma 3.1.2. Since $\Delta \mathcal{O}_a \subseteq \mathcal{O}_a$ by Lemma 4.2.1, $\pi_a \circ \Delta$ is defined on $\mathcal{O}_a$. Let $x \in \mathcal{O}_a$ and write $x = b/c$ where $b, c \in D$ have no common factor. Then, $\gcd(a, c) = 1$ as explained earlier, so $1 = ad + ce$ for some $d, e \in D$, which implies that $1 = \pi_a(1) = \pi_a(a)\pi_a(d) + \pi_a(c)\pi_a(e) = \pi_a(c)\pi_a(e)$, hence that $\pi_a(c)$ is a unit in D/I. In addition, $b = cx$, so $\pi_a(b) = \pi_a(cx) = \pi_a(c)\pi_a(x)$, so applying Δ^*, we get

$$\Delta^*\pi_a(b) = \Delta^*(\pi_a(c)\pi_a(x)) = \pi_a(c)\Delta^*\pi_a(x) + \pi_a(x)\Delta^*\pi_a(c). \tag{4.1}$$

From $b = cx$ also follows $\Delta b = c\Delta x + x\Delta c$, and applying π_a, we get

$$\pi_a(\Delta b) = \pi_a(c)\pi_a(\Delta x) + \pi_a(x)\pi_a(\Delta c). \tag{4.2}$$

But $\pi_a(\Delta b) = \pi_I(\Delta b) = \Delta^*\pi_I(b) = \Delta^*\pi_a(b)$ and $\pi_a(\Delta c) = \Delta^*\pi_a(c)$ in a similar way, so (4.2) becomes

$$\Delta^*\pi_a(b) = \pi_a(c)\pi_a(\Delta x) + \pi_a(x)\Delta^*\pi_a(c). \tag{4.3}$$

Equating (4.1) and (4.3) yields $\pi_a(c)\Delta^*\pi_a(x) = \pi_a(c)\pi_a(\Delta x)$. Since $\pi_a(c)$ is invertible in D/I, we get $\Delta^* \circ \pi_a = \pi_a \circ \Delta$. □

In the case when D is a Euclidean domain, we call $\pi_a(x)$ the *remainder of x at a*. It can be computed by the following algorithm, which has the same complexity as the extended Euclidean algorithm in D.

Remainder(x, a) (* Local remainder at a point *)

(* Given a Euclidean domain D, $a \in D \setminus \{0\}$ with $a \notin D^*$ and $x \in \mathcal{O}_a$, return $\pi_a(x)$ as an element of D. *)

$(b, c) \leftarrow$ **ExtendedEuclidean**$(a, \text{denominator}(x), 1)$
$(q, r) \leftarrow$ **PolyDivide**$(\text{numerator}(x)\ c, a)$
return r

4.3 The Order at Infinity

In the case of polynomial rings, we introduce an extra order function, called the order at infinity, which has properties similar to the order functions of the previous sections. While the usual degree function for polynomials can be used instead, the properties of the order at infinity can be generalized later to points at infinity on algebraic curves for which the degree is not defined. Let D be an integral domain of arbitrary characteristic and x an indeterminate over D throughout this section. For $a \in D[x]$, we use $\text{lc}(a)$ to denote the leading coefficient of a, *i.e.* if $a = a_0 + a_1 x + \cdots + a_n x^n$ with $a_n \neq 0$, then $\text{lc}(a) = a_n$.

Definition 4.3.1. *The* order at ∞ *is the map* $\nu_\infty : D(x) \to \mathbb{Z} \cup \{+\infty\}$ *given by* $\nu_\infty(0) = +\infty$, *and* $\nu_\infty(b/c) = \deg(c) - \deg(b)$ *for* $b, c \in D[x] \setminus \{0\}$.

Suppose that $f = b/c = d/e$ for $b, c, d, e \in D[x]$. Then $be = cd$, so $\deg(b) + \deg(e) = \deg(c) + \deg(d)$, so $\deg(c) - \deg(b) = \deg(e) - \deg(d)$, which implies that ν_∞ is well-defined on $D(x)$. We show next that ν_∞ satisfies the same properties than ν_a for an irreducible $a \in D[x]$.

Theorem 4.3.1. *Let* $f, g \in D(x)$. *Then,*

(i) $\nu_\infty(fg) = \nu_\infty(f) + \nu_\infty(g)$.
(ii) $\nu_\infty(f + g) \geq \min(\nu_\infty(f), \nu_\infty(g))$ *and equality holds if* $\nu_\infty(f) \neq \nu_\infty(g)$.
(iii) *if* $f \neq 0$ *then* $\nu_\infty(f^m) = m\nu_\infty(f)$ *for any* $m \in \mathbb{Z}$.

Proof. Let $f, g \in D(x)$ and write $f = b/c, g = d/e$ where $b, c, d, e \in D$ and $c \neq 0 \neq e$.
(i) $fg = bd/ce$ so $\nu_\infty(fg) = \deg(ce) - \deg(bd) = (\deg(c) - \deg(b)) + (\deg(e) - \deg(d)) = \nu_\infty(f) + \nu_\infty(g)$.
(ii) $f + g = (be + cd)/ce$, so $\nu_\infty(f + g) = \deg(ce) - \deg(be + cd)$. We can suppose without loss of generality that $\nu_\infty(f) \leq \nu_\infty(g)$, which implies that $\deg(c) - \deg(b) \leq \deg(e) - \deg(d)$, hence that $\deg(c) + \deg(d) \leq \deg(b) + \deg(e)$. Thus, $\deg(cd) \leq \deg(be)$ so $\deg(be + cd) \leq \deg(be)$, so $\nu_\infty(f + g) \geq \deg(ce) - \deg(be) = \deg(c) - \deg(b) = \nu_\infty(f)$. Suppose that $\nu_\infty(f) < \nu_\infty(g)$, then $\deg(cd) < \deg(be)$ as above, so $\deg(be + cd) = \deg(be)$, so $\nu_\infty(f + g) = \deg(ce) - \deg(be) = \nu_\infty(f)$.
(iii) This is trivial for $m = 1$. Suppose that it holds for $m > 0$. Then, $\nu_\infty(f^{m+1}) = \nu_\infty(f^m f) = \nu_\infty(f^m) + \nu_\infty(f) = (m + 1)\nu_\infty(f)$ so it holds

for $m + 1$. Thus (iii) holds for $m \geq 1$. For $m = 0$, $f^0 = 1 = 1/1$, so $\nu_\infty(1) = \deg(1) - \deg(1) = 0$. For $m < 0$ we have: $0 = \nu_\infty(1) = \nu_\infty(f^m f^{-m}) = \nu_\infty(f^m) - m\nu_\infty(f)$, so $\nu_\infty(f^m) = m\nu_\infty(f)$. Thus (iii) holds for any $m \in \mathbb{Z}$.
□

Since ν_∞ satisfies properties similar to ν_a, it is natural to define the notions of the localization and value map at infinity in a manner similar to what was done in the previous section at a point.

Definition 4.3.2. *We define the* localization at ∞ *to be*

$$\mathcal{O}_\infty = \{f \in D(x) \text{ such that } \nu_\infty(f) \geq 0\}.$$

Intuitively, $\mathcal{O}_\infty$, which is a local ring, is the set of all the rational functions in $D(x)$ for which the degree of the denominator is at least that of the numerator, *i.e.* which have no pole at infinity. As expected, $\mathcal{O}_\infty$ satisfies properties similar to $\mathcal{O}_a$ for an irreducible $a \in D[x]$.

Lemma 4.3.1.

(i) $\mathcal{O}_\infty$ *is a subring of* $D(x)$.
(ii)

$$f \in x^{-1}\mathcal{O}_\infty \iff \nu_\infty(f) \geq 1$$

   where $x^{-1}\mathcal{O}_\infty$ *is the ideal generated by* x^{-1} *in* $\mathcal{O}_\infty$.
(iii) $fx^{\nu_\infty(f)} \in \mathcal{O}_\infty$ *for any* $f \in D(x)^*$.

Proof. (i) Let $f, g \in \mathcal{O}_\infty$, and write $g = b/c$ for $b, c \in D[x]$. Then $\nu_\infty(f) \geq 0$ and $\nu_\infty(g) \geq 0$, so $\deg(b) \leq \deg(c)$. But $\deg(-b) = \deg(b)$, so $\nu_\infty(-g) = \nu_\infty(-b/c) \geq 0$, so $\nu_\infty(f - g) \geq 0$ and $\nu_\infty(fg) \geq 0$ by Theorem 4.3.1. Hence, $f - g \in \mathcal{O}_\infty$ and $fg \in \mathcal{O}_\infty$. In addition, $0 \in \mathcal{O}_\infty$ since $\nu_\infty(0) = +\infty$, and $1 \in \mathcal{O}_\infty$ since $\nu_\infty(1) = 0$, so $\mathcal{O}_\infty$ is a subring of $D(x)$.
(ii) Let $f \in x^{-1}\mathcal{O}_\infty$, then $f = g/x$ for some $g \in \mathcal{O}_\infty$, so $\nu_\infty(f) = \nu_\infty(g) - \nu_\infty(x) = \nu_\infty(g) + 1 \geq 1$. Conversely, let $f \in D(x)$ be such that $\nu_\infty(f) \geq 1$, and let $g = fx$. If $f = 0$, then $xf = 0 \in \mathcal{O}_\infty$, so $f \in x^{-1}\mathcal{O}_\infty$. Otherwise, $f \neq 0$ so $g \neq 0$ and we have $\nu_\infty(g) = \nu_\infty(f) + \nu_\infty(x) = \nu_\infty(f) - 1 \geq 0$, which implies that $g \in \mathcal{O}_\infty$, hence that $f = g/x \in x^{-1}\mathcal{O}_\infty$.
(iii) Let $f \in D(x)^*$. Then, $\nu_\infty\left(fx^{\nu_\infty(f)}\right) = \nu_\infty(f) - \nu_\infty(f) = 0$, so $fx^{\nu_\infty(f)} \in \mathcal{O}_\infty$.
□

Definition 4.3.3. *Let F be the quotient field of D. We define the* value at ∞ *to be the map* $\pi_\infty : \mathcal{O}_\infty \to F$ *given by:*

$$\pi_\infty(f) = \begin{cases} \mathrm{lc}(b)/\mathrm{lc}(c), & \text{if } \nu_\infty(f) = 0, \\ 0, & \text{if } \nu_\infty(f) > 0. \end{cases}$$

where $b, c \in D[x]$ *and* $f = b/c$.

Suppose that $f = b/c = d/e$ where $b, c, d, e \in D[x]$ and that $\nu_\infty(f) = 0$. Then, $be = cd$, so $\mathrm{lc}(b)\mathrm{lc}(e) = \mathrm{lc}(c)\mathrm{lc}(d)$, so $\mathrm{lc}(b)/\mathrm{lc}(c) = \mathrm{lc}(d)/\mathrm{lc}(e)$, which implies that π_∞ is well-defined on $\mathcal{O}_\infty$.

Theorem 4.3.2.

(i) $\ker(\pi_\infty) = x^{-1}\mathcal{O}_\infty$.
(ii) π_∞ is a surjective ring-homomorphism from $\mathcal{O}_\infty$ onto the quotient field F of D, hence a field-isomorphism between $\mathcal{O}_\infty/x^{-1}\mathcal{O}_\infty$ and F.

Proof. (i) Let $f \in \mathcal{O}_\infty$. If $f \in x^{-1}\mathcal{O}_\infty$, then $\nu_\infty(f) \geq 1$ by Lemma 4.3.1, so $\pi_\infty(f) = 0$ from the definition of π_∞. Conversely, suppose that $f \notin x^{-1}\mathcal{O}_\infty$, which implies that $\nu_\infty(f) = 0$, and write $f = b/c$ where $b, c \in D[x]$. The leading coefficients of b and c are never 0 by definition, so $\pi_\infty(f) \neq 0$. Hence $\ker(\pi_\infty) = x^{-1}\mathcal{O}_\infty$.
(ii) Let F be the quotient field of D and $\omega \in F$. If $\omega = 0$, then $\omega = \pi_\infty(0)$. Otherwise, write $\omega = b/c$ with $b, c \in D$ and $b \neq 0 \neq c$. Then, $\deg(b) = \deg(c) = 0$, so $\nu_\infty(b/c) = 0$, so $\pi_\infty(b/c) = b/c = \omega$. Hence, π_∞ is surjective. Taking $\omega = 1$ yields $\pi_\infty(1) = 1$. Let $f, g \in \mathcal{O}_\infty$. Then, $\nu_\infty(f) \geq 0$ and $\nu_\infty(g) \geq 0$. Suppose that $\nu_\infty(f) > 0$. Then, $\nu_\infty(fg) = \nu_\infty(f) + \nu_\infty(g) > 0$, so $\pi_\infty(fg) = 0 = \pi_\infty(f)\pi_\infty(g)$ since $\pi_\infty(f) = 0$. Similarly, $\pi_\infty(fg) = 0 = \pi_\infty(f)\pi_\infty(g)$ if $\nu_\infty(g) > 0$ so suppose that $\nu_\infty(f) = \nu_\infty(g) = 0$. Write $f = b/c$ and $g = d/e$ where $b, c, d, e \in D[x]$. Then, $\nu_\infty(fg) = 0$ by Theorem 4.3.1, so

$$\pi_\infty(fg) = \frac{\mathrm{lc}(bd)}{\mathrm{lc}(ce)} = \frac{\mathrm{lc}(b)}{\mathrm{lc}(c)}\frac{\mathrm{lc}(d)}{\mathrm{lc}(e)} = \pi_\infty(f)\pi_\infty(g)\,.$$

Suppose that $\nu_\infty(f) > 0$ and $\nu_\infty(g) > 0$. Then, $\nu_\infty(f + g) > 0$ by Theorem 4.3.1, so $\pi_\infty(f + g) = 0 = \pi_\infty(f) + \pi_\infty(g)$ since $\pi_\infty(f) = \pi_\infty(g) = 0$. We can now assume without loss of generality that $\nu_\infty(f) = 0$, *i.e.* that $\deg(b) = \deg(c)$. Suppose first that $\nu_\infty(g) > 0$. Then, $\deg(d) < \deg(e)$, so $\deg(cd) < \deg(be)$, so $\deg(be + cd) = \deg(be) = \deg(ce)$ and $\mathrm{lc}(be + cd) = \mathrm{lc}(be) = \mathrm{lc}(b)\mathrm{lc}(e)$. We also have $\nu_\infty(f + g) = \nu_\infty((be + cd)/ce) = 0$, so

$$\pi_\infty(f + g) = \frac{\mathrm{lc}(b)\mathrm{lc}(e)}{\mathrm{lc}(c)\mathrm{lc}(e)} = \frac{\mathrm{lc}(b)}{\mathrm{lc}(c)} = \pi_\infty(f) + \pi_\infty(g)$$

since $\pi_\infty(g) = 0$. Suppose finally that $\nu_\infty(g) = 0$. Then, $\deg(d) = \deg(e)$, so $\deg(cd) = \deg(be)$, so $\deg(be + cd) \leq \deg(be) = \deg(ce)$. If $\deg(be + cd) = \deg(be)$, then $\nu_\infty(f + g) = 0$ and $\mathrm{lc}(be + cd) = \mathrm{lc}(b)\mathrm{lc}(e) + \mathrm{lc}(c)\mathrm{lc}(d)$, so

$$\pi_\infty(f + g) = \frac{\mathrm{lc}(be + cd)}{\mathrm{lc}(ce)} = \frac{\mathrm{lc}(b)}{\mathrm{lc}(c)} + \frac{\mathrm{lc}(d)}{\mathrm{lc}(e)} = \pi_\infty(f) + \pi_\infty(g)\,.$$

If $\deg(be + cd) < \deg(be)$, then $\nu_\infty(f + g) > 0$ and $\mathrm{lc}(b)\mathrm{lc}(e) = \mathrm{lc}(c)\mathrm{lc}(d)$, so $\pi_\infty(f) = \pi_\infty(g)$, so $\pi_\infty(f) + \pi_\infty(g) = 0 = \pi_\infty(f + g)$. Hence, π_∞ is a ring-homomorphism. Since $\ker(\pi_\infty) = x^{-1}\mathcal{O}_\infty$ by part (i), this implies that π_∞ is a field-isomorphism between $\mathcal{O}_\infty/x^{-1}\mathcal{O}_\infty$ and the quotient field of D. $\quad\square$

ValueAtInfinity(f)   (* Value at infinity *)

(* Given a Euclidean domain D, and $f \in \mathcal{O}_\infty$, return $\pi_\infty(x)$. *)

if $f = 0$ **then return** 0
$a \leftarrow$ numerator(f), $b \leftarrow$ denominator(f)
if $\deg(b) > \deg(a)$ **then return** 0
return(lc(a)/lc(b))

4.4 Residues and the Rothstein–Trager Resultant

We present in this section the properties of the order function that are used for integration, namely the relation between the orders of a function and its derivative at a point, and the basic theory of residues in monomial extensions, up to the fundamental property of the Rothstein–Trager resultant. This relation and the various residue formulas let us connect the poles of a function to the poles of the functions that appear in its integral. Throughout this section, let K be a differential field of characteristic 0 with derivation D, and t be a monomial over K. We first define the notion of a residue at a normal polynomial.

Definition 4.4.1. *Let $p \in K[t] \setminus K$ be normal, and $\mathcal{R}_p$ be the set*

$$\mathcal{R}_p = \{f \in K(t) \text{ such that } pf \in \mathcal{O}_p\}.$$

We define the residue at p *to be the map* residue$_p : \mathcal{R}_p \to K[t]/(p)$ *given by*

$$\text{residue}_p(f) = \pi_p\left(f\frac{p}{Dp}\right).$$

Let $q \in K[t]$ be any irreducible factor of p. Then $q \nmid Dp$ since p is normal, so $1/Dp \in \mathcal{O}_q$. Since this holds for any irreducible factor of p, we have $1/Dp \in \mathcal{O}_p$. For $f \in \mathcal{R}_p$, $pf \in \mathcal{O}_p$, so $fp/Dp \in \mathcal{O}_p$, which means that residue$_p$ is well-defined. Since $\pi_p(a) = a$ for any $a \in K$, we identify K and $\pi_p(K) \subseteq K[t]/(p)$ when dealing with residues. Thus, when we say in the rest of this section that f has a residue $\alpha \in K$, we mean the residue of f is the image of an element of K by π_p.

Theorem 4.4.1. *Let $p \in K[t] \setminus K$ be normal. Then, $\mathcal{R}_p$ is a vector space over K, ker(residue$_p$) $= \mathcal{O}_p$, and residue$_p$ is a K-vector space isomorphism between $\mathcal{R}_p/\mathcal{O}_p$ and $K[t]/(p)$.*

Proof. We have $0, 1 \in \mathcal{R}_p$ since $0, p \in \mathcal{O}_p$. Let $f, g \in \mathcal{R}_p$ and $c \in K \subseteq \mathcal{O}_p$. Then, $pf, pg \in \mathcal{O}_p$, so $cpf + pg \in \mathcal{O}_p$ since $\mathcal{O}_p$ is a ring. Hence, $cf + g \in \mathcal{R}_p$, so $\mathcal{R}_p$ is a vector space over K.

Let $f \in \mathcal{O}_p$. We have $1/Dp \in \mathcal{O}_p$ as earlier, so $f/Dp \in \mathcal{O}_p$, so $pf/Dp \in p\mathcal{O}_p$, which implies that residue$_p(f) = \pi_p(pf/Dp) = 0$ by Theorem 4.2.1. Hence, $\mathcal{O}_p \subseteq$ ker(residue$_p$). Conversely, let $f \in$ ker(residue$_p$).

Then, $\pi_p(fp/Dp) = 0$, so $pf/Dp \in p\mathcal{O}_p$ by Theorem 4.2.1, which implies that $f/Dp \in \mathcal{O}_p$. But $Dp \in \mathcal{O}_p$, so $f = Dp(f/Dp) \in \mathcal{O}_p$. Hence $\ker(\text{residue}_p) = \mathcal{O}_p$.

Let $f, g \in \mathcal{R}_p$ and $c \in K$. Since π_p is a ring-homomorphism by Theorem 4.2.1 we have

$$\text{residue}_p(cf + g) = \pi_p((cf + g)\frac{p}{Dp}) = \pi_p(c)\pi_p(f\frac{p}{Dp}) + \pi_p(g\frac{p}{Dp})$$
$$= \pi_p(c)\text{residue}_p(f) + \text{residue}_p(g) .$$

But $c \in K$, so $\pi_p(c) = c$, hence $\text{residue}_p(cf + g) = c\,\text{residue}_p(f) + \text{residue}_p(g)$ in $K[t]/(p)$, so residue_p is a K-vector space homomorphism. Let $\omega \in K[t]/(p)$. Since π_p is surjective by Theorem 4.2.1, there exists $g \in \mathcal{O}_p$ such that $\pi_p(g) = \omega\pi_p(Dp)$. Let $f = g/p$. Then $pf \in \mathcal{O}_p$ so $f \in \mathcal{R}_p$, and

$$\text{residue}_p(f) = \pi_p(f\frac{p}{Dp}) = \frac{\pi_p(fp)}{\pi_p(Dp)} = \omega$$

hence residue_p is surjective. Since $\ker(\text{residue}_p) = \mathcal{O}_p$, this implies that residue_p is a K-vector space isomorphism between $\mathcal{R}_p/\mathcal{O}_p$ and $K[t]/(p)$. $\quad\square$

Example 4.4.1. Let $K = \mathbb{Q}$, t be a monomial over K with $Dt = 1$ (*i.e.* $D = d/dt$), and $p = t \in K[t]$ is normal and irreducible. We have $f = 1/t \in \mathcal{R}_p$ but $f^2 = 1/t^2 \notin \mathcal{R}_p$, so $\mathcal{R}_p$ is not a ring, even when p is normal and irreducible.

The following formula gives a useful relation between the residue at a normal polynomial and at any of its nontrivial factors.

Lemma 4.4.1. *Let $p \in K[t] \setminus K$ be normal, and $q \in K[t] \setminus K$ be a factor of p. Then, $\mathcal{R}_p \subseteq \mathcal{R}_q$ and $\text{residue}_q(f) = \pi_q(\text{residue}_p(f))$ for any $f \in \mathcal{R}_p$.*

Proof. Since $q \mid p$, we have $\mathcal{O}_p \subseteq \mathcal{O}_q$ and $\pi_q(\pi_p(g)) = \pi_q(g)$ for any $g \in \mathcal{O}_p$. Write $p = qr$ with $r \in K[t]$. Since p is normal, p is squarefree, so $\gcd(q, r) = 1$, which means that $1/r \in \mathcal{O}_q$. Let $f \in \mathcal{R}_p$. Then, $pf \in \mathcal{O}_q$, so $qf = pf/r \in \mathcal{O}_q$, which implies that $f \in \mathcal{R}_q$. Since p is normal, $\gcd(p, Dp) = 1$, so let $a, b \in K[t]$ be such that $aDp + bp = 1$. We have

$$arDq + (aDr + br)q = a(rDq + qDr) + brq = aDp + bp = 1 ,$$

so $\gcd(rDq, q) = 1$ and $\pi_q(a) = \pi_q(1/(rDq))$. Then,

$$\pi_q(\text{residue}_p(f)) = \pi_q\left(\pi_p\left(f\frac{p}{Dp}\right)\right) = \pi_q(\pi_p(fap)) = \pi_q(faqr)$$
$$= \pi_q\left(f\frac{qr}{rDq}\right) = \pi_q\left(f\frac{q}{Dq}\right) = \text{residue}_q(f) .$$

$\square$

Example 4.4.2. Let $K = \mathbb{Q}$, t be a monomial over K with $Dt = 1$ (*i.e.* $D = d/dt$), and $f = (t-2)/(t^2-1) \in K[t]$. Then,

$$\text{residue}_{t^2-1}(f) = \pi_{t^2-1}\left(\frac{t-2}{2t}\right) = \frac{1-2t}{2},$$

while

$$\text{residue}_{t-1}(f) = \pi_{t-1}\left(\frac{t-2}{t+1}\right) = -\frac{1}{2} = \pi_{t-1}\left(\frac{1-2t}{2}\right)$$

and

$$\text{residue}_{t+1}(f) = \pi_{t+1}\left(\frac{t-2}{t-1}\right) = \frac{3}{2} = \pi_{t+1}\left(\frac{1-2t}{2}\right).$$

Theorem 4.4.2. *Let $f \in K(t) \setminus \{0\}$ and $p \in K[t]$ be irreducible.*

(i)   *If p is normal, then $\nu_p(Df) = \nu_p(f) - 1$ if $\nu_p(f) \neq 0$, $\nu_p(Df) \geq 0$ if $\nu_p(f) = 0$. Furthermore,*

$$\pi_p(p^{1-\nu_p(f)}Df) = \nu_p(f)\pi_p(p^{-\nu_p(f)}f)\pi_p(Dp).$$

(ii) *$p \in S \Longrightarrow \nu_p(Df) \geq \nu_p(f)$.*
(iii) *$p \in S_1$ and $\nu_p(f) \neq 0 \Longrightarrow \nu_p(Df) = \nu_p(f)$.*

Proof. Let $p \in K[t]$ be irreducible, $f \in K(t) \setminus \{0\}$ and $n = \nu_p(f)$. Let $g = fp^{-n}$. By Lemma 4.2.1, $g \in \mathcal{O}_p$. Also,

$$Df = ngp^{n-1}Dp + p^n Dg. \tag{4.4}$$

Write $g = b/c$ where $b, c \in K[t]$ and $\gcd(b,c) = 1$. We have $\nu_p(g) = \nu_p(f) + \nu_p(p^{-n}) = n - n = 0$, so $\nu_p(b) - \nu_p(c) = 0$. But at most one of $\nu_p(b)$ and $\nu_p(c)$ can be nonzero since $\gcd(b,c) = 1$, so $\nu_p(b) = \nu_p(c) = 0$. We have

$$Dg = \frac{cDb - bDc}{c^2}$$

so $\nu_p(Dg) = \nu_p(bDc - cDb) - 2\nu_p(c) = \nu_p(bDc - cDb) \geq 0$ since $bDc - cDb \in K[t]$. By Lemma 4.2.1, this implies that $Dg \in \mathcal{O}_p$. Suppose that $n = 0$. Then $f = g$, so $Df = Dg \in \mathcal{O}_p$, so $\nu_p(Df) = \nu_p(Dg) \geq 0$. This implies that $\nu_p(pDf) > 0$, hence that $\pi_p(pDf) = 0$ by Theorem 4.2.1. This is valid regardless of whether p is normal or special, so (i) and (ii) hold when $n = 0$. Suppose now that $n \neq 0$.

(i) p is normal, so $\gcd(p, Dp) = 1$, so $\nu_p(Dp) = 0$. This implies that $Dp \in \mathcal{O}_p$ and

$$\nu_p(ngp^{n-1}Dp) = \nu_p(g) + n - 1 = n - 1 < n \leq \nu_p(p^n Dg)$$

so from (4.4) and Theorem 4.1.1 we get $\nu_p(Df) = n - 1$. We then have $p^{1-n}Df \in \mathcal{O}_p$ by Lemma 4.2.1, and from (4.4) we get $\pi_p(p^{1-n}Df) = \pi_p(ngDp + pDg)$. Since g, Dp, p and Dg are all in $\mathcal{O}_p$ and π_p is a ring-homomorphism, we have

$$\pi_p(p^{1-n}Df) = n\pi_p(g)\pi_p(Dp) + \pi_p(p)\pi_p(Dg) = n\pi_p(p^{-n}f)\pi_p(Dp)$$

since $\pi_p(p) = 0$.

(ii) $p \in \mathcal{S}$ so $p \mid Dp$, which means that $\nu_p(Dp) \geq 1$. Hence, $\nu_p(ngp^{n-1}Dp) \geq n$. Since $\nu_p(p^n Dg) = n + \nu_p(Dg) \geq n$, from (4.4) and Theorem 4.1.1 we get $\nu_p(Df) \geq n$.

(iii) Let $p \in \mathcal{S}_1$, and suppose that $n \neq 0$. Assume first that $p = t - \alpha$ for $\alpha \in K$. Then, $p_\alpha(\alpha)$ is not a logarithmic derivative of a K-radical, where $p_\alpha = (Dt - D\alpha)/(t - \alpha) = Dp/p$. Let $h = Dg + np_\alpha g$. Since $p \mid Dp$, $p_\alpha \in K[t]$, hence $p_\alpha \in \mathcal{O}_p$. In addition, $g \in \mathcal{O}_p$ and $Dg \in \mathcal{O}_p$ as seen above, so $h \in \mathcal{O}_p$. Since π_p is a ring-homomorphism, we have

$$\pi_p(h) = \pi_p(Dg + np_\alpha g) = \pi_p(Dg) + n\pi_p(g)\pi_p(p_\alpha).$$

We have $\nu_p(g) = 0$, so $g \notin p\mathcal{O}_p$ by Lemma 4.2.1, which implies that $\pi_p(g) \neq 0$ by Theorem 4.2.1. Suppose that $\pi_p(h) = 0$. Then, using the facts that (p) is a differential ideal of $K[t]$ (Lemma 3.4.3) and that $D^* \circ \pi_p = \pi_p \circ D$ (Theorem 4.2.1) where D^* is the induced derivation on $K[t]/(p)$ (Lemma 3.1.2), we get:

$$-np_\alpha(\alpha) = -n\pi_p(p_\alpha) = \frac{\pi_p(Dg)}{\pi_p(g)} = \frac{D^*\pi_p(g)}{\pi_p(g)} = \frac{D^*u}{u}$$

where $u = \pi_p(g) \in K[t]/(t-\alpha)$. But $K[t]/(t-\alpha) \simeq K$, and D^* is an extension of D by Lemma 3.4.3, so $u \in K$ and $D^*u = Du$, which implies that $p_\alpha(\alpha)$ is a logarithmic derivative of a K-radical, in contradiction with p being of the first kind. Hence $\pi_p(h) \neq 0$, so $\nu_p(h) = 0$ since $h \in \mathcal{O}_p$. From (4.4) we have $Df = (Dg + np_\alpha g)p^n = hp^n$, so $\nu_p(Df) = \nu_p(h) + \nu_p(p^n) = n$.

Let now p have arbitrary degree $m > 0$, let $\overline{K}$ be the algebraic closure of K, and $p = (t - \alpha_1) \cdots (t - \alpha_m)$ be the factorization of p in $\overline{K}[t]$. t is a monomial over $\overline{K}$ and p and the $t - \alpha_i$'s are in $\mathcal{S}_{1,\overline{K}[t]:\overline{K}}$ by Theorem 3.4.4. Then, $\nu_{t-\alpha_i}(f) = n$ for each i by Theorem 4.1.2, so $\nu_{t-\alpha_i}(Df) = n$ by the previous proof. Hence, $\nu_p(Df) = n$ by Theorem 4.1.2. $\square$

Example 4.4.3. Let $K = \mathbb{Q}$, t be a monomial over K with $Dt = 1$ (*i.e.* $D = d/dt$), $p = t \in K[t]$ is normal and irreducible, and $f = t^m + 1 \in K(t)$ for an arbitrary integer $m > 0$. We have $\nu_t(f) = 0$, but $Df = mt^{m-1}$, so $\nu_t(Df) = m - 1$. This shows that one cannot give a general upper bound on $\nu_p(Df)$ when $\nu_p(f) = 0$.

Theorem 4.4.2 has several useful consequences: $K\langle t \rangle$ must be a differential subring of $K(t)$, and we get formulas for the orders and residues of logarithmic derivatives, and for the residue at a given p.

Corollary 4.4.1. *Let $f \in K(t)$.*

(i) *f simple w.r.t $D \implies \nu_p(f) \geq -1$ for any normal irreducible $p \in K[t]$.*

(ii) *$f \in K\langle t \rangle \iff \nu_p(f) \geq 0$ for any normal irreducible $p \in K[t]$.*

(iii) *$K\langle t \rangle$ is a differential subring of $K(t)$.*

Proof. Let $f \in K(t)$ and write $f = a/b$ with $a, b \in K[t]$, $\gcd(a, b) = 1$, and $b \neq 0$. Let $p \in K[t]$ be normal irreducible.

(i) If f is simple, then b is normal hence squarefree. If $p \mid b$, then $p \nmid a$, so $\nu_p(f) = -\nu_p(b) \geq -1$ since b is squarefree.

(ii) If $f \in K\langle t \rangle$, then $b \in \mathcal{S}$, so $p \nmid b$, which implies that $\nu_p(f) = \nu_p(a) \geq 0$. Conversely, suppose that $\nu_q(f) \geq 0$ for any normal irreducible $q \in K[t]$, and let $p \in K[t]$ be a normal irreducible factor of b. Then, $p \nmid a$, so $\nu_p(f) = -\nu_p(b) < 0$ in contradiction with our hypothesis. Hence, all the irreducible factors of b are special, so $b \in \mathcal{S}$ by Theorem 3.4.1, which implies that $f \in K\langle t \rangle$.

(iii) $K\langle t \rangle$ is not empty since $K[t] \subseteq K\langle t \rangle$. Let $f, g \in K\langle t \rangle$ and $p \in K[t]$ be normal irreducible. Then, $\nu_p(f) \geq 0$ and $\nu_p(g) \geq 0$ by part (ii). We have $\nu_p(-g) = \nu_p(g)$ by Lemma 4.1.2, so $\nu_p(f - g) \geq \min(\nu_p(f), \nu_p(-g)) \geq 0$. Hence, $f - g \in K\langle t \rangle$ by part (ii). In addition, $\nu_p(fg) = \nu_p(f) + \nu_p(g) \geq 0$, so $fg \in K\langle t \rangle$, hence $K\langle t \rangle$ is a subring of $K(t)$. If $\nu_p(f) = 0$, then $\nu_p(Df) \geq 0$ by Theorem 4.4.2. Otherwise, $\nu_p(f) > 0$, so $\nu_p(Df) = \nu_p(f) - 1 \geq 0$ by Theorem 4.4.2. Thus $Df \in K\langle t \rangle$ in any case, so $K\langle t \rangle$ is a differential subring of $K(t)$. □

Corollary 4.4.2. *Let $f \in K(t) \setminus \{0\}$ and $p \in K[t]$ be irreducible. Then,*

(i) $\nu_p(Df/f) \geq -1$.
(ii) $\nu_p(Df/f) = -1 \iff \nu_p(f) \neq 0$ and p is normal.
(iii) If p is normal, then $\nu_p(Df) \neq -1$ and $\mathrm{residue}_p(Df/f) = \nu_p(f)$.

Proof. Let $p \in K[t]$ be irreducible, $f \in K(t) \setminus \{0\}$, $n = \nu_p(f)$ and $m = \nu_p(Df)$.

(i) By Theorem 4.4.2, either $m \geq n$ or $m = n - 1$, so $\nu_p(Df/f) = m - n \geq -1$ in any case.

(ii) Suppose that $n \neq 0$ and p is normal. Then, by Theorem 4.4.2, $m = n - 1$, so $\nu_p(Df/f) = -1$. Conversely, suppose that $\nu_p(Df/f) = -1$, then $m = n - 1 < n$. By Theorem 4.4.2, $m \geq n$ if either $p \in \mathcal{S}$ or $n = 0$, so p must be normal and $n \neq 0$.

(iii) Suppose that p is normal. If $n \geq 0$, then $m \geq 0$ by Theorem 4.4.2. If $n < 0$, then $m = n - 1 < -1$ by Theorem 4.4.2, so $m \neq -1$. If $n = 0$, then $\nu_p(Df/f) > -1$ by parts (i) and (ii), so $Df/f \in \mathcal{O}_p$, which implies that $\mathrm{residue}_p(Df/f) = 0$ by Theorem 4.4.1. Suppose now that $n \neq 0$. By (ii) we have $\nu_p(Df/f) = -1$, hence $\nu_p(pDf/f) = 0$ so $pDf/f \in \mathcal{O}_p$. By Theorem 4.4.2, we have $\nu_p(Df) = n - 1$ and $\pi_p(p^{1-n}Df) = n\pi_p(p^{-n}f)\pi_p(Dp)$. Since $p, Dp, p^{1-n}Df$ and $p^{-n}f$ are all in $\mathcal{O}_p$ and π_p is a field-homomorphism by Theorem 4.2.1 (p is irreducible), we have

$$
\begin{aligned}
\mathrm{residue}_p(\frac{Df}{f}) &= \pi_p(\frac{Df}{f}\frac{p}{Dp}) = \pi_p(\frac{p^{1-n}Df}{p^{-n}fDp}) \\
&= \frac{\pi_p(p^{1-n}Df)}{\pi_p(p^{-n}f)\pi_p(Dp)} = \frac{n\pi_p(p^{-n}f)\pi_p(Dp)}{\pi_p(p^{-n}f)\pi_p(Dp)} = n \, .
\end{aligned}
$$

$\square$

Lemma 4.4.2. *Let $p \in K[t]$ be normal irreducible, $g \in \mathcal{O}_p$ and $d \in K[t]$ be such that $\nu_p(d) = 1$. Then, $\mathrm{residue}_p(g/d) = \pi_p(g/Dd)$.*

Proof. Since $\nu_p(d) = 1$, $\nu_p(Dd) = 0$ by Theorem 4.4.2, so $\nu_p(g/Dd) = \nu_p(g) \geq 0$, which implies that $g/Dd \in \mathcal{O}_p$ by Lemma 4.2.1. In addition, $\nu_p(pg/d) = 1 + \nu_p(g) - 1 = \nu_p(g) \geq 0$, so $g/d \in \mathcal{R}_p$, so both $\mathrm{residue}_p(g/d)$ and $\pi_p(g/Dd)$ are defined. Write $d = pq$ for some $q \in K[t]$, and let $h = gpDq/qDpDd$. Then, $\nu_p(q) = \nu_p(d) - \nu_p(p) = 0$, $\nu_p(Dp) = 0$ since p is normal, $\nu_p(Dd) = \nu_p(d) - 1 = 0$ by Theorem 4.4.2, so $\nu_p(h) = \nu_p(g) + 1 + \nu_p(Dq) \geq 1$. This implies that $h \in p\mathcal{O}_p$ by Lemma 4.2.1, hence that $\pi_p(h) = 0$ by Theorem 4.2.1. In addition, we have

$$\frac{g}{Dd} + h = g\left(\frac{1}{Dd} + \frac{pDq}{qDpDd}\right) = g\frac{qDp + pDq}{qDpDd} = g\frac{1}{qDp} = \frac{g}{d}\frac{p}{Dp}$$

so

$$\mathrm{residue}_p\left(\frac{g}{d}\right) = \pi_p\left(\frac{g}{d}\frac{p}{Dp}\right) = \pi_p\left(h + \frac{g}{Dd}\right)$$
$$= \pi_p(h) + \pi_p\left(\frac{g}{Dd}\right) = \pi_p\left(\frac{g}{Dd}\right).$$

$\square$

Lemma 4.4.3. *Let $q \in K[t]$ be normal irreducible and $f \in K(t)$ be such that $\nu_q(f) = -1$. Write $f = p + a/d$ where $p, a, d \in K[t]$, $d \neq 0$, $\deg(a) < \deg(d)$ and $\gcd(a, d) = 1$. Then, for any $\alpha \in K$,*

$$q \mid \gcd(a - \alpha Dd, d) \iff \mathrm{residue}_q(f) = \alpha.$$

Proof. Since $\nu_q(f) = -1$, we have $\nu_q(a) = 0$ and $\nu_q(d) = 1$, so $\nu_q(Dd) = 0$ by Theorem 4.4.2. This implies that $Dd \in \mathcal{O}_q$ and that $\nu_q(1/Dd) = 0$, hence that $1/Dd \in \mathcal{O}_q$. Furthermore, $a, p \in \mathcal{O}_q$ and $f = (a + pd)/d$, so $\mathrm{residue}_q(f) = \pi_q((a + pd)/Dd) = \pi_q(a/Dd)$ by Lemma 4.4.2.

Suppose that $q \mid \gcd(a - \alpha Dd, d)$. Then,

$$0 = \pi_q(a - \alpha Dd) = \pi_q(a) - \alpha\pi_q(Dd)$$

so

$$\alpha = \frac{\pi_q(a)}{\pi_q(Dd)} = \pi_q\left(\frac{a}{Dd}\right) = \mathrm{residue}_q(f).$$

Conversely, suppose that $\alpha = \mathrm{residue}_q(f) = \pi_q(a/Dd)$. Then,

$$\pi_q(a - \alpha Dd) = \pi_q(a) - \alpha\pi_q(Dd) = \pi_q(a) - \pi_q(\frac{a}{Dd})\pi_q(Dd) = 0$$

so $q \mid a - \alpha Dd$, hence $q \mid \gcd(a - \alpha Dd, d)$. $\square$

We can now state the fundamental property of the Rothstein–Trager resultant, namely that from any simple function, one can construct a polynomial over K whose nonzero roots in K are exactly the residues of f that are in K. Note that in general, not all the residues of f are in K, expect when K is algebraically closed.

Theorem 4.4.3. *Let $f \in K(t)$ be simple w.r.t D, and write $f = p + a/d$ where $p, a, d \in K[t]$, $d \neq 0$, $\deg(a) < \deg(d)$, and $\gcd(a, d) = 1$. Let*

$$r = \text{resultant}_t(a - zDd, d) \in K[z] \tag{4.5}$$

where z is an indeterminate over K. Then, for any $\alpha \in K^*$,

$$r(\alpha) = 0 \iff \text{residue}_q(f) = \alpha \text{ for some normal irreducible } q \in K[t].$$

We call the polynomial r given by (4.5) the Rothstein–Trager resultant *of f.*

Proof. For any $\beta \in K^*$, let $r_\beta = \text{resultant}_t(a - \beta Dd, d) \in K$, and $\sigma_\beta : K[z] \to K$ be the ring homomorphism given by $\sigma_\beta(z) = \beta$ and $\sigma_\beta(x) = x$ for any $x \in K$. Define $\overline{\sigma}_\beta : K[z][t] \to K[t]$ by $\overline{\sigma}_\beta(\sum a_j t^j) = \sum \sigma_\beta(a_j) t^j$. Since $\overline{\sigma}_\beta(d) = d$, $\deg_t(\overline{\sigma}_\beta(d)) = \deg_t(d)$, so $r(\beta) = \overline{\sigma}_\beta(r) = \pm \text{lc}(d)^{m_\beta} r_\beta$ for some nonnegative integer m_β by Theorem 1.4.3.

Recall that f simple means that d is normal, hence squarefree, *i.e.* that $\nu_p(f) \geq -1$ for any normal irreducible $p \in K[t]$. Let $\alpha \in K^*$ be such that $r(\alpha) = 0$. Then, $r_\alpha = 0$, so $\deg(g) > 0$ by Corollary 1.4.2 where $g = \gcd(a - \alpha Dd, d)$. Let then $q \in K[t]$ be an irreducible factor of g. Since $q \mid d$ and f is simple, q is normal. Also, $\nu_q(d) = 1$ since d is squarefree, so $\alpha = \text{residue}_q(f)$ by Lemma 4.4.3.

Conversely suppose that $\text{residue}_q(f) = \alpha \in K^*$ for some normal irreducible $q \in K[t]$. Then, $\text{residue}_q(f) \neq 0$, so $f \notin \mathcal{O}_q$ by Theorem 4.4.1, which implies that $\nu_q(f) = -1$. Hence, $\nu_q(d) = 1$, so $q \mid \gcd(a - \alpha Dd, d)$ by Lemma 4.4.3. Therefore, $r_\alpha = 0$ by Corollary 1.4.2, so $r(\alpha) = 0$.   □

Let F be a field of characteristic 0, x be an indeterminate over F, and D be the derivation d/dx on $F(x)$. Since every irreducible $q \in F[x]$ is normal with respect to d/dx, applying the above result to $K = \overline{F}$, we see that Theorem 4.4.3 and Lemma 4.4.3 respectively prove parts (i) and (ii) of Theorem 2.4.1.

There are similar results relating the order at infinity of an element of $K(t)$ and its derivative.

Theorem 4.4.4. *Let $f \in K(t) \setminus \{0\}$. Then,*

(i) $\nu_\infty(Df) \geq \nu_\infty(f) - \max(0, \delta(t) - 1)$.
(ii)   *If t is nonlinear and $\nu_\infty(f) \neq 0$, then equality holds in (i), and*

$$\pi_\infty \left(t^{1-\delta(t)} \frac{Df}{f} \right) = -\nu_\infty(f)\lambda(t).$$

(iii) If t is nonlinear and $\nu_\infty(f) = 0$, then the strict inequality holds in (i), i.e. $\nu_\infty(Df) > 1 - \delta(t)$, and

$$\pi_\infty\left(t^{1-\delta(t)}\frac{Df}{f}\right) = 0.$$

Proof. Write $f = a/d$ where $a, d \in K[t]$, $d \neq 0$ and $\gcd(a, d) = 1$. Then, $Df = (dDa - aDd)/d^2$, so $\nu_\infty(Df) = 2\deg(d) - \deg(dDa - aDd)$. Let $m = \max(0, \delta(t) - 1)$.

(i) By Lemma 3.4.2, $\deg(Dd) \leq \deg(d) + m$ and $\deg(Da) \leq \deg(a) + m$. Hence, $\deg(dDa - aDd) \leq \deg(a) + \deg(d) + m$, so $\nu_\infty(Df) \geq \deg(d) - \deg(a) - m = \nu_\infty(f) - m$.

(ii) Suppose that t is nonlinear and $\nu_\infty(f) \neq 0$. Then $m = \delta(t) - 1$ and $Df \neq 0$ by Lemma 3.4.5. Suppose that $\deg(d) = 0$, then $\deg(a) \neq 0$ since $\nu_\infty(f) \neq 0$, so $\deg(dDa - aDd) = \deg(Da) = \deg(a) + m$ by Lemma 3.4.2, which implies that $\nu_\infty(Df) = -\deg(a) - m = \nu_\infty(f) - m$, hence that $\nu_\infty(t^{-m}Df/f) = 0$. Furthermore, $\mathrm{lc}(dDa - aDd) = \mathrm{lc}(dDa) = d\deg(a)\mathrm{lc}(a)\lambda(t)$ also by Lemma 3.4.2. Hence,

$$\pi_\infty\left(t^{-m}\frac{Df}{f}\right) = \frac{d\deg(a)\mathrm{lc}(a)\lambda(t)}{d^2}\frac{d}{\mathrm{lc}(a)} = \deg(a)\lambda(t) = -\nu_\infty(f)\lambda(t).$$

Suppose now that $\deg(a) = 0$, then $\deg(d) \neq 0$ since $\nu_\infty(f) \neq 0$, so $\deg(dDa - aDd) = \deg(Dd) = \deg(d) + m$ by Lemma 3.4.2, which implies that $\nu_\infty(Df) = \deg(d) - m = \nu_\infty(f) - m$, hence that $\nu_\infty(t^{-m}Df/f) = 0$. Furthermore, $\mathrm{lc}(dDa - aDd) = \mathrm{lc}(-aDd) = -a\deg(d)\mathrm{lc}(d)\lambda(t)$ also by Lemma 3.4.2. Hence,

$$\pi_\infty\left(t^{-m}\frac{Df}{f}\right) = -\frac{a\deg(d)\mathrm{lc}(d)\lambda(t)}{\mathrm{lc}(d)^2}\frac{\mathrm{lc}(d)}{a} = -\deg(d)\lambda(t) = -\nu_\infty(f)\lambda(t).$$

Suppose finally that $\deg(a) \neq 0$ and $\deg(d) \neq 0$. Then, by Lemma 3.4.2, the leading term of $dDa - aDd$ is

$$\mathrm{lc}(d)\deg(a)\mathrm{lc}(a)\lambda(t)t^{\deg(a)+\deg(d)+m}$$
$$-\mathrm{lc}(a)\deg(d)\mathrm{lc}(d)\lambda(t)t^{\deg(a)+\deg(d)+m}$$
$$= -\nu_\infty(f)\mathrm{lc}(a)\mathrm{lc}(d)\lambda(t)t^{\deg(a)+\deg(d)+m}. \quad (4.6)$$

Since $\nu_\infty(f) \neq 0$, this gives $\deg(dDa - aDd) = \deg(a) + \deg(d) + m$, hence $\nu_\infty(Df) = \nu_\infty(f) - m$, so $\nu_\infty(t^{-m}Df/f) = 0$. Furthermore,

$$\pi_\infty\left(t^{-m}\frac{Df}{f}\right) = -\frac{\nu_\infty(f)\mathrm{lc}(a)\mathrm{lc}(d)\lambda(t)}{\mathrm{lc}(d)^2}\frac{\mathrm{lc}(d)}{\mathrm{lc}(a)} = -\nu_\infty(f)\lambda(t).$$

(iii) Suppose that t is nonlinear and $\nu_\infty(f) = 0$. Then $m = \delta(t) - 1$ and $\deg(a) = \deg(d)$. If $Df = 0$, then $\nu_\infty(Df) = +\infty > -m$, so suppose that $Df \neq 0$. If $\deg(a) \neq 0$, then $\deg(dDa - aDd) < \deg(a) + \deg(d) + m$ by (4.6), so $\nu_\infty(Df) = 2\deg(d) - \deg(dDa - aDd) > -m$. If $\deg(a) = \deg(d) = 0$, then $f \in K$, so $Df \in K$, which implies that $\nu_\infty(Df) = 0 > -m$. Hence, $\nu_\infty(t^{-m}Df/f) > 0$, so $\pi_\infty(t^{-m}Df/f) = 0$. □

Exercises

Exercise 4.1. Let (k, D) be a differential field of characteristic 0, t a monomial over k, and $f \in k(t)$ be simple. Show that if there are $h \in k\langle t \rangle$ and $g \in k(t)$ such that $Dg = f + h$, then $f \in k[t]$.

Exercise 4.2. Let (k, D) be a differential field of characteristic 0, t a monomial over k, and $f \in k(t)$.

a) Show that if f is the logarithmic derivative of a nonzero element of $k(t)$, then f is simple and can be written as

$$f = p + \frac{a}{d}$$

where $p, a, d \in k[t]$, $\deg(p) < \max(1, \delta(t))$, $d \neq 0$, $\deg(a) < \deg(d)$, $\gcd(a, d) = 1$, and d is normal. Furthermore, all the roots in $\overline{k}$ of $r = \text{resultant}_t(a - zDd, d)$ are integers.

b) Show that if f is the logarithmic derivative of a $k(t)$-radical, then f is simple and can be written as

$$f = p + \frac{a}{d}$$

where $p, a, d \in k[t]$, $\deg(p) < \max(1, \delta(t))$, $d \neq 0$, $\deg(a) < \deg(d)$, $\gcd(a, d) = 1$, and d is normal. Furthermore, all the roots in $\overline{k}$ of $r = \text{resultant}_t(a - zDd, d)$ are rational numbers.

Exercise 4.3 (Indicial equation of a linear differential operator).
Let (k, D) be a differential field of characteristic 0, t a monomial over k, $p \in k[t]$ be normal and irreducible, and $f \in k(t)$ be such that $\nu_p(f) < 0$.

a) Show that $\nu_p(D^n f) = \nu_p(f) - n$ for any $n \in \mathbb{N}$.
b) Show that

$$\pi_p\left(p^{n-\nu_p(f)} D^n f\right) = \pi_p\left(p^{-\nu_p(f)}(Dp)^n f\right) \prod_{i=0}^{n-1} (\nu_p(f) - i)$$

for any $n \in \mathbb{N}$.

c) Let $n \in \mathbb{N}$ and $a_0, a_1, \ldots, a_n \in k(t)$ be such that $n > 0$ and $a_n \neq 0$. Let $\mu = \max_{0 \le i \le n}(i - \nu_p(a_i))$,

$$P(z) = \sum_{\substack{0 \le i \le n \\ i - \nu_p(a_i) = \mu}} \pi_p\left(p^{-\nu_p(a_i)}(Dp)^i a_i\right) \prod_{j=0}^{i-1}(z - j) \quad \in k[t]/(p)[z]$$

and $R(z) = \text{resultant}_t(p, P) \in k[z]$. Show that either

$$\nu_p\left(\sum_{i=0}^{n} a_i D^i f\right) = \nu_p(f) - \mu$$

or $P(\nu_p(f)) = R(\nu_p(f)) = 0$.

5. Integration of Transcendental Functions

Having developped the required machinery in the previous chapters, we can now describe the integration algorithm. In this chapter, we define formally the integration problem in an algebraic setting, prove the main theorem of symbolic integration (Liouville's Theorem), and describe the main part of the integration algorithm.

From now on, and without further mention, all the fields in this book are of characteristic 0. We also use the convention throughout that $\deg(0) = -\infty$.

5.1 Elementary and Liouvillian Extensions

We give in this section precise definitions of elementary functions, and of the problem of integrating functions in finite terms. Throughout this section, let k be a differential field and K a differential extension of k.

Definition 5.1.1. $t \in K$ is a primitive over k if $Dt \in k$. $t \in K^*$ is an hyperexponential over k if $Dt/t \in k$. $t \in K$ is Liouvillian over k if t is either algebraic, or a primitive or an hyperexponential over k. K is a Liouvillian extension of k if there are $t_1, \ldots, t_n$ in K such that $K = k(t_1, \ldots, t_n)$ and t_i is Liouvillian over $k(t_1, \ldots, t_{i-1})$ for i in $\{1, \ldots, n\}$.

We write $t = \int a$ when t is a primitive over k such that $Dt = a$, and $t = e^{\int a}$ when t is an hyperexponential over k such that $Dt/t = a$. Given that t is Liouvillian over k, we need to know whether t is algebraic or transcendental over k. We show that there are simple necessary and sufficient conditions that guarantee that a primitive or hyperexponential is in fact a monomial over k.

Lemma 5.1.1. *If t is a primitive over k and Dt is not the derivative of an element of k, then Dt is not the derivative of an element of any algebraic extension of k.*

Proof. Let t be a primitive over k, $a = Dt$, and suppose that a is not the derivative of an element of k. Let E be any algebraic extension of k, and suppose that $D\alpha = a$ for some $\alpha \in E$. Let Tr be the trace map from $k(\alpha)$ to k, $n = [k(\alpha) : k]$, and $b = Tr(\alpha)/n \in k$. By Theorem 3.2.4,

$$Db = \frac{1}{n}D(Tr(\alpha)) = \frac{1}{n}Tr(D\alpha) = \frac{1}{n}Tr(a) = a$$

in contradiction with $Du \neq a$ for any $u \in k$.   □

Theorem 5.1.1. *If t is a primitive over k and Dt is not the derivative of an element of k, then t is a monomial over k, $\mathrm{Const}(k(t)) = \mathrm{Const}(k)$, and $S = k$ (i.e. $S^{\mathrm{irr}} = S_1^{\mathrm{irr}} = \emptyset$). Conversely, if t is transcendental and primitive over k, and $\mathrm{Const}(k(t)) = \mathrm{Const}(k)$, then Dt is not the derivative of an element of k.*

Proof. Let t be a primitive over k, $a = Dt$, $\overline{k}$ be the algebraic closure of k, and suppose that a is not the derivative of an element of k. Then, $D\alpha \neq a$ for any $\alpha \in \overline{k}$ by Lemma 5.1.1, so t must be transcendental over k, hence it is a monomial over k. Suppose that $p \in S \setminus k$. Let then $\beta \in \overline{k}$ be a root of p. Then, $D\beta = Dt = a$ by Theorem 3.4.3, in contradiction with $D\alpha \neq a$ for any $\alpha \in \overline{k}$, so $p \in k$. Conversely, $k \subseteq S$ by definition. Let $c \in \mathrm{Const}(k(t))$. By Lemma 3.4.5, both the numerator and denominator of c must be special, hence in k, so $c \in k$, which implies that $\mathrm{Const}(k(t)) \subseteq \mathrm{Const}(k)$. The reverse inclusion is given by Lemma 3.3.1, so $\mathrm{Const}(k(t)) = \mathrm{Const}(k)$.

Conversely, let t be a transcendental primitive over k and suppose that $\mathrm{Const}(k(t)) = \mathrm{Const}(k)$. If there exists $b \in k$ such that $Dt = Db$, then $c = t - b \in \mathrm{Const}(k(t))$, so $c \in k$ in contradiction with t transcendental over k. Hence Dt is not the derivative of an element in k.   □

Theorem 5.1.2. *If t is an hyperexponential over k and Dt/t is not a logarithmic derivative of a k-radical, then t is a monomial over k, $\mathrm{Const}(k(t)) = \mathrm{Const}(k)$, and $S^{\mathrm{irr}} = S_1^{\mathrm{irr}} = \{t\}$. Conversely, if t is transcendental and hyperexponential over k, and $\mathrm{Const}(k(t)) = \mathrm{Const}(k)$, then Dt/t is not a logarithmic derivative of a k-radical.*

Proof. Let t be an hyperexponential over k, $a = Dt/t$, $\overline{k}$ be the algebraic closure of k, and suppose that a is not a logarithmic derivative of a k-radical. We have $Dt/t = a$ and a is not a logarithmic derivative of a $\overline{k}$-radical by Lemma 3.4.8, so t must be transcendental over k, hence it is a monomial over k since $Dt = at$.

Let $p = bt^m$ for $b \in k$ and $m \geq 0$. Then, $Dp = (Db + mab)t^m$, so $p \mid Dp$, which means that $p \in S$. Let now $p \in S^{\mathrm{irr}}$ and suppose that p has a nonzero root $\beta \in \overline{k}^*$. Then, $D\beta/\beta = Dt/t = a$ by Theorem 3.4.3, in contradiction with $D\alpha/\alpha \neq a$ for any $\alpha \in \overline{k}^*$. Hence the only root of p in $\overline{k}$ is 0, so $p = t$.

We have $S_1^{\mathrm{irr}} \subseteq S^{\mathrm{irr}}$ by definition. Conversely, let $p \in S^{\mathrm{irr}}$. Then $p = t$, so the only root of p in $\overline{k}$ is $\beta = 0$. We have $p_\beta = p_0 = Dt/t = a$, which is not a logarithmic derivative of a k-radical, so $p \in S_1^{\mathrm{irr}}$, which implies that $S_1^{\mathrm{irr}} = S^{\mathrm{irr}}$.

Let $c \in \mathrm{Const}(k(t))$. By Lemma 3.4.5, both the numerator and denominator of c must be special, hence $c = bt^q$ for $b \in k$ and $q \in \mathbb{Z}$. Suppose that $b \neq 0$ and $q \neq 0$. Then, $0 = Dc = (Db + qab)t^q$, so $Db/b = qa$, which implies that a

is a logarithmic derivative of a k-radical, in contradiction with our hypothesis. Hence, $b = 0$ or $q = 0$, so $c \in k$, which implies that $\text{Const}(k(t)) \subseteq \text{Const}(k)$. The reverse inclusion is given by Lemma 3.3.1, so $\text{Const}(k(t)) = \text{Const}(k)$.

Conversely, let t be a transcendental hyperexponential over k and suppose that $\text{Const}(k(t)) = \text{Const}(k)$. If there exist $b \in k^*$ and an integer $n \neq 0$ such that $nDt/t = Db/b$, then $c = t^n/b \in \text{Const}(k(t))$, so $c \in k$ in contradiction with t transcendental over k. Hence Dt/t is not a logarithmic derivative of a k-radical.   □

In practice, we only consider primitives and hyperexponentials that satisfy the hypotheses of Theorems 5.1.1 or 5.1.2. As we have seen, such primitives and hyperexponentials are monomials that satisfy the extra condition $\text{Const}(k(t)) = \text{Const}(k)$. Those monomials are traditionally called *Liouvillian monomials* in the literature.

Definition 5.1.2. $t \in K$ *is a* Liouvillian monomial over k *if t is transcendental and Liouvillian over k and* $\text{Const}(k(t)) = \text{Const}(k)$.

One should be careful that our definition of monomial in Chap. 3 does not require $\text{Const}(k(t)) = \text{Const}(k)$, so it is possible for a monomial in the sense of Chap. 3 to be Liouvillian over k and yet *not* a Liouvillian monomial in the sense of Definition 5.1.2 (for example $\log(2)$ over $\mathbb{Q}$). Theorems 5.1.1 and 5.1.2 can be seen as necessary and sufficient conditions for a primitive or hyperexponential to be a Liouvillian monomial. Furthermore, those theorems describe all the special polynomials in such extensions, and they are all of the first kind. We also have:

$$k\langle t \rangle = \begin{cases} k[t], & \text{if } Dt \in k, \\ k[t, t^{-1}], & \text{if } Dt/t \in k. \end{cases} \tag{5.1}$$

The fact that k and $k(t)$ have the same field of constants allows us to refine the relationship between the degree of a polynomial and its derivative in a Liouvillian monomial extension, and to strenghten Theorem 4.4.4.

Lemma 5.1.2. *Let t be a Liouvillian monomial over k, $f \in k(t)$ be such that $Df \neq 0$, and write $f = p/q$ where $p, q \in k[t]$ and q is monic. If $\nu_\infty(f) = 0$, then $\nu_\infty(Df) \geq 0$. Otherwise, $\nu_\infty(f) \neq 0$ and*

$$\nu_\infty(Df) = \begin{cases} \nu_\infty(f), & \text{if } Dt/t \in k \text{ or } D(\text{lc}(p)) \neq 0, \\ \nu_\infty(f) + 1, & \text{if } Dt \in k \text{ and } D(\text{lc}(p)) = 0. \end{cases}$$

Proof. If $\nu_\infty(f) = 0$, then $\nu_\infty(Df) \geq 0$ by Theorem 4.4.4, so suppose from now on that $\nu_\infty(f) \neq 0$. Then, $n - m \neq 0$ where $n = \deg(p)$ and $m = \deg(q)$. We have

$$Df = \frac{qDp - pDq}{q^2}$$

hence $\nu_\infty(Df) = 2m - \deg(qDp - pDq)$, so we need to compute $\deg(qDp - pDq)$. Write $p = bt^n + r$ and $q = t^m + s$ where $b \in k^*$ and $r, s \in k[t]$ satisfy

$\deg(r) < n$ and $\deg(s) < m$. We treat the primitive and hyperexponential cases separately.

Primitive case: Suppose that $Dt = a \in k$. Then,

$$Dp = (Db)t^n + nabt^{n-1} + Dr \tag{5.2}$$

and

$$Dq = mat^{m-1} + Ds$$

so $\deg(Dq) < m$ since $\deg(Ds) < m$ by Lemma 3.4.2.
Suppose first that $Db \neq 0$. Then, $\deg(Dp) = n$ since $\deg(Dr) < n$ by Lemma 3.4.2, so $\deg(qDp) = m + n$ and $\deg(pDq) < m + n$, which implies that $\deg(qDp - pDq) = m + n$, hence that

$$\nu_\infty(Df) = 2m - (m + n) = m - n = \nu_\infty(f).$$

Suppose now that $Db = 0$, and write $r = ct^{n-1} + u$ and $s = dt^{m-1} + v$, where $c, d \in k$ and $u, v \in k[t]$ satisfy $\deg(u) < n - 1$ and $\deg(v) < m - 1$. We then have

$$
\begin{aligned}
qDp - pDq &= (Dc + nab)t^{n+m-1} + (n-1)act^{n+m-2} + t^m Du \\
&\quad + (dt^{m-1} + v)Dp - b(Dd + ma)t^{n+m-1} \\
&\quad - (m-1)abdt^{n+m-2} - bt^n Dv - (ct^{n-1} + u)Dq \\
&= (Dc - bDd + (n-m)ab)\, t^{n+m-1} \\
&\quad + ((n-1)c - (m-1)bd)\, at^{n+m-2} \\
&\quad + (dt^{m-1} + v)Dp + t^m Du - bt^n Dv - (ct^{n-1} + u)Dq.
\end{aligned}
$$

Since $n - m \neq 0$ and $b \neq 0$, $c - bd + (n-m)bt \notin k$, so $D(c - bd + (n-m)bt) \neq 0$ since $\mathrm{Const}(k(t)) = \mathrm{Const}(k)$. But

$$D(c - bd + (n-m)bt) = Dc - bDd + (n-m)ab$$

since $b \in \mathrm{Const}(k)$, hence $Dc - bDd + (n-m)ab \neq 0$. In addition, (5.2) and $Db = 0$ imply that $\deg(Dp) < n$, and Lemma 3.4.2 imply that $\deg(Du) < n - 1$ and $\deg(Dv) < m - 1$. Hence, $(dt^{m-1} + v)Dp$, $t^m Du$, $bt^n Dv$ and $(ct^{n-1} + u)Dq$ all have degrees strictly smaller than $n + m - 1$, which implies that $\deg(qDp - pDq) = n + m - 1$, hence that $\nu_\infty(Df) = 2m - (n+m-1) = m - n + 1 = \nu_\infty(f) + 1$.

Hyperexponential case: Suppose that $Dt/t = a \in k$. Then,

$$
\begin{aligned}
qDp - pDq &= (Db + nab)t^{n+m} + t^m Dr + sDp - bmat^{n+m} - bt^n Ds - rDq \\
&= (Db + (n-m)ab)\, t^{n+m} + (sDp - rDq + t^m Dr - bt^n Ds).
\end{aligned}
$$

Since $n - m \neq 0$ and $b \neq 0$, $bt^{n-m} \notin k$, so $D(bt^{n-m}) \neq 0$ since $\mathrm{Const}(k(t)) = \mathrm{Const}(k)$. But $D(bt^{n-m}) = (Db + (n-m)ab)\, t^{n-m}$, so $Db + (n-m)ab \neq 0$. In addition, $\deg(Dp) \leq n$, $\deg(Dq) \leq m$, $\deg(Dr) < n$ and $\deg(Ds) < m$ by Lemma 3.4.2, so sDp, rDq, $t^m Dr$ and $bt^n Ds$ all have degrees strictly smaller than $n + m$, which implies that $\deg(qDp - pDq) = n + m$, hence that $\nu_\infty(Df) = 2m - (n+m) = m - n = \nu_\infty(f)$.   □

Note that when applied to polynomials $p \in k[t]$ when t is a Liouvillian monomial over k, Lemma 5.1.2 implies that

$$\deg(Dp) = \begin{cases} \deg(p), & \text{if } Dt/t \in k \text{ or } D(\mathrm{lc}(p)) \neq 0, \\ \deg(p) - 1, & \text{if } Dt \in k \text{ and } D(\mathrm{lc}(p)) = 0 \end{cases}$$

whenever $Dp \neq 0$, and we often use it in this context in the sequel.

We now introduce the particular Liouvillian extensions that define the integration in finite terms problem, namely the elementary extensions.

Definition 5.1.3. $t \in K$ *is a* logarithm *over* k *if* $Dt = Db/b$ *for some* $b \in k^*$. $t \in K^*$ *is an* exponential *over* k *if* $Dt/t = Db$ *for some* $b \in k$. $t \in K$ *is* elementary *over* k *if* t *is either algebraic, or a logarithm or an exponential over* k. $t \in K$ *is an* elementary monomial *over* k *if* t *is transcendental and elementary over* k, *and* $\mathrm{Const}(k(t)) = \mathrm{Const}(k)$.

We write $t = \log(b)$ when t is a logarithm over k such that $Dt = Db/b$, and $t = e^b$ when t is an exponential over k such that $Dt/t = b$. Since logarithms are primitives and exponentials are hyperexponentials, elementary monomials are Liouvillian monomials and all the results of this section apply to them.

Definition 5.1.4. K *is an* elementary extension *of* k *if there are* $t_1, \ldots, t_n$ *in* K *such that* $K = k(t_1, \ldots, t_n)$ *and* t_i *is elementary over* $k(t_1, \ldots, t_{i-1})$ *for* i *in* $\{1, \ldots, n\}$. *We say that* $f \in k$ *has an* elementary integral *over* k *if there exists an elementary extension* E *of* k *and* $g \in E$ *such that* $Dg = f$. *An* elementary function *is any element of any elementary extension of* $(\mathbb{C}(x), d/dx)$.

We can now define precisely the *problem of integration in closed form*: given a differential field k and an integrand $f \in k$, to decide in a finite number of steps whether f has an elementary integral over k, and to compute one if it has any. Note that there is a difference between having an elementary integral over k and having an elementary antiderivative: consider $k = \mathbb{C}(x, t_1, t_2)$ where x, t_1, t_2 are indeterminates over $\mathbb{C}$, with the derivation D given by $Dx = 1$, $Dt_1 = t_1$ and $Dt_2 = t_1/x$ (*i.e.* $t_1 = e^x$ and $t_2 = \mathrm{Ei}(x)$). Then,

$$\int \frac{e^x \mathrm{Ei}(x)}{x}\, dx = \frac{\mathrm{Ei}(x)^2}{2} \in k$$

so $e^x \mathrm{Ei}(x)/x$ has an elementary integral over k even though its integral is not an elementary function. The two notions coincide only when k itself is a field of elementary functions.

Remark that the elementary functions of Definition 5.1.4 include all the usual elementary functions of analysis, since the trigonometric functions and their inverses can be rewritten in terms of complex exponential and logarithms by the usual formulas derived from Euler's formula $e^{f\sqrt{-1}} = \cos(f) + \sin(f)\sqrt{-1}$. Those transformations have the computational inconvenience that they introduce $\sqrt{-1}$, and it turns out that they can be avoided when integrating real trigonometric functions (Sections 5.8 and 5.10).

5.2 Outline and Scope of the Integration Algorithm

We outline in this section the integration algorithm so that the structure of the remaining sections and chapters will be easier to follow. Given an integrand $f(x)dx$, we first need to construct a differential field containing f, and the integration algorithm we describe requires that f be contained in a differential field of the form $K = C(t_1, t_2, \ldots, t_n)$ where $C = \mathrm{Const}(K)$, $Dt_1 = 1$ (i.e. $t_1 = x$ is the integration variable), and each t_i is a monomial over $C(t_1, \ldots, t_{i-1})$. If the formula for $f(x)$ contains only Liouvillian operations, this requirement can be checked by integrating recursively the argument of each primitive or hyperexponential before adjoining it[1], and verifying using Theorem 5.1.1 or Theorem 5.1.2 that it is a Liouvillian monomial. Another alternative, which is in general more efficient, is to apply the algorithms that are derived from the various structure theorems, whenever they are applicable (Chap. 9).

Example 5.2.1. Consider

$$\int \log(x) \log(x+1) \log(2x^2 + 2x) dx \,.$$

We construct the differential field $K = \mathbb{Q}(x, t_1, t_2, t_3)$ with

$$Dx = 1, \quad Dt_1 = \frac{1}{x}, \quad Dt_2 = \frac{1}{x+1} \quad \text{and} \quad Dt_3 = \frac{2x+1}{x^2+x} \,.$$

As we construct K, we integrate at each step and make the following verifications:

- $\int dx \notin \mathbb{Q}$, so x is a Liouvillian monomial over $\mathbb{Q}$;
- $\int dx/x \notin \mathbb{Q}(x)$, so t_1 is a Liouvillian monomial over $\mathbb{Q}(x)$;
- $\int dx/(x+1) \notin \mathbb{Q}(x, t_1)$ so t_2 is a Liouvillian monomial over $\mathbb{Q}(x, t_1)$;
-
$$\int \frac{2x+1}{x^2+x} dx = t_1 + t_2 \in \mathbb{Q}(x, t_1, t_2)$$

  so t_3 is not a Liouvillian monomial over $\mathbb{Q}(x, t_1, t_2)$, and K is isomorphic as a differential field to $\mathbb{Q}(c)(x, t_1, t_2)$ where $c = t_3 - t_1 - t_2 \in \mathrm{Const}(K)$.
- Alternatively, applying the Risch structure Theorem (Corollary 9.3.1), we find that the linear equation (9.8) for $a = 2x^2 + 2x$ becomes

$$\frac{r_1}{x} + \frac{r+2}{x+1} = \frac{2x+1}{x^2+x}$$

  which has the rational solution $r_1 = r_2 = 1$. This implies that Dt_3 is the derivative of an element of K and that $c = t_3 - t_1 - t_2 \in \mathrm{Const}(K)$.

---

[1] A simpler version of the integration algorithm can be used for those verifications, see Sect. 5.12

Example 5.2.2. Consider

$$\int \left(e^{2x} + e^{x+\log(x)/2}\right) dx.$$

We construct the differential field $K = \mathbb{Q}(x, t_1, t_2, t_3)$ with

$$Dx = 1, \quad Dt_1 = 2t_1, \quad Dt_2 = \frac{1}{x} \quad \text{and} \quad Dt_3 = \left(1 + \frac{1}{2x}\right) t_3.$$

As we construct K we integrate at each step and make the following verifications:

- $\int dx \notin \mathbb{Q}$, so x is a Liouvillian monomial over $\mathbb{Q}$;
- $\int 2dx \neq \log(v)/n$ for any $v \in \mathbb{Q}(x)$ and $n \in \mathbb{Z}$, so 2 is not the logarithmic derivative of a $\mathbb{Q}(x)$-radical, which implies that t_1 is a Liouvillian monomial over $\mathbb{Q}(x)$;
- $\int dx/x \notin \mathbb{Q}(x, t_1)$, so t_2 is a Liouvillian monomial over $\mathbb{Q}(x, t_1)$;
-
$$\int \left(1 + \frac{1}{2x}\right) dx = \frac{1}{2}\log(xt_1)$$

  so $1 + 1/(2x)$ is the logarithmic derivative of a $\mathbb{Q}(x, t_1, t_2)$-radical, so t_3 is not a Liouvillian monomial over $\mathbb{Q}(x, t_1, t_2)$, and K is isomorphic as a differential field to $\mathbb{Q}\left(x, t_1, t_2, \sqrt{xt_1}\right)$.
- Alternatively, applying the Risch structure Theorem (Corollary 9.3.1), we find that the linear equation (9.9) for $b = x + t_2/2$ becomes

$$\frac{r_2}{x} + 2r_1 = 1 + \frac{1}{2x}$$

  which has the rational solution $r_1 = r_2 = 1/2$. This implies that Dt_3/t_3 is the logarithmic derivative of a K-radical, and that $c = t_3^2/(xt_1) \in \text{Const}(K)$.

Note that the requirement that each t_i be a monomial eliminates expressions containing algebraic functions from the algorithm presented here. Although the problem of integrating elementary functions containing algebraic functions is also decidable, the algorithms used in the algebraic function case are beyond the scope of this book [8, 9, 13, 14, 26, 58, 59, 61, 76].

Once we have a tower of monomials $K = C(t_1, \ldots, t_n)$, the algorithms of this chapter reduce the problem of integrating an element of K to various integration-related problems involving elements of $C(t_1, \ldots, t_{n-1})$, thereby eliminating the monomial t_n. Since the reduced problems involve integrands in a tower of smaller transcendence degree over C, we can use the algorithm recursively on them, and termination is ensured. In order to avoid writing the full tower of extensions throughout this book, we write $K = k(t)$ where $k = C(t_1, \ldots, t_{n-1})$ and $t = t_n$ is a monomial over k, and the task of the algorithms of this chapter is to reduce integrating a given element of $k(t)$ to integration-related problems over k. If t is elementary over k, then having an elementary integral over $k(t)$ is equivalent to having an elementary integral over k, so the algorithms we present in this book provide a complete decision procedure for the problem of deciding whether an element of a purely transcendental elementary extension of $(C(x), d/dx)$ has an elementary integral over $C(x)$. For more general functions, when t is not elementary over k, it can be proven that if t is either an hyperexponential monomial or nonlinear monomial over k with $S_1^{irr} = S^{irr}$, then having an elementary integral over $k(t)$ is equivalent to having an elementary integral over k (Exercise 5.5), so the algorithm is complete for integrands built from transcendental logarithms, arc-tangents, hyperexponentials and tangents. The only obstruction to a complete algorithm for Liouvillian integrands is the case where t is a nonelementary primitive over k: even though we can reduce the problem to an integrand in k, the problem becomes however to determine whether $f \in k$ has an elementary integral over $k(t)$, and although there are algorithms for special types of primitive monomials [5, 20, 21, 38, 39, 79], this problem has not been solved for general monomials (Exercise 5.5 f)). As will be seen from numerous examples in this book, the algorithm can still be used successfully on many integrands involving nonelementary monomials. It cannot however always provide a proof on nonexistence of an elementary integral over $k(t)$ when t is a nonelementary primitive over k. The reduction from $k(t)$ to t is also incomplete for general nonlinear monomials, but is complete for tangents and hyperbolic tangents.

The general line of the integration algorithm is to perform successive reductions, which all transform the integrand to a "simpler" one, until the remaining integrand is in k (Fig. 5.1):

- The *Hermite reduction* (Sect. 5.3), which can be applied to arbitrary monomials, transforms a general integrand to the sum of a simple and a reduced integrand;
- The *polynomial reduction* (section 5.4), which can be applied to nonlinear monomials, reduces the degree of the polynomial part of an integrand;
- The *residue criterion* (Sect. 5.6), which can be applied to arbitrary monomials, either proves that an integrand does not have an elementary integral over $k(t)$, or transforms it to a reduced integrand (*i.e.* an integrand in $k\langle t \rangle$);
- Reduced integrands are integrated by specific algorithms for each case of Liouvillian or hypertangent monomial (Sect. 5.8, 5.9 and 5.10). Those algorithms either prove that there is no elementary integral over $k(t)$, or reduce the problem to various integration-related problems over k. Algorithms for solving those related problems are described in Chap. 6, 7 and 8.

Except for the last part, the various reductions are applicable to arbitrary monomial extensions.

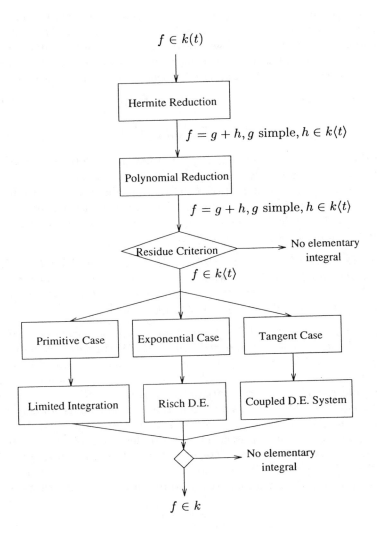

Fig. 5.1. General outline of the integration algorithm

5.3 The Hermite Reduction

We have seen in Sect. 2.2 that the Hermite reduction rewrites any rational function as the sum of a derivative and a rational function with a squarefree denominator. In this section, we show that the Hermite reduction can be applied to the normal part of any element of a monomial extension. Let (k, D) be a differential field and t a monomial over k for the next two sections.

Definition 5.3.1. *For $f \in k(t)$, we define the* polar multiplicity *of f to be*

$$\mu(f) = - \min_{p \in k[t] \setminus k} (\nu_p(f)).$$

Note that $\mu(0) = -\infty$ and that $\mu(f) \geq 0$ for any $f \neq 0$, since in that case there is always some polynomial $p \in k[t]$ for which $\nu_p(f) = 0$. Also, the minimum in the above definition can be taken over all the irreducible or squarefree factors of the denominator of f. It is easy to see that for $f \neq 0$, $\mu(f)$ is exactly the highest power appearing in any squarefree factorization of the denominator of f (Exercise 5.1).

Theorem 5.3.1. *Let $f \in k(t)$. Using only the extended Euclidean algorithm in $k[t]$, one can find $g, h, r \in k(t)$ such that h is simple, r is reduced, and $f = Dg + h + r$. Furthermore, the denominators of g, h and r divide the denominator of f, and either $g = 0$ or $\mu(g) < \mu(f)$.*

Proof. Let $f = f_p + f_s + f_n$ be the canonical representation of f, and write $f_n = a/d$ with $a, d \in k[t]$ and $\gcd(a, d) = 1$. We proceed by induction on $m = \mu(f_n)$. Let $d = d_1 d_2^2 \cdots d_m^m$ be a squarefree factorization of d. If $m \leq 1$, then either $f_n = 0$ or d is normal. In both cases, f_n is simple, so $g = 0$, $h = f_n$ and $r = f_p + f_s \in k\langle t \rangle$ satisfy the theorem.

Otherwise, $m > 1$, so assume that the theorem holds for any nonzero $g = g_p + g_n + g_s$ with $\mu(g_n) < m$, and let $v = d_m$ and $u = d/v^m$. Since every squarefree factor of d is normal by the definition of the canonical representation, v is normal, so $\gcd(Dv, v) = 1$. In addition, $\gcd(u, v) = 1$ by the definition of a squarefree factorization, so $\gcd(uDv, v) = 1$. Hence, we can use the extended Euclidean algorithm to find $b, c \in k[t]$ such that

$$\frac{a}{1 - m} = b u \, Dv + c v.$$

Multiplying both sides by $(1 - m)/(uv^m)$ gives

$$f_n = \frac{a}{uv^m} = \frac{(1 - m)bDv}{v^m} + \frac{(1 - m)c}{uv^{m-1}}$$

so, adding and subtracting Db/v^{m-1} to the right hand side, we get

$$f_n = \left(\frac{Db}{v^{m-1}} - \frac{(m-1)bDv}{v^m} \right) + \frac{(1 - m)c - uDb}{uv^{m-1}} = Dg_0 + w$$

where $g_0 = b/v^{m-1}$ and $w = ((1-m)c - uDb)/(uv^{m-1})$. Since the denominator of w divides uv^{m-1}, w has no special part, so let $w = w_p + w_n$ be the canonical representation of w. Since $\mu(w) \leq m - 1$, we have $\mu(w_n) \leq m - 1$, so by induction we can find g_1, h_1 and r_1 in $k(t)$ such that $w_n = Dg_1 + h + r_1$, h is simple, r_1 is reduced, the denominators of g_1, h and r_1 divide uv^{m-1}, and $\mu(g_1) < \mu(w)$ if $g_1 \neq 0$. Let then $g = g_0 + g_1$ and $r = f_p + w_p + f_s + r_1$, and write e for the denominator of f. Note that $d \mid e$ by the definition of the canonical representation. The denominator of g_1 divides uv^{m-1} and $g_0 = b/v^{m-1}$, so the denominator of g divides d hence e. The denominator of h divides uv^{m-1}, so it divides d hence e. The denominator of w divides d and the denominator of r_1 divides uv^{m-1}, so the denominator of r divides e. In addition, f_p, w_p, f_s and r_1 are in $k\langle t \rangle$, which is a subring of $k(t)$ by Corollary 4.4.1, so $r \in k\langle t \rangle$. Finally, we have

$$\begin{aligned} f = f_p + f_s + f_n &= f_p + f_s + Dg_0 + w \\ &= f_p + f_s + Dg_0 + w_p + Dg_1 + h + r_1 = Dg + h + r \end{aligned}$$

which proves the theorem. $\qquad\square$

Although we have used the quadratic version of the Hermite reduction in the above proof, the other versions are also valid in monomial extensions (Exercise 5.2). Instead of splitting a rational function into a derivative and a simple rational function, the Hermite reduction splits any element of $k(t)$ into a derivative, a simple and a reduced element. Thus, it reduces any integration problem to integrands that are the sum of a simple and a reduced element.

HermiteReduce(f, D) (* Hermite Reduction – quadratic version *)

(* Given a derivation D on $k(t)$ and $f \in k(t)$, return $g, h, r \in k(t)$ such that $f = Dg + h + r$, h is simple and r is reduced. *)

$(f_p, f_s, f_n) \leftarrow$ **CanonicalRepresentation**(f, D)
$(a, d) \leftarrow (\text{numerator}(f_n), \text{denominator}(f_n))$ (* d is monic *)
$(d_1, \ldots, d_m) \leftarrow$ **SquareFree**(d)
$g \leftarrow 0$
for $i \leftarrow 2$ **to** m such that $\deg(d_i) > 0$ **do**
$\quad v \leftarrow d_i$
$\quad u \leftarrow d/v^i$
$\quad$ **for** $j \leftarrow i - 1$ **to** 1 **step** -1 **do**
$\quad\quad (b, c) \leftarrow$ **ExtendedEuclidean**$(u\,Dv, v, -a/j)$
$\quad\quad g \leftarrow g + b/v^j$
$\quad\quad a \leftarrow -jc - u\,Db$
$\quad d \leftarrow uv$
$(q, r) \leftarrow$ **PolyDivide**(a, uv)
return$(g, r/(uv), q + f_p + f_s)$

Example 5.3.1. Let $k = \mathbb{Q}(x)$ with $D = d/dx$, and let t be a monomial over k satisfying $Dt = 1 + t^2$, *i.e.* $t = \tan(x)$, and consider

$$f = \frac{x - \tan(x)}{\tan(x)^2} = \frac{x - t}{t^2} \in k(t).$$

Since f has no polynomial part and t is normal in $k[t]$, the canonical representation of f is $(f_p, f_s, f_n) = (0, 0, f)$ so we get $a = x - t$ and $d = t^2 = d_2^2$ where $d_2 = t$. We then have:

i	v	u	j	b	c	a
2	t	1	1	$-x$	$xt + 1$	$-xt$

and $a/uv = -xt/t = -x$, so the Hermite reduction returns $(-x/t, 0, -x)$, which means that

$$\int \frac{x - \tan(x)}{\tan(x)^2}\, dx = -\frac{x}{\tan(x)} - \int x\, dx$$

and the remaining integrand is in $k\langle t\rangle$.

The Hermite reduction can also be iterated, yielding a decomposition of f into a sum of higher-order derivatives of reduced and simple elements of $k(t)$ (Exercise 5.3).

5.4 The Polynomial Reduction

In the case of nonlinear monomials, another reduction allows us to rewrite any polynomial in $k[t]$ as the sum of a derivative and a polynomial of degree less than $\delta(t)$.

Theorem 5.4.1. *If t is a nonlinear monomial, then for any $p \in k[t]$, we can find $q, r \in k[t]$ such that $p = Dq + r$ and $\deg(r) < \delta(t)$.*

Proof. We proceed by induction on $n = \deg(p)$. If $n < \delta(t)$, then $q = 0$ and $r = p$ satisfy the theorem. Otherwise $n \geq \delta(t)$ so assume that the theorem holds for any $a \in k[t]$ with $\deg(a) < n$. Let

$$c = \frac{\operatorname{lc}(p)}{(n - \delta(t) + 1)\lambda(t)} \in k,$$

$q_0 = ct^{n-\delta(t)+1}$, and $r_0 = p - Dq_0$. Since t is nonlinear and $\deg(q_0) > 0$, Lemma 3.4.2 implies that $\deg(Dq_0) = \deg(q_0) + \delta(t) - 1 = n$, and that the leading coefficient of Dq_0 is $(n - \delta(t) + 1)\, c\, \lambda(t) = \operatorname{lc}(p)$. Hence, $\deg(r_0) < n$, so by induction we can find $q_1, r \in k[t]$ such that $r_0 = Dq_1 + r$ and $\deg(r) < \delta(t)$. Therefore,

$$p = Dq_0 + r_0 = Dq_0 + Dq_1 + r = Dq + r$$

where $q = q_0 + q_1 \in k[t]$. $\square$

PolynomialReduce(p, D) (* Polynomial Reduction *)

(* Given a derivation D on $k(t)$ and $p \in k[t]$ where t is a nonlinear monomial over k, return $q, r \in k[t]$ such that $p = Dq + r$, and $\deg(r) < \delta(t)$. *)

if $\deg(p) < \delta(t)$ **then return**$(0, p)$
$m \leftarrow \deg(p) - \delta(t) + 1$
$q_0 \leftarrow (\operatorname{lc}(p)/(m\lambda(t)))\, t^m$
$(q, r) \leftarrow$ **PolynomialReduce**$(p - Dq_0, D)$
return$(q_0 + q, r)$

Example 5.4.1. Let $k = \mathbb{Q}(x)$ with $D = d/dx$, and let t be a monomial over k satisfying $Dt = 1 + t^2$, *i.e.* $t = \tan(x)$, and consider

$$p = 1 + x\tan(x) + \tan(x)^2 = 1 + xt + t^2 \in k[t].$$

We have $\delta(t) = 2$, $\lambda(t) = 1$, and applying **PolynomialReduce**, we get $m = \deg(p) - 1 = 1$, $q_0 = t$, $Dq_0 = 1 + t^2$, so $p - Dq_0 = xt$, which has degree 1. Thus,

$$\int (1 + x\tan(x) + \tan(x)^2)\, dx = \tan(x) + \int x\tan(x)dx$$

and it will be proven later that the remaining integral is not an elementary function.

If $S \neq k$, *i.e.* $S^{\mathrm{irr}} \neq \emptyset$, then any nontrivial element of S can be used to eliminate the term of degree $\delta(t) - 1$ from a polynomial.

Theorem 5.4.2. *Suppose that t is a nonlinear monomial. Let $p \in k[t]$ with $\deg(p) < \delta(t)$, $a \in k$ be the coefficient of $t^{\delta(t)-1}$ in p, and $c = a/\lambda(t)$. Then,*

$$\deg\left(p - \frac{c}{\deg(q)}\frac{Dq}{q}\right) < \delta(t) - 1$$

for any $q \in S \setminus k$.

Proof. Let $q \in S \setminus k$, then $Dq/q \in k[t]$ and by Lemma 3.4.2, $\deg(Dq/q) = \deg(Dq) - \deg(q) = \delta(t) - 1$, and the leading coefficient of Dq is $\deg(q)\operatorname{lc}(q)\lambda(t)$. Hence,

$$\operatorname{lc}\left(\frac{c}{\deg(q)}\frac{Dq}{q}\right) = \frac{c}{\deg(q)}\frac{\deg(q)\operatorname{lc}(q)\lambda(t)}{\operatorname{lc}(q)} = c\lambda(t) = a$$

which implies that the degree of $p - c/\deg(q)\, Dq/q$ is at most $\delta(t) - 2$. $\square$

5.5 Liouville's Theorem

Given a differential field K and an integrand $f \in K$, if an elementary integral is found, it can be easily proven correct by differentiation. Furthermore, there are usually several ways to find elementary integrals when they exist. Proving that f has no elementary integral is however quite a different problem, since we need results that connect the existence of an elementary integral to a special form of the integrand. The first such result is Laplace's principle [41], which states roughly that we can simplify the integration problem by allowing only new logarithms to appear linearly in the integral, all the other functions must be in the integrand already[2]. Liouville was the first to state and prove a precise theorem from this observation, first in the case of algebraic integrands [43, 44], then for more general integrands [45]. See Chap. IX of [47] for the fascinating history of Liouville's Theorem in the 19th century. This theorem has become the main tool used in proving that no elementary integral exists for a given function. Furthermore, since it provides an explicit class of elementary extensions to search for an integral, it forms the basis of the integration algorithm. While Liouville used analytic arguments, it is now possible to prove it algebraically in the context of differential fields. Algebraic techiques were first used by Ostrowski [54], who presented a modern proof of Liouville's Theorem, together with an algorithm that reduces integrating in $k(t)$ to integrating in k when t is a primitive monomial over k. The first complete algebraic proof of Liouville's Theorem was then published by Rosenlicht [64] and the first proof of the strong version of Liouville's Theorem by Risch, who published it together with a complete integration algorithm for purely transcendental elementary functions [60]. We follow both of them here, first presenting essentially Rosenlicht's proof of the weak Liouville Theorem, and then progressively removing the restrictions on the constant fields, obtaining Risch's proof of the strong Liouville Theorem. We remark that Liouville's Theorem has been extended in various directions [17, 56, 66, 71], but those extensions go beyond the scope of this book. Integration algorithms that yield nonelementary integrals [20, 21, 38, 39] are based on such extensions [71].

Theorem 5.5.1 (Liouville's Theorem). *Let K be a differential field and $f \in K$. If there exist an elementary extension E of K with $\mathrm{Const}(E) = \mathrm{Const}(K)$ and $g \in E$ such that $Dg = f$, then there are $v \in K$, $u_1, \ldots, u_n \in K^*$ and $c_1, \ldots, c_n \in \mathrm{Const}(K)$ such that*

$$f = Dv + \sum_{i=1}^{n} c_i \frac{Du_i}{u_i}. \tag{5.3}$$

---

[2] "... la différentiation laissant subsister les quantités exponentielles et radicales, et ne faisant disparaitre les quantités logarithmiques qu'autant qu'elles ont multipliées par des constantes, on doit en conclure que l'intégrale d'une fonction différentielle ne peut contenir d'autres quantités exponentielles et radicales que celles qui sont contenues dans cette fonction..."

Proof. Write $C = \text{Const}(K)$ and let E be an elementary extension of K with $\text{Const}(E) = C$ and $g \in E$ be such that $Dg = f$. Then, there are $t_1, \ldots, t_m \in E$ such that $E = K(t_1, \ldots, t_m)$ and each t_i is elementary over $K(t_1, \ldots, t_{i-1})$. We proceed by induction on m. For $m = 0$, we have $E = K$, so letting $v = g \in K$, we get $f = Dv$, which is of the form (5.3) with $n = 0$. Suppose now that $m > 0$ and that the theorem holds for any elementary extension generated by $m - 1$ elements. Let $t = t_1$ and $F = K(t)$. Since $K \subseteq F \subseteq E$, then $C \subseteq \text{Const}(F) \subseteq \text{Const}(E) = C$, so $\text{Const}(F) = C$. In addition, $f \in F$, and $E = F(t_2, \ldots, t_m)$ is an elementary extension of F generated by $m - 1$ elements, so by induction there are $v \in F$, $u_1, \ldots, u_n \in F^*$ and $c_1, \ldots, c_n \in C$ such that

$$f = Dv + \sum_{i=1}^{n} c_i \frac{Du_i}{u_i}. \tag{5.4}$$

Case 1: t transcendental over K. Then, since $\text{Const}(F) = C$, t is Liouvillian monomial over K by Theorems 5.1.1 and 5.1.2. Let $p \in K[t]$ be normal and irreducible. We have $\nu_p(Du_i/u_i) \geq -1$ by Corollary 4.4.2, hence $\nu_p(\sum_{i=1}^{n} c_i Du_i/u_i) \geq -1$ by Theorem 4.1.1. Suppose that $\nu_p(v) < 0$. Then, $\nu_p(Dv) = \nu_p(v) - 1 < -1$ by Theorem 4.4.2, so $\nu_p(f) = \min(\nu_p(Dv), -1) < -1$ by Theorem 4.1.1, in contradiction with $f \in K$. Hence $\nu_p(v) \geq 0$, so, since this holds for any normal irreducible p, $v \in K\langle t \rangle$. Hence, $Dv \in K\langle t \rangle$ by Corollary 4.4.1. Write now $u_i = w_i \prod_{j=1}^{n_i} p_{ij}^{e_{ij}}$ where $w_i \in K^*$, each $p_{ij} \in K[t]$ is monic irreducible, and the e_{ij}'s are integers. Then, using the logarithmic derivative identity and grouping together all the terms involving the same p_{ij}, we get

$$f = Dv + \sum_{i=1}^{n} c_i \frac{Dw_i}{w_i} + \sum_{j=1}^{N} d_j \frac{Dq_j}{q_j} \tag{5.5}$$

where the q_j's are in $K[t]$, monic, irreducible and coprime. Write

$$g = \sum_{i=1}^{n} c_i \frac{Dw_i}{w_i} \in K, \quad h = \sum_{j=1}^{N} d_j \frac{Dq_j}{q_j},$$

and suppose that one of the q_j's, say q_k, is normal. We have $\nu_{q_k}(q_k) = 1$ and $\nu_{q_k}(q_j) = 0$ for $j \neq k$, so $\nu_{q_k}(d_k Dq_k/q_k) = -1$ and $\nu_{q_k}(d_j Dq_j/q_j) = 0$ by Corollary 4.4.2. This implies that $\nu_{q_k}(\sum_{j \neq k} d_j Dq_j/q_j) \geq 0$, hence that $\nu_{q_k}(h) = -1$. But q_k is normal and $Dv \in K\langle t \rangle$, hence $\nu_{q_k}(Dv) \geq 0$, so $\nu_{q_k}(f) = -1$, in contradiction with $f \in K$. Hence all the q_j's in equation (5.5) are special.

Case 1a: t is a logarithm over K. Then, $Dt = Da/a$ for some $a \in K^*$, and every irreducible $p \in K[t]$ is normal by Theorem 5.1.1, so $N = 0$ in equation (5.5) and $v, Dv \in K[t]$. From (5.5) we get $Dv = f - g \in K$. By Lemma 5.1.2, this implies that or $v = ct + b$ where $b, c \in K$ and $Dc = 0$ (otherwise $\deg(Dv) \geq 1$). Hence,

$$f = Db + c\frac{Da}{a} + \sum_{i=1}^{n} c_i \frac{Dw_i}{w_i}$$

which is of the form (5.3).

Case 1b: t is an exponential over K. Then, $Dt/t = Da$ for some $a \in K$, and the only special monic irreducible $p \in K[t]$ is $p = t$ by Theorem 5.1.2, so $N = 1$ in equation (5.5) and $q_1 = t$ (with d_1 possibly 0). Hence, $d_1 Dq_1/q_1 = d_1 Dt/t = d_1 Da$, so $f = Dw + g$ where $w = v + d_1 a \in K\langle t \rangle$. Suppose that $\nu_t(w) < 0$, then $\nu_t(Dw) = \nu_t(w) < 0$ by Theorem 4.4.2 since $t \in S^{\mathrm{irr}}$, so $\nu_t(f) < 0$ in contradiction with $f \in K$. Hence, $\nu_t(w) \geq 0$ so $w \in K[t]$. By Lemma 5.1.2, $\nu_\infty(Dw) = \nu_\infty(w)$, so $\deg(Dw) = \deg(w)$, which implies that $\deg(w) = 0$ since $f = Dw + g \in K$. Hence $w \in K$ and

$$f = Dw + \sum_{i=1}^{n} c_i \frac{Dw_i}{w_i}$$

which is of the form (5.3).

Case 2: t algebraic over K. Let $Tr : F \to K$ and $N : F \to K$ be the trace and norm maps from F to K and $d = [F : K]$. Applying Tr to both sides of equation (5.4) we get:

$$Tr(f) = Tr(Dv + \sum_{i=1}^{n} c_i \frac{Du_i}{u_i}) = Tr(Dv) + \sum_{i=1}^{n} c_i\, Tr(\frac{Du_i}{u_i})$$

since Tr is K-linear and the c_i's are in K. We have $Tr(f) = df$ since $f \in K$, and

$$Tr(Dv) = D(Tr(v)) \quad \text{and} \quad Tr\left(\frac{Du_i}{u_i}\right) = \frac{DN(u_i)}{N(u_i)}$$

by Theorem 3.2.4, so

$$f = Dw + \sum_{i=1}^{n} \frac{c_i}{d} \frac{Dw_i}{w_i}$$

which is of the form (5.3) with $w = Tr(v)/d \in K$ and $w_i = N(u_i) \in K^*$. □

Of course, in practice we may have to adjoin new constants in order to compute integrals, as we have seen in Chap. 2. We first show that new transcendental constants are not necessary in order to express an elementary integral.

Theorem 5.5.2. *Let K be a differential field with algebraically closed constant field and $f \in K$. If there exist an elementary extension E of K and $g \in E$ such that $Dg = f$, then there are $v \in K$, $u_1, \ldots, u_n \in K^*$ and $c_1, \ldots, c_n \in \mathrm{Const}(K)$ such that*

$$f = Dv + \sum_{i=1}^{n} c_i \frac{Du_i}{u_i}.$$

Proof. Suppose that there exist an elementary extension E of K and $g \in E$ such that $Dg = f$. Write $\text{Const}(K) = C$, $\text{Const}(E) = C(a_1, \ldots, a_m)$ for some constants $a_1, \ldots, a_m$ in E, and let $F = K(a_1, \ldots, a_m)$. Since $C(a_1, \ldots, a_m) \subseteq F \subseteq E$, $C(a_1, \ldots, a_m) \subseteq \text{Const}(F) \subseteq \text{Const}(E)$, so F and E have the same constant subfield. In addition, $f \in F$ and E is elementary over F, so by Theorem 5.5.1, there are $v \in F$, $u_1, \ldots, u_m \in F^*$ and $c_1, \ldots, c_n \in \text{Const}(F)$ such that

$$f = Dv + \sum_{i=1}^{n} c_i \frac{Du_i}{u_i}. \tag{5.6}$$

Let $X_1, \ldots, X_m$ be independent indeterminates over K. Since the elements of F are rational functions in $a_1, \ldots, a_n$, we can write

$$v = \frac{p(a_1, \ldots, a_m)}{q(a_1, \ldots, a_m)}, \quad c_i = \frac{r_i(a_1, \ldots, a_m)}{s_i(a_1, \ldots, a_m)} \quad \text{and} \quad u_i = \frac{p_i(a_1, \ldots, a_m)}{q_i(a_1, \ldots, a_m)} \tag{5.7}$$

where p, q, p_i, q_i are in $K[X_1, \ldots, X_m]$, and r_i, s_i are in $C[X_1, \ldots, X_m]$. In addition, $g(a_1, \ldots, a_m) \neq 0$, where

$$g = q \left(\prod_{i=1}^{n} s_i \right) \left(\prod_{i=1}^{n} p_i \right) \left(\prod_{i=1}^{n} q_i \right) \in K[X_1, \ldots, X_m].$$

Replacing v, $c_1, \ldots, c_m$ and $u_1, \ldots, u_m$ by the fractions (5.7) in (5.6), and clearing denominators, we obtain a polynomial $f \in K[X_1, \ldots, X_m]$ such that $f(a_1, \ldots, a_m) = 0$. By Lemma 3.3.6 applied to g and $S = \{f\}$, there are $b_1, \ldots, b_m \in C$ such that $g(b_1, \ldots, b_m) \neq 0$ and $f(b_1, \ldots, b_m) = 0$. But this implies that

$$f = Dw + \sum_{i=1}^{n} d_i \frac{Dw_i}{w_i}$$

where

$$w = \frac{p(b_1, \ldots, b_m)}{q(b_1, \ldots, b_m)}, \quad d_i = \frac{r_i(b_1, \ldots, b_m)}{s_i(b_1, \ldots, b_m)} \quad \text{and} \quad w_i = \frac{p_i(b_1, \ldots, b_m)}{q_i(b_1, \ldots, b_m)}.$$

Since $p, q, p_i, q_i \in K[X_1, \ldots, X_m]$ and $r_i, s_i \in C[X_1, \ldots, X_m]$, we get $w \in K$, $w_1, \ldots, w_n \in K^*$ and $d_1, \ldots, d_n \in C$, which proves the theorem. $\qquad \square$

We can finally remove all the constant restrictions in Liouville's Theorem, showing that for arbitrary constant subfields, v in (5.3) can be taken in K, and the u_i's can be taken in $K(c_1, \ldots, c_n)$.

Theorem 5.5.3 (Liouville's Theorem – Strong version). *Let K be a differential field, $C = \text{Const}(K)$, and $f \in K$. If there exist an elementary extension E of K and $g \in E$ such that $Dg = f$, then there are $v \in K$, $c_1, \ldots, c_n \in \overline{C}$, and $u_1, \ldots, u_n \in K(c_1, \ldots, c_n)^*$ such that*

$$f = Dv + \sum_{i=1}^{n} c_i \frac{Du_i}{u_i}.$$

Proof. Suppose that there exist an elementary extension E of K and $g \in E$ such that $Dg = f$. Since $\overline{C}K$ is algebraic over K, $\mathrm{Const}(\overline{C}K) = \overline{C} \cap \overline{C}K = \overline{C}$ by Corollary 3.3.1. Hence, $\overline{C}K$ has an algebraically closed constant subfield, $f \in \overline{C}K$, $g \in \overline{C}E$, which is an elementary extension of $\overline{C}K$, so by Theorem 5.5.2, there are $v \in \overline{C}K$, $u_1, \ldots, u_n \in (\overline{C}K)^*$ and $c_1, \ldots, c_n \in \overline{C}$ such that

$$f = Dv + \sum_{i=1}^{n} c_i \frac{Du_i}{u_i}.$$

$F = K(v, u_1, \ldots, u_n, c_1, \ldots, c_n)$ is finite algebraic over K, so let $Tr_K^F : F \to K$ be the trace from F to K, $\overline{K}$ be the algebraic closure of K and $\sigma_1, \ldots, \sigma_m$ be the distinct embeddings of F in $\overline{K}$ over K. Each σ_j can be extended to a field automorphism of $\overline{K}$ over K, and since Tr_K^F and each σ_j commute with D by Theorem 3.2.4, we have

$$mf = \sum_{j=1}^{m} f^{\sigma_j} = Tr_K^F(Dv) + \sum_{j=1}^{m} \sum_{i=1}^{n} c_i^{\sigma_j} \frac{Du_i^{\sigma_j}}{u_i^{\sigma_j}}$$

so

$$f = Dw + \sum_{j=1}^{m} \sum_{i=1}^{n} d_{ij} \frac{Dw_{ij}}{w_{ij}}$$

with

$$w = \frac{1}{m} Tr_K^F(v) \in K, \quad d_{ij} = \frac{1}{m} c_i^{\sigma_j} \in \overline{K} \quad \text{and} \quad w_{ij} = u_i^{\sigma_j} \in \overline{K}^*.$$

In addition, $\mathrm{Const}(\overline{K}) = \overline{C} \cap \overline{K} = \overline{C}$ by Corollary 3.3.1, and $Dd_{ij} = D(c_i^{\sigma_j}/m) = (Dc_i)^{\sigma_j}/m = 0$, so $d_{ij} \in \overline{C}$ for each i and j. Let now $L = K(d_{11}, \ldots, d_{mn})$ and $M = L(w_{11}, \ldots, w_{mn})$. Since L is algebraic over K, $\overline{K}$ is the algebraic closure of L. Since M is finite algebraic over L, let $Tr_L^M : M \to L$ and $N : M \to L$ be the trace and norm maps from M to L. Since $d_{ij} \in L$ and Tr_L^M is L-linear, we have

$$Tr_L^M \left(d_{ij} \frac{Dw_{ij}}{w_{ij}} \right) = d_{ij} Tr_L^M \left(\frac{Dw_{ij}}{w_{ij}} \right) = d_{ij} \frac{DN(w_{ij})}{N(w_{ij})}$$

by Theorem 3.2.4, so

$$kf = Tr_L^M(f) = Tr_L^M(Dw) + Tr_L^M \left(\sum_{j=1}^{m} \sum_{i=1}^{n} d_{ij} \frac{Dw_{ij}}{w_{ij}} \right)$$

$$= kDw + \sum_{j=1}^{m} \sum_{i=1}^{n} d_{ij} \frac{DN(w_{ij})}{N(w_{ij})}$$

hence

$$f = Dw + \sum_{j=1}^{m} \sum_{i=1}^{n} \frac{d_{ij}}{k} \frac{Dz_{ij}}{z_{ij}}$$

which is of the form (5.3) with $w \in K$, $d_{ij} \in \overline{C}$ and $z_{ij} = N(w_{ij})$ in $K(d_{11}, \ldots, d_{mn})^*$. $\qquad \square$

5.6 The Residue Criterion

Now that Liouville's Theorem gives us a way of proving that a function has no elementary integral over a given field, we can complete the integration algorithm. For the rest of this chapter, let (k, D) be a differential field and t a monomial over k. From the Hermite reduction, we can assume without loss of generality that the integrand is given as the sum of a simple and a reduced element of $k(t)$.

We have seen in Sect. 2.4 that the Rothstein–Trager algorithm expresses the integral of a simple rational function with no polynomial part as a sum of logarithms. In this section, we show that this algorithm can be generalized to any monomial extension, where it will either prove that a function has no elementary integral, or reduce the problem to integrating elements of $k\langle t \rangle$. Rothstein had already generalized this algorithm to elementary transcendental extensions in his dissertation [68].

Lemma 5.6.1. *Let $f \in k(t)$ be simple. If there are $h \in k\langle t \rangle$, an algebraic extension E of $\mathrm{Const}(k)$, $v \in k(t)$, $c_1, \ldots, c_n \in E$, and $u_1, \ldots, u_n \in Ek(t)$ such that*

$$f + h = Dv + \sum_{i=1}^{n} c_i \frac{Du_i}{u_i}$$

then

$$\mathrm{residue}_p(f) = \sum_{i=1}^{n} c_i \nu_p(u_i)$$

for any normal irreducible $p \in Ek[t]$.

Proof. Let $f \in k(t)$ be simple, and suppose that there are $h \in k\langle t \rangle$, an algebraic extension E of $\mathrm{Const}(k)$, $v \in k(t)$, $c_1, \ldots, c_n \in E$, and $u_1, \ldots, u_n \in Ek(t)$ such that $f + h = Dv + \sum_{i=1}^{n} c_i D(u_i)/u_i$. Note that $f + h$ is simple since $h \in k\langle t \rangle$. Let $p \in Ek[t]$ be normal and irreducible. Then, for each i, $\nu_p(Du_i/u_i) \geq -1$ and $\mathrm{residue}_p(Du_i/u_i) = \nu_p(u_i)$ by Corollary 4.4.2. Suppose that $\nu_p(v) < 0$. Then $\nu_p(Dv) = \nu_p(v) - 1 < -1$ by Theorem 4.4.2, which implies that $\nu_p(f + h) < -1$ in contradiction with $f + h$ being simple. Hence $\nu_p(v) \geq 0$, so $\nu_p(Dv) \geq 0$, which implies that $\mathrm{residue}_p(Dv) = 0$. Furthermore, $\nu_p(h) \geq 0$, so $\mathrm{residue}_p(h) = 0$. Since $\mathrm{residue}_p$ is Ek-linear, we get

$$
\begin{aligned}
\mathrm{residue}_p(f) &= \mathrm{residue}_p(f) + \mathrm{residue}_p(h) = \mathrm{residue}_p(f + h) \\
&= \mathrm{residue}_p(Dv) + \sum_{i=1}^{n} c_i \, \mathrm{residue}_p\left(\frac{Du_i}{u_i}\right) = \sum_{i=1}^{n} c_i \, \nu_p(u_i).
\end{aligned}
$$

$\square$

Lemma 5.6.2. *Suppose that $\mathrm{Const}(k)$ is algebraically closed and let $f \in k(t)$ be simple. If there exists $h \in k\langle t \rangle$ such that $f + h$ has an elementary integral over $k(t)$, then $\mathrm{residue}_p(f) \in \mathrm{Const}(k)$ for any normal irreducible $p \in k[t]$.*

Proof. Let $C = \text{Const}(k)$, and suppose C is algebraically closed and that $f + h$ has an elementary integral over $k(t)$ where $f \in k(t)$ is simple and $h \in k\langle t \rangle$. By Theorem 5.5.1, there are $v, u_1, \ldots, u_n \in k$ and $c_1, \ldots, c_n \in C$ such that

$$f + h = Dv + \sum_{i=1}^{n} c_i \frac{Du_i}{u_i}.$$

Let $p \in k[t]$ be normal and irreducible. By Lemma 5.6.1 we have

$$\text{residue}_p(f) = \sum_{i=1}^{n} c_i \, \nu_p(u_i) \in C.$$

$\square$

Example 5.6.1. Let $k = \mathbb{Q}$, t be a monomial over k with $Dt = 1$ (*i.e.* $D = d/dt$), and

$$f = \frac{2t - 2}{t^2 + 1} \in k[t].$$

Then, f has an elementary integral over $k(t)$:

$$\int \frac{2t - 2}{t^2 + 1} \, dt = (1 + \sqrt{-1}) \log(1 + t\sqrt{-1}) + (1 - \sqrt{-1}) \log(1 - t\sqrt{-1}).$$

On the other hand, $t^2 + 1$ is irreducible over $\mathbb{Q}$, but

$$\text{residue}_{t^2+1}(f) = \pi_{t^2+1}\left(\frac{2t - 2}{2t}\right) = t + 1$$

which is not a constant. This shows that the hypothesis that the constant field of k be algebraically closed is required in Lemma 5.6.2. If we replace $\mathbb{Q}$ by $\mathbb{C}$, then $t^2 + 1 = (t - \sqrt{-1})(t + \sqrt{-1})$,

$$\text{residue}_{t-\sqrt{-1}}(f) = \pi_{t-\sqrt{-1}}\left(\frac{2t - 2}{t + \sqrt{-1}}\right) = 1 + \sqrt{-1}$$

and

$$\text{residue}_{t+\sqrt{-1}}(f) = \pi_{t+\sqrt{-1}}\left(\frac{2t - 2}{t - \sqrt{-1}}\right) = 1 - \sqrt{-1}$$

which are constants. This shows that the hypothesis that p be irreducible is also required in Lemmas 5.6.1 and 5.6.2.

Theorem 5.6.1. *Let $f \in k(t)$ be simple, and write $f = p + a/d$ where $p, a, d \in k[t]$, $d \neq 0$, $\deg(a) < \deg(d)$, and $\gcd(a, d) = 1$. Let z be an indeterminate over k,*

$$r = \text{resultant}_t(a - zDd, d) \in k[z],$$

$r = r_s r_n$ be a splitting factorization of r w.r.t. the coefficient lifting κ_D of D to $k[z]$, and

$$g = \sum_{r_s(\alpha)=0} \alpha \frac{Dg_\alpha}{g_\alpha} \tag{5.8}$$

where $g_\alpha = \gcd(a - \alpha Dd, d) \in k(\alpha)[t]$ and the sum is taken over all the distinct roots of r_s. Then,

(i) $g \in k(t)$, the denominator of g divides d, and $f - g$ is simple.

(ii) If there exists $h \in k\langle t \rangle$ such that $f + h$ has an elementary integral over $k(t)$, then $r_n \in k$ and $f - g \in k[t]$.

(iii) If there are $h \in k\langle t \rangle$, an algebraic extension E of $\mathrm{Const}(k)$, $v \in k(t)$, $c_1, \ldots, c_n \in E$, and $u_1, \ldots, u_n \in Ek(t)$ such that

$$f + h = Dv + \sum_{i=1}^{n} c_i \frac{Du_i}{u_i}$$

then r_s factors linearly over E.

Proof. (i) Let $r_s = c r_1^{e_1} \cdots r_n^{e_n}$ be the irreducible factorization of r_s in $k[z]$. Then, g can be rewritten as

$$g = \sum_{i=1}^{n} \sum_{r_i(\alpha)=0} \alpha \frac{Dg_\alpha}{g_\alpha}.$$

For each i, let k_i be $k(t)$ extended by all the roots of r_i, and α_i be a given root of r_i. Since k_i is a finitely generated algebraic extension of $k(t)$, the field automorphisms of k_i over $k(t)$ commute with D by Theorem 3.2.4, so we get

$$g = \sum_{i=1}^{n} Tr_i \left(\alpha_i \frac{Dg_{\alpha_i}}{g_{\alpha_i}} \right)$$

by Theorem 3.2.4 where Tr_i is the trace map from $k(t)(\alpha_i)$ to $k(t)$. Hence, $g \in k(t)$. Furthermore, since $g_\alpha \mid d$ for each root α of r_s, $\mathrm{lcm}_{r_s(\alpha)=0}(g_\alpha) \mid d$, so the denominator of g also divides d. Hence the denominator of $f - g$ divides d, which implies that $f - g$ is simple since d is normal.

(ii) Suppose that $f + h$ has an elementary integral over $k(t)$ for some $h \in k\langle t \rangle$, and let $\overline{k}$ be the algebraic closure of k. By Corollary 3.4.1, t is a monomial over $\overline{k}$, and simple (resp. reduced) elements of $k(t)$ remain simple (resp. reduced) when viewed as elements of $\overline{k}(t)$. Furthermore $f + h$ has an elementary integral over $\overline{k}(t)$, so we work with $\overline{k}(t)$ in the rest of this proof. Let $\alpha \in \overline{k}$ be any root of r. If $\alpha = 0$, then $D\alpha = 0$. Otherwise $\alpha \neq 0$ and $\alpha = \mathrm{residue}_q(f)$ for some normal irreducible $q \in \overline{k}[x]$ by Theorem 4.4.3, hence $D\alpha = 0$ by Lemma 5.6.2. Thus $r_s(\alpha) = 0$ in both cases by Theorem 3.5.2, so $r_n(\alpha) \neq 0$ since $\gcd(r_n, r_s) = 1$. Since this holds for all the roots of r, we have $r_n \in k$.

For any $\alpha \in \overline{k}$, write $g_\alpha = \gcd(d, a - \alpha Dd)$. Note that all the irreducible factors of g_α must be normal, since $g_\alpha \mid d$, which is normal. Let $\alpha, \beta \in \overline{k}$, and $q \in \overline{k}[t]$ be a normal irreducible common factor of g_α and g_β. Then

$\alpha = \text{residue}_q(a/d) = \beta$ by Lemma 4.4.3, so $\gcd(g_\alpha, g_\beta) = 1$ when $\alpha \neq \beta$. Let now $q \in \overline{k}[t]$ be irreducible and normal, and $\beta = \text{residue}_q(f)$. If $\beta = 0$, then q does not divide d, so q does not divide any g_α, which implies that $\nu_q(g) \geq 0$, hence that $\text{residue}_q(g) = 0 = \text{residue}_q(f - g)$. If $\beta \neq 0$, then $r(\beta) = 0$ by Theorem 4.4.3, and $q \mid g_\beta$ by Lemma 4.4.3, so $r_s(\beta) = 0$ since $r_n \in k$. Since d is squarefree, g_β is squarefree, so $\nu_q(g_\beta) = 1$. By Theorem 4.4.1, residue_q is $\overline{k}$-linear, so we get

$$\text{residue}_q(f - g) = \beta - \sum_{r_s(\alpha)=0} \alpha \, \text{residue}_q\left(\frac{D g_\alpha}{g_\alpha}\right) = \beta - \sum_{r_s(\alpha)=0} \alpha \, \nu_q(g_\alpha)$$

by Corollary 4.4.2. Since $\nu_q(g_\alpha) = 0$ for $\alpha \neq \beta$, this gives $\text{residue}_q(s) = \beta - \beta = 0$. Since this holds for any normal irreducible $q \in \overline{k}[t]$ and $f - g$ is simple, we have $f - g \in \overline{k}[t]$, hence $f - g \in k[t]$.

(iii) Suppose that there are $h \in k\langle t \rangle$, an algebraic extension E of $\text{Const}(k)$, $v \in k(t)$, $c_1, \ldots, c_n \in E$, and $u_1, \ldots, u_n \in Ek(t)$ such that

$$f + h = Dv + \sum_{i=1}^{n} c_i \frac{D u_i}{u_i}. \tag{5.9}$$

Let $\overline{k}$ be the algebraic closure of k. As explained in part (ii), we can replace $k(t)$ by $\overline{k}(t)$ and view (5.9) as an equality in $\overline{k}(t)$. Let $\alpha \in \overline{k}$ be any root of r_s. By Theorem 4.4.3, $\alpha = \text{residue}_p(f)$ for some normal irreducible $p \in \overline{k}[t]$, so by Lemma 5.6.1

$$\alpha = \text{residue}_p(f) = \sum_{i=1}^{n} c_i \nu_p(u_i) \in E.$$

Hence, E contains all the roots of r_s in $\overline{k}$, so r_s factors linearly over E. □

Note that since the roots of r_s are all constants by Theorem 3.5.2, g as given by (5.8) always has an elementary integral, namely

$$\int g = \sum_{r_s(\alpha)=0} \alpha \log(\gcd(d, a - \alpha D d))$$

which is the Rothstein–Trager formula in the case of rational functions. Part (iii) of Theorem 5.6.1 applied to the rational function case proves part (iii) of Theorem 2.4.1, thereby completing the proof of that theorem. As in the rational function case, a prime factorization $r_s = u \, s_1^{e_1} \cdots s_m^{e_m}$ is required, as well as a gcd computation in $k(\alpha_i)[t]$ for each i, where α_i is a root of s_i. There is no need however to compute the splitting field of r_s. Furthermore, the monic part of r_s always has constant coefficients.

ResidueReduce(f, D) (* Rothstein–Trager resultant reduction *)

(* Given a derivation D on $k(t)$ and $f \in k(t)$ simple, return g elementary over $k(t)$ and a Boolean $b \in \{0, 1\}$ such that $f - Dg \in k[t]$ if $b = 1$, or $f + h$ and $f + h - Dg$ do not have an elementary integral over $k(t)$ for any $h \in k\langle t \rangle$ if $b = 0$. *)

$d \leftarrow$ denominator(f)
$(p, a) \leftarrow$ **PolyDivide**(numerator$(f), d$)   (* $f = p + a/d$ *)
$z \leftarrow$ a new indeterminate over $k(t)$
$r \leftarrow$ resultant$_t(d, a - zDd)$
$(r_n, r_s) \leftarrow$ **SplitFactor**(r, κ_D)
$u\, s_1^{e_1} \cdots s_m^{e_m} \leftarrow$ **factor**(r_s) (* factorization into irreducibles *)
for $i \leftarrow 1$ **to** m **do**
  $\alpha \leftarrow \alpha \mid s_i(\alpha) = 0$
  $g_i \leftarrow$ gcd$(d, a - \alpha Dd)$ (* algebraic gcd computation *)
if $\prod_{i=1}^{m} n_i \in k$ **then** $b \leftarrow 1$ **else** $b \leftarrow 0$
return$(\sum_{i=1}^{m} \sum_{\alpha \mid s_i(\alpha) = 0} \alpha \log(g_i), b)$

Example 5.6.2. Consider

$$\int \frac{2\log(x)^2 - \log(x) - x^2}{\log(x)^3 - x^2\log(x)}\, dx \,.$$

Let $k = \mathbb{Q}(x)$ with $D = d/dx$, and let t be a monomial over k satisfying $Dt = 1/x$, *i.e.* $t = \log(x)$. Our integrand is then

$$f = \frac{2t^2 - t - x^2}{t^3 - x^2 t} \in k(t)$$

which is simple since $t^3 - x^2 t$ is squarefree. We get

$$d = t^3 - x^2 t, \quad p = 0, \quad a = 2t^2 - t - x^2$$

and

$$\begin{aligned}
r &= \text{resultant}_t \left((t^3 - x^2 t, \frac{2x - 3z}{x} t^2 + (2xz - 1)t + x(z - x) \right) \\
&= 4x^3(1 - x^2) \left(z^3 - xz^2 - \frac{1}{4}z + \frac{x}{4} \right)
\end{aligned}$$

which is squarefree. Then,

$$\kappa_D r = -x^2 (4(5x^2 + 3)z^3 + 8x(3x^2 - 2)z^2 + (5x^2 - 3)z - 2x(3x^2 - 2))$$

so the splitting factorization of r w.r.t. κ_D is

$$r_s = \gcd(r, \kappa_D r) = x^2 \left(z^2 - \frac{1}{4} \right)$$

and

$$r_n = \frac{r}{r_s} = -4x(x^2 - 1)(z - x) \notin k.$$

Hence, f does not have an elementary integral. Proceeding further we get

$$g_1 = \gcd\left(t^3 + x^2 t, \frac{2x - 3\alpha}{x} t^2 + (2x\alpha - 1)t + x(\alpha - x)\right) = t + 2\alpha x$$

where $\alpha^2 - 1/4 = 0$, so

$$g = \sum_{\alpha | \alpha^2 - 1/4 = 0} \alpha \log(t + 2\alpha x) = \frac{1}{2} \log(t + x) - \frac{1}{2} \log(t - x).$$

Computing $f - Dg$ we find

$$\begin{aligned}
\int \frac{2\log(x)^2 - \log(x) - x^2}{\log(x)^3 - x^2 \log(x)} \, dx &= \frac{1}{2} \log\left(\frac{\log(x) + x}{\log(x) - x}\right) + \int \frac{dx}{\log(x)} \\
&= \frac{1}{2} \log\left(\frac{\log(x) + x}{\log(x) - x}\right) + \mathrm{Li}(x)
\end{aligned}$$

where $\mathrm{Li}(x)$ is the logarithmic integral, which has been proven to be nonelementary since $r_n \notin k$.

With the notation as in Theorem 5.6.1, we have $\gcd(r_s, r_n) = 1$, so any root α of r_s with multiplicity n is also a root of r with multiplicity n. Since $\gcd(a, d) = \gcd(d, Dd) = 1$ and $\deg(a) < \deg(d)$, we can apply Theorem 2.5.1 with $A = a$, $B = Dd$ and $C = d$, and we get that for any root α of r of multiplicity $i > 0$,

$$\gcd(d, a - \alpha Dd) = \mathrm{pp}_t(R_m)(\alpha, t)$$

where $\deg_t(R_m) = i$ and R_m is in the subresultant PRS of d and $a - zDd$ if $\deg(Dd) \leq \deg(d)$, or of $a - zDd$ and d if $\deg(Dd) > \deg(d)$. Thus, the Lazard–Rioboo–Trager algorithm is applicable in arbitrary monomial extensions, and it is not necessary to compute the prime factorization of r_s, or the g_α's appearing in (5.8), we can use the various remainders appearing in the subresultant PRS instead. As in the case of rational functions, we use a squarefree factorization of $r_s = \prod_{i=1}^n q_i^i$ to split the sum appearing in (5.8) into several summands, each indexed by the roots of q_i. We can also avoid computing $\mathrm{pp}_t(R_m)$, ensuring instead that its leading coefficient is coprime with the corresponding q_i. And since multiplying any g_α in (5.8) by an arbitrary nonzero element of $k(\alpha)$ does not change the conclusion of Theorem 5.6.1, we can make $\mathrm{pp}_t(R_m)(\alpha, t)$ monic in order to simplify the answer. This last step requires inverting an element of $k[\alpha]$ and is optional. As in the rational function case, it turns out that the leading coefficients of the $\mathrm{pp}_t(R_m)(\alpha, t)$'s are always invertible in $k[\alpha]$ (Exercise 2.7).

ResidueReduce(f, D)
(* Lazard–Rioboo–Rothstein–Trager resultant reduction *)

(* Given a derivation D on $k(t)$ and $f \in k(t)$ simple, return g elementary over $k(t)$ and a Boolean $b \in \{0, 1\}$ such that $f - Dg \in k[t]$ if $b = 1$, or $f + h$ and $f + h - Dg$ do not have an elementary integral over $k(t)$ for any $h \in k\langle t \rangle$ if $b = 0$. *)

$d \leftarrow$ denominator(f)
$(p, a) \leftarrow$ **PolyDivide**(numerator$(f), d)$ (* $f = p + a/d$ *)
$z \leftarrow$ a new indeterminate over $k(t)$
if $\deg(Dd) \le \deg(d)$
then $(r, (R_0, R_1, \ldots, R_q, 0)) \leftarrow$ **SubResultant**$_x(d, a - zDd)$
else $(r, (R_0, R_1, \ldots, R_q, 0)) \leftarrow$ **SubResultant**$_x(a - zDd, d)$
$((n_1, \ldots, n_n), (s_1, \ldots, s_n)) \leftarrow$ **SplitSquarefreeFactor**(r, κ_D)
for $i \leftarrow 1$ **to** n such that $\deg(s_i) > 0$ **do**
  if $i = \deg(d)$ **then** $S_i \leftarrow d$
  else
      $S_i \leftarrow R_m$ where $\deg_t(R_m) = i$, $1 \le m < q$
      $(A_1, \ldots, A_s) \leftarrow$ **SquareFree**(lc$_t(S_i)$)
      for $j \leftarrow 1$ **to** s **do** $S_i \leftarrow S_i / \gcd_z(A_j, s_i)^j$ (* exact quotient *)
if $\prod_{i=1}^{n} n_i \in k$ **then** $b \leftarrow 1$ **else** $b \leftarrow 0$
return$(\sum_{i=1}^{m} \sum_{\alpha | s_i(\alpha) = 0} \alpha \log(S_i(\alpha, t)), b)$

Example 5.6.3. Consider the same integrand as in example 5.6.2. We have $\deg(Dd) < \deg(d)$ and the subresultant PRS of d and $a - zDd$ is

i	R_i
0	$t^3 - x^2 t$
1	$(2 - 3z/x)t^2 + (2xz - 1)t + x(z - x)$
2	$(4x^2 - 6)z^2 + 3xz - 2x^2 + 1)t + x(z - x)(2xz - 1)$
3	$4x^3(1 - x^2)\left(z^3 - xz^2 - \frac{1}{4}z + \frac{1}{4}x\right)$

The Rothstein–Trager resultant is $r = R_3$, and its split–squarefree factorization w.r.t. κ_D is

$$s_1 = \gcd(r, \kappa_D r) = x^2\left(z^2 - \frac{1}{4}\right), \qquad n_1 = \frac{r}{s_1} = -4x(x^2 - 1)(z - x) \notin k.$$

Hence, f does not have an elementary integral. Proceeding further we find that s_1 is squarefree, and the remainder of degree 1 in t in the PRS is

$$R_2 = ((4x^2 - 6)z^2 + 3xz - 2x^2 + 1)t + x(z - x)(2xz - 1).$$

Since

$$\gcd(\mathrm{lc}_t(R_2), s_1) = \gcd\left((4x^2 - 6)z^2 + 3xz - 2x^2 + 1, x^2\left(z^2 - \frac{1}{4}\right)\right) = 1,$$

$S_1 = R_2$. Evaluating for z at a root α of $z^2 - 1/4 = 0$ we get

$$S_1(\alpha, t) = -\frac{1}{2}((2x^2 - 6\alpha x + 1)t + 4\alpha x^3 - 3x^2 + 2\alpha x)$$

so

$$
\begin{aligned}
g &= \sum_{\alpha | \alpha^2 - 1/4 = 0} \alpha \log\left(-\frac{1}{2}((2x^2 - 6\alpha x + 1)t + x(4\alpha x^2 - 3x + 2\alpha))\right) \\
&= \frac{1}{2} \log\left(-\frac{(2x^2 - 3x + 1)(t + x)}{2}\right) - \frac{1}{2} \log\left(-\frac{(2x^2 + 3x + 1)(t - x)}{2}\right).
\end{aligned}
$$

Computing $f - Dg$ we find

$$
\begin{aligned}
\int \frac{2\log(x)^2 - \log(x) - x^2}{\log(x)^3 - x^2 \log(x)}\, dx &= \frac{1}{2}\log\left(\frac{(2x^2 - 3x + 1)(\log(x) + x)}{(2x^2 + 3x + 1)(\log(x) - x)}\right) \\
&\quad + \int\left(\frac{1}{\log(x)} - \frac{6x^2 - 3}{4x^4 - 5x^2 + 1}\right) dx
\end{aligned}
$$

where the remaining integral has been proven to be nonelementary. In fact, it is the integral of a rational function plus the logarithmic integral $\mathrm{Li}(x)$. If we had decided to make $S_1(\alpha, t)$ monic, we would have obtained

$$S_1(\alpha, x) = -\frac{1}{2}(2x^2 - 6\alpha x + 1)(t + 2\alpha x)$$

so the integral is then the same as in example 5.6.2.

5.7 Integration of Reduced Functions

From the results of the previous sections, we are left with the problem of integrating reduced elements of a monomial extension. We use a specialized version of Liouville's Theorem for such elements.

Theorem 5.7.1. *Let k be a differential field, t be a monomial over k, $C = \mathrm{Const}(k(t))$, and $f \in k\langle t\rangle$. If there exist an elementary extension E of $k(t)$ and $g \in E$ such that $Dg = f$, then there are $v \in k\langle t\rangle$, $c_1, \ldots, c_n \in \overline{C}$, and $u_1, \ldots, u_n \in \mathcal{S}_{k(c_1, \ldots, c_n)[t]:k(c_1, \ldots, c_n)}$ such that*

$$f = Dv + \sum_{i=1}^{n} c_i \frac{Du_i}{u_i}.$$

Proof. Suppose that there exist an elementary extension E of $k(t)$ and $g \in E$ such that $Dg = f$. Then, by Theorem 5.5.3, there are $v \in k(t)$, $c_1, \ldots, c_n \in \overline{C}$, and $u_1, \ldots, u_n \in k(c_1, \ldots, c_n)(t)$ such that $f = Dv + \sum_{i=1}^{n} c_i D(u_i)/u_i$. Write $g = \sum_{i=1}^{n} c_i D(u_i)/u_i$. Since $g = f - Dv$, it follows that $g \in k(t)$. Let $p \in k[t]$ be normal and irreducible, and $q \in k(c_1, \ldots, c_n)[t]$ be any irreducible factor of

p over $k(c_1, \ldots, c_n)$. Then, $\nu_p(f) \geq 0$ by Corollary 4.4.1, and $\nu_q(c_i Du_i / u_i) \geq -1$ for each i by Corollary 4.4.2, so $\nu_q(g) \geq -1$. Since this holds for any irreducible factor q of p and $g \in k(t)$, Theorem 4.1.2 implies that $\nu_p(g) \geq -1$. Suppose that $\nu_p(v) < 0$. Then, $\nu_p(Dv) = \nu_p(v) - 1 < -1$ by Theorem 4.4.2, which implies that $\nu_p(Dv + g) < -1$, hence that $\nu_p(f) < -1$, in contradiction with f reduced. Hence, $\nu_p(v) \geq 0$ for all normal irreducible $p \in k[t]$, which means that $v \in k\langle t \rangle$ and $Dv \in k\langle t \rangle$ by Corollary 4.4.1.

Write now $u_i = w_i \prod_{j=1}^{n_i} p_{ij}^{e_{ij}}$ where $w_i \in k(c_1, \ldots, c_n)$, each p_{ij} is a monic irreducible element of $k(c_1, \ldots, c_n)[t]$, and the e_{ij}'s are integers. Then, using the logarithmic derivative identity and grouping together all the terms involving the same p_{ij}, we get

$$f = Dv + \sum_{i=1}^{n} c_i \frac{Dw_i}{w_i} + \sum_{j=1}^{N} d_j \frac{Dq_j}{q_j} \tag{5.10}$$

where the q_j's are in $k(c_1, \ldots, c_n)[t]$, monic, irreducible and coprime. Each w_i is special since it is in $k(c_1, \ldots, c_n)$. Suppose that q_s is normal for some s. Then, Lemma 5.6.1 applied to (5.10) implies that

$$\text{residue}_{q_s}(f) = \sum_{i=1}^{n} c_i \nu_{q_s}(w_i) + \sum_{j=1}^{n} d_j \nu_{q_s}(q_j) \, .$$

But $\text{residue}_{q_s}(f) = 0$ since $f \in k\langle t \rangle$, and $\nu_{q_s}(w_i) = 0$ since $w_i \in k(c_1, \ldots, c_n)$, and $\nu_{q_s}(q_j) = 0$ for $j \neq s$ since the q_j's are coprime. Hence, $0 = d_s \nu_{q_s}(q_s) = d_s$, so $d_s = 0$ whenever q_s is normal. Keeping only the nonzero summands in (5.10), we get that each q_j is special, which proves the theorem. $\square$

In the case of nonlinear monomials, we have seen that we can always rewrite a polynomial $p \in k[t]$ as the sum of a derivative and a polynomial of degree less than $\delta(t)$. We then have an analogue of the residue criterion that either proves that such a reduced function does not have an elementary integral, or eliminates the term of degree $\delta(t) - 1$ from its polynomial part.

Theorem 5.7.2. *Suppose that t is a nonlinear monomial. Let $f \in k\langle t \rangle$ and write $f = p + a/d$ where $p, a, d \in k[t]$, $d \neq 0$, $\deg(p) < \delta(t)$ and $\deg(a) < \deg(d)$. Let $b \in k$ be the coefficient of $t^{\delta(t)-1}$ in p, and $c = b/\lambda(t)$. If f has an elementary integral over $k(t)$ then $Dc = 0$.*

Proof. Let $C = \text{Const}(k)$. Replacing C by its algebraic closure, we can assume without loss of generality that C is algebraically closed. Suppose that f has an elementary integral over $k(t)$. Then, by Theorem 5.7.1, there are $v \in k\langle t \rangle$, $c_1, \ldots, c_n \in C$, and $u_1, \ldots, u_n \in S$ such that

$$f = Dv + \sum_{i=1}^{n} c_i \frac{Du_i}{u_i} \, . \tag{5.11}$$

By Theorem 4.4.4, $\nu_\infty(Du_i/u_i) \geq -m$ and $\pi_\infty(t^{-m}Du_i/u_i) = -\nu_\infty(u_i)\lambda(t)$ for each i where $m = \delta(t) - 1$. Furthermore, $\nu_\infty(a/d) > 0$ since $\deg(a) < \deg(d)$, so $\nu_\infty(f) \geq -\deg(p) \geq -m$ and $\pi_\infty(a/d) = \pi_\infty(t^{-m}a/d) = 0$, which implies that $\pi_\infty(t^{-m}f) = b$. Suppose that $\nu_\infty(v) < 0$, then $\nu_\infty(Dv) < -m$ by Theorem 4.4.4, so $\nu_\infty(Dv + \sum_{i=1}^n c_i Du_i/u_i) < -m$, in contradiction with $\nu_\infty(f) \geq -m$. Hence $\nu_\infty(v) \geq 0$. If $\nu_\infty(v) > 0$, then $\nu_\infty(Dv) > -m$ by Theorem 4.4.4. Otherwise, $\nu_\infty(v) = 0$ and $\nu_\infty(Dv) > -m$ also by Theorem 4.4.4. Hence $\nu_\infty(t^{-m}Dv) > 0$ in any case, so $\pi_\infty(t^{-m}Dv) = 0$. Multiplying both sides of (5.11) by t^{-m} and applying π_∞, we get

$$b = \pi_\infty(t^{-m}f) = \sum_{i=1}^n c_i \pi_\infty\left(t^{-m}\frac{Du_i}{u_i}\right) = -\sum_{i=1}^n c_i \nu_\infty(u_i)\lambda(t)$$

hence $c = b/\lambda(t) = -\sum_{i=1}^n c_i \nu_\infty(u_i)$, so $Dc = 0$.   □

If c is a constant, then Theorem 5.4.2 implies that

$$f - D\left(\frac{c}{\deg(q)}\log(q)\right)$$

has degree at most $\delta(t) - 2$ for any $q \in \mathcal{S} \setminus k$, so in the case of nonlinear monomials, we are left with reduced integrands with polynomial parts of degree at most $\delta(t) - 2$, provided that we know at least one nontrivial special polynomial. If we know that there are no nontrivial special polynomials, then integrating reduced elements of such nonlinear extensions is in fact easier, and an algorithm for that purpose will be presented in Sect. 5.11.

We have now all the necessary tools to complete the integration algorithm. In the following sections, we give algorithms that, given an integrand f in $k(t)$ for a monomial t, either prove that f has no elementary integral over $k(t)$, or compute an elementary extension E of $k(t)$ and an element $g \in E$ such that $f - Dg \in k$. This process eliminates t from the integrand, thus reducing the problem to integrating an element of k, which can be done recursively, i.e. the algorithms of this chapter can be applied to elements of k until we are left with constants to integrate. Note that when t itself is not elementary over k, then the problems of deciding whether an element of k has an elementary integral over k or over $k(t)$ are fundamentally different, so our algorithms will produce proofs of nonintegrability only if the integrand is itself an elementary function. They can be applied however to much larger classes of functions.

It turns out that it will also be necessary to assume that some related problems are solvable for elements of k. Those problems depend on the kind of monomial we are dealing with, so we need to handle the various cases separately at this point. Algorithms for all those related problems will be presented in later chapters.

5.8 The Primitive Case

In the case of primitive monomials over a differential field k, the related problem we need to solve over k is the *limited integration problem*: recall that the problem of integration in closed form is, given $f \in k$ to determine whether there exist an elementary extension E of k and $g \in E$ such that $\text{Const}(E)$ is algebraic over $\text{Const}(k)$ and $Dg = f$. Let $w_1, \ldots, w_n \in k$ be fixed. The *problem of limited integration with respect to* $w_1, \ldots, w_n$ is: given $f \in k$, determine whether there are $g \in k$ and $c_1, \ldots, c_n \in \text{Const}(k)$ such that $Dg = f - c_1 w_1 - \ldots - c_n w_n$, and to compute g and the c_i's if they exist. It is very similar to the problem of integration in closed form, except that the specific differential extension $k(\int w_1, \ldots, \int w_n)$ is provided for the integral. We present in this section an algorithm that, with appropriate assumptions on k, integrates elements of $k(t)$ when t is a primitive monomial over k. We first describe an algorithm for integrating elements of $k[t]$.

Theorem 5.8.1. *Let k be a differential field and t a primitive over k. If the problem of limited integration w.r.t. Dt is decidable for elements of k, and Dt is not the derivative of an element of k, then for any $p \in k[t]$ we can either prove that p has no elementary integral over $k(t)$, or compute $q \in k[t]$ such that $p - Dq \in k$.*

Proof. We proceed by induction on $m = \deg(p)$. If $m = 0$, then $p \in k$ and $q = 0$ satisfies the theorem, so suppose that $m > 0$ and that the theorem holds for any polynomial of degree less than m. Since Dt is not the derivative of an element of k, t is a monomial over k, $\text{Const}(k(t)) = \text{Const}(k)$, and $S = k$ by Theorem 5.1.1. Thus, Theorem 5.7.1 says that if p has an elementary integral over $k(t)$, then there are $v \in k[t]$, $c_1, \ldots, c_n \in \overline{C}$ and $u_1, \ldots, u_n \in k(c_1, \ldots, c_n)$ such that

$$p = Dv + \sum_{i=1}^{n} c_i \frac{Du_i}{u_i} \tag{5.12}$$

where $C = \text{Const}(k)$. $K = k(c_1, \ldots, c_n)$ is an algebraic extension of k, so t is transcendental over K. Furthermore, Dt is not the derivative of an element of K by Lemma 5.1.1, so t is a monomial over K and $\text{Const}(K(t)) = \text{Const}(K)$. Equating degrees in (5.12) we get $\deg(Dv) = \deg(p) = m > 0$, so $\deg(v) < m + 1$ by Lemma 5.1.2, so write $p = at^m + s$ and $v = ct^{m+1} + bt^m + w$ where $a, b, c \in k$, $s, w \in k[t]$, $\deg(s) < m$ and $\deg(w) < m$. Equating the coefficients of t^{m+1} and t^m in (5.12) we get $Dc = 0$ and

$$a = Db + (m + 1) c \, Dt. \tag{5.13}$$

Since we can solve the problem of limited integration w.r.t. Dt for elements of k and $a \in k$, we can either prove that (5.13) has no solution $b \in k, c \in \text{Const}(k)$, or find such a solution. If it has no solution, then (5.12) has no solution so p has no elementary integral over $k(t)$. If we have a solution b, c, letting $q_0 = ct^{m+1} + bt^m$, we get

$$p - Dq_0 = (at^m + s) - ((m+1)cDt + Db)\, t^m - (mbDt)t^{m-1} = s - (mbDt)t^{m-1}$$

hence $\deg(p - Dq_0) < m$. By induction we can either prove that $p - Dq_0$ has no elementary integral over $k(t)$, in which case p has no elementary integral over $k(t)$, or we get $q_1 \in k[t]$ such that $p - Dq_0 - Dq_1 \in k$, which implies that $p - Dq \in k$ where $q = q_0 + q_1$. □

IntegratePrimitivePolynomial(p, D)
(* Integration of polynomials in a primitive extension *)

(* Given a is a primitive monomial t over k, and $p \in k[t]$, return $q \in k[t]$ and a Boolean $\beta \in \{0, 1\}$ such that $p - Dq \in k$ if $\beta = 1$, or $p - Dq$ does not have an elementary integral over $k(t)$ if $\beta = 0$. *)

if $p \in k$ **then return**$(0, 1)$
$a \leftarrow \mathrm{lc}(p)$
(* LimitedIntegrate will be given in Chap. 7 *)
$(b, c) \leftarrow$ **LimitedIntegrate**(a, Dt, D) (* $a = Db + cDt$ *)
if $(b, c) =$ "no solution" **then return**$(0, 0)$
$m \leftarrow \deg(p)$
$q_0 \leftarrow ct^{m+1}/(m+1) + bt^m$
$(q, \beta) \leftarrow$ **IntegratePrimitivePolynomial**$(p - Dq_0, D)$
return$(q + q_0, \beta)$

Example 5.8.1. Consider

$$\int \left(\left(\log(x) + \frac{1}{\log(x)} \right) \mathrm{Li}(x) - \frac{x}{\log(x)} \right) dx$$

where $\mathrm{Li}(x) = \int dx / \log(x)$ is the logarithmic integral. Let $k = \mathbb{Q}(x, t_0)$ with $D = d/dx$, where t_0 is a monomial over $\mathbb{Q}(x)$ satisfying $Dt_0 = 1/x$, i.e. $t_0 = \log(x)$, and let t be a monomial over k satisfying $Dt = 1/t_0$, i.e. $t = \mathrm{Li}(x)$. Our integrand is then

$$p = \left(t_0 + \frac{1}{t_0} \right) t - \frac{x}{t_0} \in k[t].$$

We get

1. $a = \mathrm{lc}(p) = t_0 + 1/t_0$
2.

$$\left(t_0 + \frac{1}{t_0} \right) - \frac{1}{t_0} = t_0 = \log(x) = \frac{d}{dx} \left(x\log(x) - x \right) = D(xt_0 - x)$$

so $(b, c) =$ **LimitedIntegrate**$(t_0 + 1/t_0, 1/t_0, D) = (xt_0 - x, 1)$
3. $q_0 = ct^2/2 + bt = t^2/2 + (xt_0 - x)t$

4. $p - Dq_0 = -x \in k$ so the call **IntegratePrimitivePolynomial**$(-x, D)$
returns $(q, \beta) = (0, 1)$.

Hence,

$$\int \left(\left(\log(x) + \frac{1}{\log(x)} \right) \mathrm{Li}(x) - \frac{x}{\log(x)} \right) dx$$

$$= \frac{\mathrm{Li}(x)^2}{2} + (x \log(x) - x)\mathrm{Li}(x) - \int x dx$$

$$= \frac{\mathrm{Li}(x)^2}{2} + (x \log(x) - x)\mathrm{Li}(x) - \frac{x^2}{2} .$$

Putting all the pieces together, we get an algorithm for integrating elements of $k(t)$.

Theorem 5.8.2. *Let k be a differential field and t a primitive over k. If the problem of limited integration w.r.t. Dt is decidable for elements of k, and Dt is not the derivative of an element of k, then for any $f \in k(t)$ we can either prove that f has no elementary integral over $k(t)$, or compute an elementary extension E of $k(t)$ and $g \in E$ such that $f - Dg \in k$.*

Proof. Suppose that Dt is not the derivative of an element of k, then t is a monomial over k and $\mathrm{Const}(k(t)) = \mathrm{Const}(k)$ by Theorem 5.1.1. Let $f \in k(t)$. By Theorem 5.3.1, we can compute $g_1, h, r \in k(t)$ such that $f = Dg_1 + h + r$, h is simple and r is reduced. From h, which is simple, we compute $g_2 \in k(t)$ given by (5.8) in Theorem 5.6.1. Note that $g_0 = g_1 + \int g_2$ lies in some elementary extension of $k(t)$. Let $p = h - g_2$ and $q = p + r$, then $f = Dg_0 + q$ so f has an elementary integral over $k(t)$ if and only if q has one. If $p \notin k[t]$, then $p + r$ does not have an elementary integral over $k(t)$ by Theorem 5.6.1, so f does not have an elementary integral over $k(t)$. Suppose now that $p \in k[t]$. We have $k\langle t \rangle = k[t]$ by (5.1), so $r \in k[t]$, hence $q \in k[t]$. By Theorem 5.8.1 we can either prove that q has no elementary integral over $k(t)$, in which case f has no elementary integral over $k(t)$, or compute $s \in k[t]$ such that $q - Ds \in k$, in which case $f - Dg \in k$ where $g = g_0 + s$.   □

IntegratePrimitive(f, D) (* Integration of primitive functions *)

(* Given a is a primitive monomial t over k, and $f \in k(t)$, return g
elementary over $k(t)$ and $\beta \in \{0, 1\}$ such that $f - Dg \in k$ if $\beta = 1$, or
$f - Dg$ does not have an elementary integral over $k(t)$ if $\beta = 0$. *)

$(g_1, h, r) \leftarrow$ **HermiteReduce**(f, D)
$(g_2, \beta) \leftarrow$ **ResidueReduce**(h, D)
if $\beta = 0$ then return$(g_1 + g_2, 0)$
$(q, \beta) \leftarrow$ **IntegratePrimitivePolynomial**$(h - Dg_2 + r, D)$
return$(g_1 + g_2 + q, \beta)$

5.9 The Hyperexponential Case

In the case of hyperexponential monomials over a differential field k, the related problem we need to solve over k is the *Risch differential equation problem*: given $f, g \in k$, determine whether there exists $y \in k$ such that

$$Dy + fy = g \tag{5.14}$$

and to compute y if it exists. It may happen in general that (5.14) has more than one solution in k, so we first need to examine when this can happen.

Lemma 5.9.1. *Let (K, D) be a differential field and $\alpha \in K$. If there are $y, z \in K$ such that $y \neq z$ and $Dy + \alpha y = Dz + \alpha z$, then $\alpha = Du/u$ for some $u \in K^*$.*

Proof. Let $u = 1/(y - z) \in K^*$. Then,

$$Du - \alpha u = -\frac{Dy - Dz}{(y - z)^2} - \frac{\alpha}{y - z} = \frac{(Dz + \alpha z) - (Dy + \alpha y)}{(y - z)^2} = 0.$$

$\square$

We present in this section an algorithm that, with appropriate assumptions on k, integrates elements of $k(t)$ when t is a hyperexponential monomial over k. We first describe an algorithm for integrating elements of $k\langle t \rangle$.

Theorem 5.9.1. *Let k be a differential field and t an hyperexponential over k. If we can solve Risch differential equations over k, and Dt/t is not a logarithmic derivative of a k-radical, then for any $p \in k\langle t \rangle$ we can either prove that p has no elementary integral over $k(t)$, or compute $q \in k\langle t \rangle$ such that $p - Dq \in k$.*

Proof. Since Dt/t is not a logarithmic derivative of a k-radical, t is a monomial over k, $\mathrm{Const}(k(t)) = \mathrm{Const}(k)$, and $\mathcal{S}^{\mathrm{irr}} = \{t\}$ by Theorem 5.1.2. Thus $k\langle t \rangle = k[t, t^{-1}]$ by (5.1), and Theorem 5.7.1 says that if p has an elementary integral over $k(t)$, then there are $v \in k\langle t \rangle$, $c_1, \ldots, c_n \in \overline{C}$, $b_1, \ldots, b_n \in k(c_1, \ldots, c_n)$, and $m_1, \ldots, m_n \in \mathbb{Z}$ such that

$$p = Dv + \sum_{i=1}^{n} c_i \frac{Db_i t^{m_i}}{b_i t^{m_i}} = Dv + \frac{Dt}{t} \sum_{i=1}^{n} m_i c_i + \sum_{i=1}^{n} c_i \frac{Db_i}{b_i} \tag{5.15}$$

where $C = \mathrm{Const}(k)$. $K = k(c_1, \ldots, c_n)$ is an algebraic extension of k, so t is transcendental over K. Furthermore, Dt/t is not a logarithmic derivative of a K-radical by Lemma 3.4.8, so t is a monomial over K and $\mathrm{Const}(K(t)) = \mathrm{Const}(K)$. Since $p, v \in k[t, t^{-1}]$, write $p = \sum_{i=m}^{M} a_i t^i$ and $v = \sum_{i=r}^{R} v_i t^i$ where $a_i, v_i \in k$, $m, M, r, R \in \mathbb{Z}$, $m \leq M$ and $r \leq R$. Let $p_1 = \sum_{i=m}^{0} a_i t^i$. If $M = 0$, then $p - Dq_0 = p_1$ where $q_0 = 0 \in k\langle t \rangle$. If $M > 0$, then $\nu_\infty(p) = -M < 0$, which implies that $\nu_\infty(Dv) = -M < 0$, so $\nu_\infty(v) = -M$ by Lemma 5.1.2, hence $R = M$. Equating the coefficients of $t, \ldots, t^M$ in (5.15) we get

$$a_i = Dv_i + i\frac{Dt}{t}v_i \qquad \text{for } 1 \le i \le M.\qquad(5.16)$$

Since we can solve Risch differential equations over k and $a_i, Dt/t \in k$, we can either prove that (5.16) has no solution $v_i \in k$, or find such a solution[3] If it has no solution for some i, then (5.15) has no solution so p has no elementary integral over $k(t)$. If we have solutions v_i for $1 \le i \le M$, letting $q_0 = v_1 t + \ldots v_M t^M$, we get

$$p - Dq_0 = \sum_{i=1}^{M} a_i t^i + \sum_{i=m}^{0} a_i t^i - \sum_{i=1}^{M}\left(Dv_i + i\frac{Dt}{t}v_i\right)t^i = \sum_{i=m}^{0} a_i t^i = p_1.$$

If $m = 0$, then $p_1 \in k$ so $q = q_0$ satisfies the theorem. If $m < 0$, then $\nu_t(p_1) = -m < 0$, which implies that $\nu_t(Dv) = -m < 0$, so $\nu_t(v) = -m$ by Theorem 4.4.2 (since $t \in \mathcal{S}^{\text{irr}}$), hence $r = m$. Equating the coefficients of $t^{-1}, \ldots, t^{-m}$ in (5.15) we get

$$a_i = Dv_i + i\frac{Dt}{t}v_i \qquad \text{for } m \le i \le -1.\qquad(5.17)$$

Since we can solve Risch differential equations over k and $a_i, Dt/t \in k$, we can either prove that (5.17) has no solution $v_i \in k$, or find such a solution. If it has no solution for some i, then (5.15) has no solution, so p_1 and p have no elementary integrals over $k(t)$. If we have solutions v_i for $m \le i \le -1$, letting $q_1 = v_{-1}t^{-1} + \ldots v_{-m}t^{-m}$ and $q = q_0 + q_1 \in k\langle t\rangle$, we get

$$p - Dq = p_1 - Dq_1 = \sum_{i=m}^{-1} a_i t^i + a_0 - \sum_{i=m}^{-1}\left(Dv_i + i\frac{Dt}{t}v_i\right)t^i = a_0 \in k.$$

$\square$

IntegrateHyperexponentialPolynomial(p, D)
(* Integration of hyperexponential polynomials *)

  (* Given an hyperexponential monomial t over k and $p \in k[t, t^{-1}]$ return $q \in k[t, t^{-1}]$ and a Boolean $\beta \in \{0, 1\}$ such that $p - Dq \in k$ if $\beta = 1$, or $p - Dq$ does not have an elementary integral over $k(t)$ if $\beta = 0$. *)

  $q \leftarrow 0, \beta \leftarrow 1$
  for $i \leftarrow \nu_t(p)$ **to** $-\nu_\infty(p)$ such that $i \ne 0$ **do**
    $a \leftarrow$ **coefficient**(p, t^i)
    (* RischDE will be given in Chap. 6 *)
    $v \leftarrow$ **RischDE**$(iDt/t, a)$ (* $a = Dv + ivDt/t$ *)
    if $v = $ "no solution" **then** $\beta \leftarrow 0$ **else** $q \leftarrow q + vt^i$
  return(q, β)

---

[3] Although this fact is not needed by the algorithm, we remark that Lemma 5.9.1 implies that (5.16) has at most one solution in k.

Example 5.9.1. Consider

$$\int \left(\left(\tan(x)^3 + (x+1)\tan(x)^2 + \tan(x) + x + 2 \right) e^{\tan(x)} + \frac{1}{x^2+1} \right) dx .$$

Let $k = \mathbb{Q}(x, t_0)$ with $D = d/dx$, where t_0 is a monomial over $\mathbb{Q}(x)$ satisfying $Dt_0 = 1 + t_0^2$, i.e. $t_0 = \tan(x)$, and let t be a monomial over k satisfying $Dt = (1 + t_0^2)t$, i.e. $t = e^{\tan(x)}$. Our integrand is then

$$p = \left(t_0^3 + (x+1)t_0^2 + t_0 + x + 2 \right) t + \frac{1}{x^2+1} \in k[t] .$$

We get

1. $q = 0$, $\beta = 1$
2. $\nu_t(p) = -\nu_\infty(p) = 1$
3. $i = 1$
4. $a = \mathrm{lc}(p) = t_0^3 + (x+1)t_0^2 + t_0 + x + 2$
5. $D(t_0 + x) + (1 + t_0^2)(t_0 + x) = a$, so $v = \mathbf{RischDE}(1 + t_0^2, a) = t_0 + x$
6. $q = vt = (t_0 + x)t$
7. $p - Dq = 1/(x^2 + 1)$.

Hence,

$$\int \left(\left(\tan(x)^3 + (x+1)\tan(x)^2 + \tan(x) + x + 2 \right) e^{\tan(x)} + \frac{1}{x^2+1} \right) dx$$

$$= (\tan(x) + x)e^{\tan(x)} + \int \frac{dx}{x^2+1}$$

$$= (\tan(x) + x)e^{\tan(x)} + \arctan(x) .$$

Putting all the pieces together, we get an algorithm for integrating elements of $k(t)$.

Theorem 5.9.2. *Let k be a differential field and t an hyperexponential over k. If we can solve Risch differential equations over k, and Dt/t is not a logarithmic derivative of a k-radical, then for any $f \in k(t)$ we can either prove that f has no elementary integral over $k(t)$, or compute an elementary extension E of $k(t)$ and $g \in E$ such that $f - Dg \in k$.*

Proof. Suppose that Dt/t is not a logarithmic derivative of a k-radical, then t is a monomial over k and $\mathrm{Const}(k(t)) = \mathrm{Const}(k)$ by Theorem 5.1.2. Let $f \in k(t)$. By Theorem 5.3.1, we can compute $g_1, h, r \in k(t)$ such that $f = Dg_1 + h + r$, h is simple and r is reduced. From h, which is simple, we compute $g_2 \in k(t)$ given by (5.8) in Theorem 5.6.1. Note that $g_0 = g_1 + \int g_2$ lies in some elementary extension of $k(t)$. Let $p = h - g_2$ and $q = p + r$, then $f = Dg_0 + q$ so f has an elementary integral over $k(t)$ if and only if q has one. If $p \notin k[t]$, then $p + r$ does not have an elementary integral over $k(t)$ by Theorem 5.6.1, so f does not have an elementary integral over $k(t)$. Suppose now that $p \in k[t]$. We have $k\langle t \rangle = k[t, t^{-1}]$ by (5.1), so $r \in k[t, t^{-1}]$, hence $q \in k[t, t^{-1}]$. By Theorem 5.9.1 we can either prove that q has no elementary integral over $k(t)$, in which case f has no elementary integral over $k(t)$, or compute $s \in k[t, t^{-1}]$ such that $q - Ds \in k$, in which case $f - Dg \in k$ where $g = g_0 + s$. $\qquad\square$

IntegrateHyperexponential(f, D)
(* Integration of hyperexponential functions *)

(* Given an hyperexponential monomial t over k and $f \in k(t)$, return g elementary over $k(t)$ and a Boolean $\beta \in \{0, 1\}$ such that $f - Dg \in k$ if $\beta = 1$, or $f - Dg$ does not have an elementary integral over $k(t)$ if $\beta = 0$. *)

$(g_1, h, r) \leftarrow$ **HermiteReduce**(f, D)
$(g_2, \beta) \leftarrow$ **ResidueReduce**(h, D)
if $\beta = 0$ **then return**$(g_1 + g_2, 0)$
$(q, \beta) \leftarrow$ **IntegrateHyperexponentialPolynomial**$(h - Dg_2 + r, D)$
return$(g_1 + g_2 + q, \beta)$

5.10 The Hypertangent Case

Tangents and trigonometric functions can be integrated by transforming them to complex logarithms and exponentials, but the theory of monomial extensions allows us to integrate them directly without introducing the algebraic number $\sqrt{-1}$. We start by defining tangent monomials and computing the special polynomials. Let k be a differential field and K a differential extension of k.

Definition 5.10.1. *Let $t \in K$ be such that $t^2 + 1 \neq 0$. t is a hypertangent over k if $Dt/(t^2+1) \in k$. t is a tangent over k if $Dt/(t^2+1) = Db$ for some $b \in k$. t is a hypertangent (resp. tangent) monomial over k if t is a hypertangent (resp. tangent) over k, transcendental over k, and $\mathrm{Const}(k(t)) = \mathrm{Const}(k)$.*

We write $t = \tan(\int a)$ when t is a hypertangent over k such that $Dt/(t^2+1) = a$, and $t = \tan(b)$ when t is a tangent over k such that $Dt/(t^2 + 1) = Db$.

Lemma 5.10.1. *Let (F, D) be a differential field containing $\sqrt{-1}$, $a \in F$ be such that $a^2 + 1 \neq 0$, and $b = (\sqrt{-1} - a)/(\sqrt{-1} + a)$. Then, $b \neq 0$ and*

$$\frac{Db}{b} = 2\sqrt{-1}\,\frac{Da}{a^2 + 1}.$$

Proof. $b \neq 0$ since $a^2 + 1 \neq 0$, and we have

$$\frac{Db}{b} = D\left(\frac{\sqrt{-1}-a}{\sqrt{-1}+a}\right)\frac{\sqrt{-1}+a}{\sqrt{-1}-a}$$

$$= -2\sqrt{-1}\frac{Da}{(\sqrt{-1}+a)^2}\frac{\sqrt{-1}+a}{\sqrt{-1}-a} = 2\sqrt{-1}\frac{Da}{1+a^2}.$$

□

Theorem 5.10.1. *If t is an hypertangent over k and $\sqrt{-1}Dt/(t^2+1)$ is not a logarithmic derivative of a $k(\sqrt{-1})$-radical, then t is a monomial over k, $\mathrm{Const}(k(t)) = \mathrm{Const}(k)$, and any $p \in S^{irr}$ divides t^2+1 in $k[t]$. Furthermore, $S_1^{irr} = S^{irr}$. Conversely, if t is transcendental and hypertangent over k, and $\mathrm{Const}(k(t)) = \mathrm{Const}(k)$, then $\sqrt{-1}Dt/(t^2+1)$ is not a logarithmic derivative of a $k(\sqrt{-1})$-radical.*

Proof. Let t be an hypertangent over k, $a = Dt/(t^2+1)$, and suppose that $a\sqrt{-1}$ is not a logarithmic derivative of a $k(\sqrt{-1})$-radical. Let $\theta = \frac{\sqrt{-1}-t}{\sqrt{-1}+t} \in k(\sqrt{-1})(t)$. By Lemma 5.10.1, we have

$$\frac{D\theta}{\theta} = 2\sqrt{-1}\frac{Dt}{1+t^2} = 2a\sqrt{-1} \in k(\sqrt{-1})$$

so θ is hyperexponential over $k(\sqrt{-1})$. Since $a\sqrt{-1}$ is not a logarithmic derivative of a $k(\sqrt{-1})$-radical, $2a\sqrt{-1}$ is not one either, so by Theorem 5.1.2, θ is a monomial over $k(\sqrt{-1})$, and $\mathrm{Const}(k(\sqrt{-1})(\theta)) = \mathrm{Const}(k(\sqrt{-1}))$. But $t = \sqrt{-1}(\theta-1)/(\theta+1)$, so t is transcendental over $k(\sqrt{-1})$, hence a monomial over k since $Dt = a + at^2$. Furthermore, $k(\sqrt{-1})(\theta) = k(\sqrt{-1})(t)$, so

$$\mathrm{Const}(k(\sqrt{-1})(t)) = \mathrm{Const}(k(\sqrt{-1})(\theta)) = \mathrm{Const}(k(\sqrt{-1})) = \overline{C} \cap k(\sqrt{-1})$$

by Corollary 3.3.1 where $\overline{C}$ is the algebraic closure of $\mathrm{Const}(k)$. This implies that $\mathrm{Const}(k(t)) \subseteq \overline{C} \cap k(\sqrt{-1}) \cap k(t) \subseteq k$ since t is transcendental over k. Hence, $\mathrm{Const}(k(t)) \subseteq \mathrm{Const}(k)$. The reverse inclusion is given by Lemma 3.3.1, so $\mathrm{Const}(k(t)) = \mathrm{Const}(k)$, which implies that $\mathrm{Const}(\overline{k}(t)) = \mathrm{Const}(\overline{k})$ by Lemma 3.3.3.

We have $D(t^2+1) = 2tDt = 2at(t^2+1)$ so $t^2+1 \in S$, hence any factor of t^2+1 is special by Theorem 3.4.1. Suppose now that $p \in S$, and let $\beta \in \overline{k}$ be any root of p. $D\beta = a\beta^2 + a$ by Theorem 3.4.3, so

$$D\left(\frac{t-\beta}{\beta t+1}\right) = a\frac{(t^2-\beta^2)(\beta t+1) - (t-\beta)(t\beta^2 + t + \beta t^2 + \beta)}{(\beta t+1)^2}$$

$$= a(t-\beta)\frac{(\beta t^2 + t + \beta^2 t + \beta) - (t\beta^2 + t + \beta t^2 + \beta)}{(\beta t+1)^2} = 0$$

which implies that $c = (t-\beta)/(\beta t+1) \in \mathrm{Const}(\overline{k}(t)) \subseteq \overline{k}$. Since t is transcendental over $\overline{k}$, $(c\beta - 1)t + (c+\beta) = 0$ implies that $c\beta - 1 = c + \beta = 0$, so $\beta^2 + 1 = 0$. Since this holds for every root of p, this implies that every irreducible factor of p divides t^2+1 in $k[t]$.

We have $\mathcal{S}_1^{\mathrm{irr}} \subseteq \mathcal{S}^{\mathrm{irr}}$ by definition. Conversely, let $p \in \mathcal{S}^{\mathrm{irr}}$. Then p divides $t^2 + 1$, so all the roots of p in $\overline{k}$ satisfy $\beta^2 = -1$. Hence,

$$p_\beta = \frac{Dt - D\beta}{t - \beta} = a\frac{t^2 + 1}{t - \beta} = a(t + \beta)$$

which implies that $p_\beta(\beta) = 2a\beta = \pm 2\sqrt{-1}a$, which is not a logarithmic derivative of a $k(\sqrt{-1})$-radical, hence not a logarithmic derivative of a $k(\beta)$-radical. Thus, $p \in \mathcal{S}_1^{\mathrm{irr}}$ which implies that $\mathcal{S}_1^{\mathrm{irr}} = \mathcal{S}^{\mathrm{irr}}$.

Conversely, let t be a transcendental hypertangent over k and suppose that $\mathrm{Const}(k(t)) = \mathrm{Const}(k)$. Then, $\mathrm{Const}(\overline{k}(t)) = \mathrm{Const}(\overline{k})$ by Lemma 3.3.3. If there exist $b \in k(\sqrt{-1})^*$ and an integer $n > 0$ such that

$$n\sqrt{-1}\frac{Dt}{t^2 + 1} = \frac{Db}{b}$$

then, taking

$$\theta = \frac{\sqrt{-1} - t}{\sqrt{-1} + t} \quad \text{and} \quad c = \frac{\theta^n}{b^2} \in k(\sqrt{-1})(t)$$

we get

$$\frac{Dc}{c} = n\frac{D\theta}{\theta} - 2\frac{Db}{b} = 2n\sqrt{-1}\frac{Dt}{t^2 + 1} - 2\frac{Db}{b} = 0$$

so $c \in \mathrm{Const}(\overline{k}(t)) \subseteq \overline{k}$ in contradiction with t transcendental over k. Hence, $\sqrt{-1}Dt/(t^2 + 1)$ is not a logarithmic derivative of a $k(\sqrt{-1})$-radical.   □

As a consequence, we have

$$k\langle t \rangle = \{f \in k(t) \text{ such that } (t^2 + 1)^n f \in k[t] \text{ for some integer } n \geq 0\}$$

when t is a hypertangent monomial over k. We now present an algorithm that, with appropriate assumptions on k, integrates elements of $k(t)$ when t is a hypertangent monomial over k. Note first that if the polynomial $X^2 + 1$ factors over k, then $\sqrt{-1} \in k$, so $k(t) = k(\theta)$ where $\theta = (\sqrt{-1} - t)/(\sqrt{-1} + t)$ is a hyperexponential monomial over k. Hence we can use the algorithm for integrating elements of hyperexponential extensions in this case, so we can assume for the rest of this section that $X^2 + 1$ is irreducible over k, in other words that $\sqrt{-1} \notin k$. Since hypertangents are nonlinear monomials, integrating elements of $k[t]$ is straightforward.

Theorem 5.10.2. *Let k be a differential field not containing $\sqrt{-1}$, and t an hypertangent over k. If $\sqrt{-1}Dt/(t^2 + 1)$ is not a logarithmic derivative of a $k(\sqrt{-1})$-radical, then for any $p \in k[t]$ we can compute $q \in k[t]$ and $c \in k$ such that*

$$p - Dq - c\frac{D(t^2 + 1)}{t^2 + 1} \in k.$$

Furthermore, if $Dc \neq 0$, then p has no elementary integral over k.

Proof. Let $\alpha = Dt/(t^2 + 1) \in k$. Since $\alpha\sqrt{-1}$ is not a logarithmic derivative of a $k(\sqrt{-1})$-radical, t is a monomial over k, $\mathrm{Const}(k\langle t \rangle) = \mathrm{Const}(k)$, and all the special irreducible polynomials divide $t^2 + 1$ in $k[t]$ by Theorem 5.10.1. Since $\sqrt{-1} \notin k$, $t^2 + 1$ is irreducible over k, so $\mathcal{S}^{\mathrm{irr}} = \{t^2 + 1\}$. Since $\delta(t) = 2$, Theorem 5.4.1 shows how to compute $q, r \in k[t]$ such that $p - Dq = r$ and $\deg(r) \leq 1$. Write $r = at + b$ where $a, b \in k$, and let $c = a/(2\alpha) \in k$. Since $h = t^2 + 1 \in \mathcal{S}$, Theorem 5.4.2 says that $\deg(r - cDh/h) < 1$, hence that

$$p - Dq - c\,\frac{D(t^2 + 1)}{t^2 + 1} \in k\,.$$

Suppose now that $Dc \neq 0$, and that r has an elementary integral over $k(t)$. Then, by Theorem 5.7.1, there are $v \in k\langle t \rangle$, $c_1, \ldots, c_n \in \overline{C}$, $b_1, \ldots, b_n \in k(c_1, \ldots, c_n)$, and $m_1, \ldots, m_n \in \mathbb{Z}$ such that

$$at + b = Dv + \sum_{i=1}^{n} c_i \frac{Db_i(t^2 + 1)^{m_i}}{b_i(t^2 + 1)^{m_i}} = Dv + 2t\alpha \sum_{i=1}^{n} m_i c_i + \sum_{i=1}^{n} c_i \frac{Db_i}{b_i}\,. \quad (5.18)$$

If $\nu_\infty(v) < 0$, then $\nu_\infty(Dv) = \nu_\infty(v) - 1 < -1$ by Theorem 4.4.4, in contradiction with (5.18), hence $\nu_\infty(v) \geq 0$, which implies that $\nu_\infty(Dv) \geq 0$ by Theorem 4.4.4. Let $c = a/(2\alpha) \in k$. Equating the coefficients of t in (5.18), we get $a = 2\alpha \sum_{i=1}^{n} m_i c_i$, so

$$c = \frac{a}{2\alpha} = \sum_{i=1}^{n} m_i c_i \in \mathrm{Const}(k)$$

in contradiction with $Dc \neq 0$. Hence (5.18) has no solution if $Dc \neq 0$, which implies that r, and hence p, have no elementary integral over $k(t)$. $\square$

IntegrateHypertangentPolynomial(p, D)
(* Integration of hypertangent polynomials *)

(* Given a differential field k such that $\sqrt{-1} \notin k$, a hypertangent monomial t over k and $p \in k[t]$, return $q \in k[t]$ and $c \in k$ such that $p - Dq - cD(t^2 + 1)/(t^2 + 1) \in k$ and $p - Dq$ does not have an elementary integral over $k(t)$ if $Dc \neq 0$. *)

$(q, r) \leftarrow$ **PolynomialReduce**(p, D) $\qquad\qquad$ (* $\deg(r) \leq 1$ *)
$\alpha \leftarrow Dt/(t^2 + 1)$
$c \leftarrow$ **coefficient**$(r, t)/(2\alpha)$
return(q, c)

Example 5.10.1. Consider

$$\int \left(\tan(x)^2 + x\tan(x) + 1 \right) dx$$

Let $k = \mathbb{Q}(x)$ with $D = d/dx$, and and let t be a monomial over k satisfying $Dt = 1 + t^2$, *i.e.* $t = \tan(x)$. Our integrand is then

$$p = t^2 + xt + 1 \in k[t].$$

We get

1. $(q, r) = \textbf{PolynomialReduce}(t^2 + xt + 1) = (t, xt)$
2. $\alpha = Dt/(t^2 + 1) = 1$
3. $c = x/2$.

Since $Dc = 1/2 \neq 0$, we conclude that

$$\int \left(\tan(x)^2 + x\tan(x) + 1\right) dx = \tan(x) + \int x\tan(x)dx$$

and the latter integral is not an elementary function.

For reduced elements in an hypertangent extension, the related problem we need to solve over k is the *coupled differential system problem*: given $f_1, f_2, g_1, g_2 \in k$, determine whether there are $y_1, y_2 \in k$ such that

$$\begin{pmatrix} Dy_1 \\ Dy_2 \end{pmatrix} + \begin{pmatrix} f_1 & -f_2 \\ f_2 & f_1 \end{pmatrix} \begin{pmatrix} y_1 \\ y_2 \end{pmatrix} = \begin{pmatrix} g_1 \\ g_2 \end{pmatrix}$$

and to compute y_1 and y_2 if they exist.

Theorem 5.10.3. *Let k be a differential field not containing $\sqrt{-1}$, and t an hypertangent over k. If we can solve coupled differential systems over k, and $\sqrt{-1}Dt/(t^2 + 1)$ is not a logarithmic derivative of a $k(\sqrt{-1})$-radical, then for any $p \in k\langle t \rangle$ we can either prove that p has no elementary integral over $k(t)$, or compute $q \in k\langle t \rangle$ such that $p - Dq \in k[t]$.*

Proof. Let $\alpha = Dt/(t^2 + 1) \in k$. Since $\alpha\sqrt{-1}$ is not a logarithmic derivative of a $k(\sqrt{-1})$-radical, t is a monomial over k, $\text{Const}(k(t)) = \text{Const}(k)$, and all the special irreducible polynomials divide $t^2 + 1$ in $k[t]$ by Theorem 5.10.1. Since $\sqrt{-1} \notin k$, $t^2 + 1$ is irreducible over k, so $S^{\text{irr}} = S_1^{\text{irr}} = \{t^2 + 1\}$. Thus, Theorem 5.7.1 says that if p has an elementary integral over $k(t)$, then there are $v \in k\langle t \rangle$, $c_1, \ldots, c_n \in \overline{C}$, $b_1, \ldots, b_n \in k(c_1, \ldots, c_n)$, and $m_1, \ldots, m_n \in \mathbb{Z}$ such that

$$\begin{aligned} p &= Dv + \sum_{i=1}^{n} c_i \frac{Db_i(t^2 + 1)^{m_i}}{b_i(t^2 + 1)^{m_i}} \\ &= Dv + 2t\alpha \sum_{i=1}^{n} m_i c_i + \sum_{i=1}^{n} c_i \frac{Db_i}{b_i} = Dv + w \end{aligned} \tag{5.19}$$

where $C = \text{Const}(k)$, and

$$w = 2t\alpha \sum_{i=1}^{n} m_i c_i + \sum_{i=1}^{n} c_i \frac{Db_i}{b_i} \in k(c_1, \ldots, c_n)[t].$$

$K = k(c_1, \ldots, c_n)$ is an algebraic extension of k, so t is transcendental over K. Furthermore, $\alpha\sqrt{-1}$ is not a logarithmic derivative of a $K(\sqrt{-1})$-radical by Lemma 3.4.8, so t is a monomial over K and $\text{Const}(K(t)) = \text{Const}(K)$. We proceed by induction on $-\nu_{t^2+1}(p)$. If $\nu_{t^2+1}(p) \geq 0$, then $p - Dq \in k[t]$ where $q = 0 \in k\langle t \rangle$, so suppose that $m = -\nu_{t^2+1}(p) > 0$ and that the theorem holds for all $h \in k\langle t \rangle$ with $-\nu_{t^2+1}(h) < m$. Since $p \in k\langle t \rangle$ and $m = -\nu_{t^2+1}(p) > 0$, we have $p = r/(t^2+1)^m$ where $r \in k[t]$ and $\gcd(r, t^2+1) = 1$. Since $\nu_{t^2+1}(p) = -m < 0$, (5.19 implies that $\nu_{t^2+1}(Dv) = -m < 0$, hence that $\nu_{t^2+1}(v) = -m$ by Theorem 4.4.2, since $t^2 + 1 \in \mathcal{S}_1$. Thus, $v = s/(t^2+1)^m$ where $s \in k[t]$ and $\gcd(s, t^2 + 1) = 1$. Dividing r and s by $t^2 + 1$, we get $r = r_0(t^2 + 1) + at + b$ and $s = s_0(t^2 + 1) + ct + d$, where $r_0, s_0 \in k[t]$, $a, b, c, d \in k$, $at + b \neq 0$, and $ct + d \neq 0$. From (5.19), we get

$$\frac{at + b}{(t^2 + 1)^m} + \frac{r_0}{(t^2 + 1)^{m-1}} = D\left(\frac{ct + d}{(t^2 + 1)^m} + \frac{s_0}{(t^2 + 1)^{m-1}}\right) + w$$

$$= \frac{tDc + c\alpha(t^2 + 1) + Dd}{(t^2 + 1)^m} - \frac{2m\alpha t(t^2 + 1)(ct + d)}{(t^2 + 1)^{m+1}} + Dw_0 + w$$

$$= \frac{tDc + Dd}{(t^2 + 1)^m} - 2m\alpha\frac{ct^2 + dt}{(t^2 + 1)^m} + c\alpha\frac{1}{(t^2 + 1)^{m-1}} + Dw_0 + w$$

$$= \frac{tDc + Dd}{(t^2 + 1)^m} - 2m\alpha\frac{dt - c}{(t^2 + 1)^m} + c\alpha\frac{1 - 2m}{(t^2 + 1)^{m-1}} + Dw_0 + w$$

where $w_0 = s_0/(t^2 + 1)^{m-1}$. Since $\nu_{t^2+1}(w_0) > -m$, $\nu_{t^2+1}(Dw_0) > -m$ by Theorem 4.4.2, so, equating the coefficients of $(t^2 + 1)^{-m}$ we get

$$at + b = (Dc - 2m\alpha d)t + Dd + 2m\alpha c$$

which implies that

$$\begin{pmatrix} Dc \\ Dd \end{pmatrix} + \begin{pmatrix} 0 & -2m\alpha \\ 2m\alpha & 0 \end{pmatrix} \begin{pmatrix} c \\ d \end{pmatrix} = \begin{pmatrix} a \\ b \end{pmatrix} \tag{5.20}$$

Since we can solve coupled differential systems over k and $a, b, \alpha \in k$, we can either prove that (5.20) has no solution $c, d \in k$, or find such a solution. If it has no solution in k, then (5.19) has no solution, so p has no elementary integral over $k(t)$. If we have a solution $c, d \in k$, letting $q_0 = (ct + d)/(t^2 + 1)^m \in k\langle t \rangle$, we get

$$p - Dq_0 = \frac{u}{(t^2 + 1)^{m-1}}$$

for some $u \in k[t]$, so $\nu_{t^2+1}(p - Dq_0) > -m$. By induction we can either prove that $p - Dq_0$ has no elementary integral over $k(t)$, in which case p has no elementary integral over $k(t)$, or we get $q_1 \in k\langle t \rangle$ such that $p - Dq_0 - Dq_1 \in k[t]$, which implies that $p - Dq \in k[t]$ where $q = q_0 + q_1$. $\qquad\square$

IntegrateHypertangentReduced(p, D)
(* Integration of hypertangent reduced elements *)

(* Given a differential field k such that $\sqrt{-1} \notin k$, a hypertangent monomial t over k and $p \in k\langle t \rangle$, return $q \in k\langle t \rangle$ and a Boolean $\beta \in \{0, 1\}$ such that $p - Dq \in k[t]$ if $\beta = 1$, or $p - Dq$ does not have an elementary integral over $k(t)$ if $\beta = 0$. *)

$m \leftarrow -\nu_{t^2+1}(p)$
if $m \leq 0$ **then return**$(0, 1)$
$h \leftarrow (t^2 + 1)^m p$ (* $h \in k[t]$ *)
$(q, r) \leftarrow$ **PolyDivide**$(h, t^2 + 1)$ (* $h = (t^2 + 1)q + r, \deg(r) \leq 1$ *)
$a \leftarrow$ **coefficient**(r, t), $b \leftarrow r - at$ (* $r = at + b$ *)
(* CoupledDESystem will be given in Chap. 8 *)
$(c, d) \leftarrow$ **CoupledDESystem**$(0, 2mDt/(t^2 + 1), a, b)$
(* $Dc - 2mDt/(t^2 + 1)d = a, Dd + 2mDt/(t^2 + 1)c = b$ *)
if $(c, d) =$ "no solution" **then return**$(0, 0)$
$q_0 \leftarrow (ct + d)/(t^2 + 1)^m$
$(q, \beta) \leftarrow$ **IntegrateHypertangentReduced**$(p - Dq_0, D)$
return$(q + q_0, \beta)$

Example 5.10.2. Consider

$$\int \frac{\sin(x)}{x}\, dx \,.$$

Let $k = \mathbb{Q}(x)$ with $D = d/dx$, and and let t be a monomial over k satisfying $Dt = (1 + t^2)/2$, *i.e.* $t = \tan(x/2)$. Using the classical half-angle formula, our integrand is then

$$p = \frac{\sin(x)}{x} = \frac{2\tan(x/2)}{x(\tan(x/2)^2 + 1)} = \frac{2t/x}{t^2 + 1} \in k\langle t \rangle \,.$$

We get $Dt/(t^2 + 1) = 1/2$ and

1. $m = -\nu_{t^2+1}(p) = 1$
2. $h = p(t^2 + 1) = 2t/x$
3. $(q, r) =$ **PolyDivide**$(2t/x, t^2 + 1) = (0, 2t/x)$, so $(a, b) = (2/x, 0)$
4. Since

$$\begin{pmatrix} Dc \\ Dd \end{pmatrix} + \begin{pmatrix} 0 & -1 \\ 1 & 0 \end{pmatrix} \begin{pmatrix} c \\ d \end{pmatrix} = \begin{pmatrix} 2/x \\ 0 \end{pmatrix}$$

has no solution in $\mathbb{Q}(x)$, **CoupledDESystem**$(0, 1, 2/x, 0)$ returns "no solution".

Hence,

$$\int \frac{\sin(x)}{x}\, dx$$

is not an elementary function.

Example 5.10.3. Consider

$$\int \frac{\tan(x)^5 + \tan(x)^3 + x^2 \tan(x) + 1}{(\tan(x)^2 + 1)^3} \, dx \, .$$

Let $k = \mathbb{Q}(x)$ with $D = d/dx$, and and let t be a monomial over k satisfying $Dt = 1 + t^2$, *i.e.* $t = \tan(x)$. Our integrand is then

$$p = \frac{t^5 + t^3 + x^2 t + 1}{(t^2 + 1)^3} \in k\langle t \rangle \, .$$

We get $Dt/(t^2 + 1) = 1$ and

1. $m = -\nu_{t^2+1}(p) = 3$
2. $h = p(t^2 + 1)^3 = t^5 + t^3 + x^2 t + 1$
3. $(q, r) = \mathbf{PolyDivide}(h, t^2 + 1) = (t^3, x^2 t + 1)$, so $(a, b) = (x^2, 1)$
4. Since

$$\begin{pmatrix} Dc \\ Dd \end{pmatrix} + \begin{pmatrix} 0 & -6 \\ 6 & 0 \end{pmatrix} \begin{pmatrix} c \\ d \end{pmatrix} = \begin{pmatrix} x^2 \\ 1 \end{pmatrix}$$

   has the solution $c = x/18 + 1/6$ and $d = 1/108 - x^2/6$ in $\mathbb{Q}(x)$,
   $(c, d) = \mathbf{CoupledDESystem}(0, 6, x^2, 1) = (x/18 + 1/6, 1/108 - x^2/6)$.

5.

$$q_0 = \frac{ct + d}{(t^2 + 1)^3} = \frac{(1 + x/3)\, t - (x^2 - 1/18)}{6\,(t^2 + 1)^3} \, ,$$

$$p - Dq_0 = \frac{t^3 + 5x/18 + 15/18}{(t^2 + 1)^2}$$

6. Recursively calling $(q, \beta) = \mathbf{IntegrateHypertangentReduced}(p - Dq_0)$, we get $\beta = 1$ and

$$q = \frac{5\,(1 + x/3)\, t + 77/12}{24\,(t^2 + 1)^2} + \frac{5\,(1 + x/3)\, t - 43/6}{16\,(t^2 + 1)}$$

7.

$$p - D(q + q_0) = \frac{5}{16}\left(1 + \frac{x}{3}\right)$$

Hence,

$$\int \frac{\tan(x)^5 + \tan(x)^3 + x^2 \tan(x) + 1}{(\tan(x)^2 + 1)^3} \, dx =$$

$$\frac{(1 + x/3)\tan(x) - (x^2 - 1/18)}{6\,(\tan(x)^2 + 1)^3} + \frac{5\,(1 + x/3)\tan(x) + 77/12}{24\,(\tan(x)^2 + 1)^2}$$

$$+ \frac{5\,(1 + x/3)\tan(x) - 43/6}{16\,(\tan(x)^2 + 1)} + \frac{5}{16}\int\left(1 + \frac{x}{3}\right) dx$$

and the remaining integral is of course $x + x^2/6$.

Putting all the pieces together, we get an algorithm for integrating elements of $k(t)$.

Theorem 5.10.4. *Let k be a differential field not containing $\sqrt{-1}$, and t an hypertangent over k. If we can solve coupled differential systems over k, and $\sqrt{-1}Dt/(t^2+1)$ is not a logarithmic derivative of a $k(\sqrt{-1})$-radical, then for any $f \in k(t)$ we can either prove that f has no elementary integral over $k(t)$, or compute an elementary extension E of $k(t)$ and $g \in E$ such that $f - Dg \in k$.*

Proof. Suppose that $\sqrt{-1}Dt/(t^2+1)$ is not a logarithmic derivative of a $k(\sqrt{-1})$-radical, then t is a monomial over k and $\mathrm{Const}(k(t)) = \mathrm{Const}(k)$ by Theorem 5.10.1. Let $f \in k(t)$. By Theorem 5.3.1, we can compute $g_1, h, r \in k(t)$ such that $f = Dg_1 + h + r$, h is simple and r is reduced. From h, which is simple, we compute $g_2 \in k(t)$ given by (5.8) in Theorem 5.6.1. Note that $g_0 = g_1 + \int g_2$ lies in some elementary extension of $k(t)$. Let $p = h - g_2$ and $q = p + r$, then $f = Dg_0 + q$ so f has an elementary integral over $k(t)$ if and only if q has one. If $p \notin k[t]$, then $p + r$ does not have an elementary integral over $k(t)$ by Theorem 5.6.1, so f does not have an elementary integral over $k(t)$. Suppose now that $p \in k[t]$. Then $p \in k\langle t\rangle$ so $q \in k\langle t\rangle$. By Theorem 5.10.3 we can either prove that q has no elementary integral over $k(t)$, in which case f has no elementary integral over $k(t)$, or compute $s \in k\langle t\rangle$ such that $u = q - Ds \in k[t]$, in which case by Theorem 5.10.2, we compute $v \in k[t]$ and $c \in \mathrm{Const}(k)$ such that $u - Dv - cD(t^2+1)/(t^2+1) \in k$. If $Dc \neq 0$, then u, and hence f, have no elementary integral over $k(t)$, otherwise $Dc = 0$ so $f - Dg \in k$ where $g = g_0 + s + v + c\int D(t^2+1)/(t^2+1)$ lies in some elementary extension of $k(t)$. $\square$

IntegrateHypertangent(f, D) (* Integration of hypertangent functions *)

(* Given a differential field k such that $\sqrt{-1} \notin k$, a hypertangent monomial t over k and $f \in k(t)$, return g elementary over $k(t)$ and a Boolean $\beta \in \{0,1\}$ such that $f - Dg \in k$ if $\beta = 1$, or $f - Dg$ does not have an elementary integral over $k(t)$ if $\beta = 0$. *)

$(g_1, h, r) \leftarrow$ **HermiteReduce**(f, D)
$(g_2, \beta) \leftarrow$ **ResidueReduce**(h, D)
if $\beta = 0$ **then return**$(g_1 + g_2, 0)$
$p \leftarrow h - Dg_2 + r$
$(q_1, \beta) \leftarrow$ **IntegrateHypertangentReduced**(p, D)
if $\beta = 0$ **then return**$(g_1 + g_2 + q_1, 0)$
$(q_2, c) \leftarrow$ **IntegrateHypertangentPolynomial**$(p - Dq_1, D)$
if $Dc = 0$ **then return**$(g_1 + g_2 + q_1 + q_2 + c\log(t^2+1), 1)$
else return$(g_1 + g_2 + q_1 + q_2, 0)$

5.11 The Nonlinear Case with no Specials

In the case of nonlinear monomials over a differential field k, we have seen that we can reduce the problem to integrating reduced elements of the form $p + a/d$ where $p, a \in k[t]$, $d \in \mathcal{S} \setminus \{0\}$, $\deg(p) < \delta(t)$ and $\deg(a) < \deg(d)$. Furthermore, Theorem 5.7.2 provides a criterion for nonintegrability, and if an element of $\mathcal{S} \setminus k$ is known, allows us to reduce the problem to $\deg(p) < \delta(t) - 1$. We address in this section the case $\mathcal{S} = k$, $i.e.$ $\mathcal{S}^{\mathrm{irr}} = \emptyset$, which corresponds to interesting classes of functions as will be illustrated in the examples. Note that if $\mathcal{S}^{\mathrm{irr}} = \emptyset$, then $k\langle t \rangle = k[t]$, so as a result of the polynomial reduction (Sect. 5.4), we consider integrands of the form $p \in k[t]$ with $\deg(p) < \delta(t)$. It turns out that if such elements are integrable, then they must be in k.

Corollary 5.11.1. *Suppose that t is a nonlinear monomial and that $\mathcal{S}^{\mathrm{irr}} = \emptyset$. Let $p \in k[t]$ be such that $\deg(p) < \delta(t)$. If p has an elementary integral over $k(t)$, then $p \in k$.*

Proof. Let $C = \mathrm{Const}(k(t))$, $p \in k[t]$ be such that $\deg(p) < \delta(t)$, and suppose that p has an elementary integral over $k(t)$. By Theorem 5.7.1 there are $v \in k[t]$, $c_1, \ldots, c_n \in \overline{C}$ and $u_1, \ldots, u_n \in \mathcal{S}_{k(c_1, \ldots, c_n)[t]:k(c_1, \ldots, c_n)}$ such that $p = Dv + g$ where $g = \sum_{i=1}^{n} c_i D(u_i)/u_i$. Note that $g = p - Dv \in k[t]$. Since $\mathcal{S}^{\mathrm{irr}}_{k[t]:k} = \emptyset$, it follows that $\mathcal{S}^{\mathrm{irr}}_{k(c_1, \ldots, c_n)[t]:k(c_1, \ldots, c_n)} = \emptyset$ (Exercise 3.5), hence that $\mathcal{S}_{k(c_1, \ldots, c_n)[t]:k(c_1, \ldots, c_n)} = k(c_1, \ldots, c_n)$. This implies that $g \in k(c_1, \ldots, c_n)$. Since $g \in k[t]$, we get that $g \in k$. Suppose that $\deg(v) \geq 1$, then,

$$\deg(p) = \deg(Dv + g) = \deg(Dv) = \deg(v) + \delta(t) - 1 \geq \delta(t)$$

in contradiction with $\deg(p) < \delta(t)$. Hence, $v \in k$, so $p = Dv + g \in k$. $\qquad\square$

This provides a complete algorithm for integrating elements of $k(t)$.

Theorem 5.11.1. *Let k be a differential field and t be a nonlinear monomial over k be such that $\mathcal{S}^{\mathrm{irr}} = \emptyset$. Then, for any $f \in k(t)$ we can either prove that f has no elementary integral over $k(t)$, or compute an elementary extension E of $k(t)$ and $g \in E$ such that $f - Dg \in k$.*

Proof. Suppose that t is a nonlinear monomial over k and that $\mathcal{S}^{\mathrm{irr}} = \emptyset$. Then, $\mathrm{Const}(k(t)) = \mathrm{Const}(k)$ by Lemma 3.4.5. Let $f \in k(t)$. By Theorem 5.3.1, we can compute $g_1, h, r \in k(t)$ such that $f = Dg_1 + h + r$, h is simple and r is reduced. From h, which is simple, we compute $g_2 \in k(t)$ given by (5.8) in Theorem 5.6.1. Note that $g_0 = g_1 + \int g_2$ lies in some elementary extension of $k(t)$. Let $p = h - g_2$ and $q = p + r$, then $f = Dg_0 + q$ so f has an elementary integral over $k(t)$ if and only if q has one. If $p \notin k[t]$, then $p + r$ does not have an elementary integral over $k(t)$ by Theorem 5.6.1, so f does not have an elementary integral over $k(t)$. Suppose now that $p \in k[t]$. We have $k\langle t \rangle = k[t]$ by (5.1), so $r \in k[t]$, hence $q \in k[t]$. By Theorem 5.4.1 we compute $q_1, q_2 \in k[t]$ such that $q = Dq_1 + q_2$ and $\deg(q_2) < \delta(t)$. We now have $f - Dg = q_2$ where $g = g_0 + g_1$. If $q_2 \in k$, then the theorem is proven, otherwise $0 < \deg(q_2) < \delta(t)$, so q_2, and therefore f, have no elementary integral over k by Corollary 5.11.1. $\square$

IntegrateNonLinearNoSpecial(f, D)
(* Integration of nonlinear monomials with no specials *)

(* Given a is a nonlinear monomial t over k with $\mathcal{S}^{\mathrm{irr}} = \emptyset$, and $f \in k(t)$, return g elementary over $k(t)$ and a Boolean $\beta \in \{0,1\}$ such that $f - Dg \in k$ if $\beta = 1$, or $f - Dg$ does not have an elementary integral over $k(t)$ if $\beta = 0$. *)

$(g_1, h, r) \leftarrow$ **HermiteReduce**(f, D)
$(g_2, \beta) \leftarrow$ **ResidueReduce**(h, D)
if $\beta = 0$ then **return**$(g_1 + g_2, 0)$
$(q_1, q_2) \leftarrow$ **PolynomialReduce**$(h - Dg_2 + r, D)$
if $q_2 \in k$ then $\beta \leftarrow 1$ else $\beta \leftarrow 0$
return$(g_1 + g_2 + q_1, \beta)$

Example 5.11.1. Let $\nu \in \mathbb{Z}$ be any integer and consider

$$\int \frac{J_{\nu+1}(x)}{J_\nu(x)} \, dx$$

where $J_\nu(x)$ is the Bessel function of the first kind of order ν. From

$$\frac{dJ_\nu(x)}{dx} = -J_{\nu+1}(x) + \frac{\nu}{x} J_\nu(x)$$

we get

$$\int \frac{J_{\nu+1}(x)}{J_\nu(x)} \, dx = \int \nu \frac{dx}{x} - \int \frac{dJ_\nu(x)/dx}{J_\nu(x)} \, dx = \nu \log(x) - \int \phi_\nu(x) dx$$

where $\phi_\nu(x)$ is the logarithmic derivative of $J_\nu(x)$. Since $J_\nu(x)$ is a solution of the Bessel equation

$$y''(x) + \frac{1}{x} y'(x) + \left(1 - \frac{\nu^2}{x^2}\right) y(x) = 0 \qquad (5.21)$$

it follows that $\phi_\nu(x)$ is a solution of the Riccati equation

$$y'(x) + y(x)^2 + \frac{1}{x} y(x) + \left(1 - \frac{\nu^2}{x^2}\right) = 0. \qquad (5.22)$$

Let $k = \mathbb{Q}(x)$ with $D = d/dx$, and let t be a monomial over k satisfying $Dt = -t^2 - t/x - (1 - \nu^2/x^2)$, i.e. $t = \phi_\nu(x)$. It can be proven that $\mathcal{S}^{\mathrm{irr}} = \emptyset$ in this extension[4] so Corollary 5.11.1 implies that t has no elementary integral over k, hence that

$$\int \frac{J_{\nu+1}(x)}{J_\nu(x)} \, dx = \nu \log(x) - \int \phi_\nu(x) dx$$

where the remaining integral is not elementary over $\mathbb{Q}(x, \phi_\nu(x))$.

Example 5.11.2. Let $\nu \in \mathbb{C}$ be any complex number and consider

$$\int \frac{x^2 \phi_\nu^5 + x\phi_\nu^4 - \nu^2 \phi_\nu^3 - x(x^2+1)\phi_\nu^2 - (x^2 - \nu^2)\phi_\nu - x^5/4}{x^2\phi_\nu^4 + x^2(x^2+2)\phi_\nu^2 + x^2 + x^4 + x^6/4} \, dx$$

where $\phi_\nu(x)$ is the logarithmic derivative of $J_\nu(x)$, the Bessel function of the first kind of order ν. Let $k = \mathbb{Q}(x)$ with $D = d/dx$, and let t be a monomial over k satisfying $Dt = -t^2 - t/x - (1 - \nu^2/x^2)$, i.e. $t = \phi_\nu(x)$. Our integrand is then

$$f = \frac{x^2 t^5 + xt^4 - \nu^2 t^3 - x(x^2+1)t^2 - (x^2 - \nu^2)t - x^5/4}{x^2 t^4 + x^2(x^2+2)t^2 + x^2 + x^4 + x^6/4}$$

and we get

1. Calling $(g_1, h, r) = $ **HermiteReduce**(f, D) we get

$$g_1 = -\frac{1 + x^2/4}{t^2 + 1 + x^2/2}, \quad h = -\frac{(\nu^2 + x^4/2)t + x^3 + x}{x^2 t^2 + x^2 + x^4/2}, \quad \text{and } r = t + \frac{1}{x}.$$

2. Calling $(g_2, \beta) = $ **ResidueReduce**(h, D) we get $\beta = 1$ and

$$g_2 = -\frac{1}{2} \log\left(t^2 + 1 + \frac{x^2}{2}\right).$$

3. We have $h - Dg_2 + r = 0$, so $(q_1, q_2) = (0, 0)$.

Hence $f = Dg_1 + Dg_2$, which means that

$$\int \frac{x^2 \phi_\nu^5 + x\phi_\nu^4 - \nu^2 \phi_\nu^3 - x(x^2+1)\phi_\nu^2 - (x^2 - \nu^2)\phi_\nu - x^5/4}{x^2\phi_\nu^4 + x^2(x^2+2)\phi_\nu^2 + x^2 + x^4 + x^6/4} \, dx =$$

$$-\frac{1 + x^2/4}{\phi_\nu(x)^2 + 1 + x^2/2} - \frac{1}{2} \log\left(\phi_\nu(x)^2 + 1 + x^2/2\right).$$

Note that the above integral is valid regardless of whether $\mathcal{S}^{\text{irr}}$ is empty.

The above examples used Bessel functions, but in fact the algorithm of this section can be applied whenever the integrand can be expressed in terms of the logarithmic derivative of a function defined by a second-order linear ordinary differential equation. If the defining equation is known not to have solutions in quadratures (for example for Airy functions), then $\mathcal{S}^{\text{irr}} = \emptyset$, as explained in note 4 of this chapter.

---

[4] The fact that (5.21) has no solutions in quadratures for $\nu \in \mathbb{Z}$ (its Galois group is $SL_2(\mathbb{C})$) implies that (5.22) has no algebraic function solution, hence no solution in $\bar{k}$. Theorem 3.4.3 then implies that $\mathcal{S}^{\text{irr}} = \emptyset$.

5.12 In–Field Integration

We outline in this section minor variants of the integration algorithm that are used for deciding whether an element of $k(t)$ is either a

- derivative of an element of $k(t)$,
- logarithmic derivative of an element of $k(t)$,
- logarithmic derivative of a $k(t)$-radical.

As we have seen in Sect. 5.2, such procedures are needed when building the tower of fields containing the integrand. Furthermore, they will be needed at various points by the algorithms of the remaining chapters, in particular when bounding orders and degrees.

Note that the structure Theorems of Chap. 9 provide efficient alternatives to the use of modified integration algorithms, and in some cases the only complete algorithms for recognizing logarithmic derivatives.

Recognizing Derivatives

The first problem is, given $f \in k(t)$, to determine whether there exists $u \in k(t)$ such that $Du = f$, and to compute such an u if it exists. We first perform the Hermite reduction on f, obtaining $g \in k(t)$, a simple $h \in k(t)$, and $r \in k\langle t\rangle$ such that $f = Dg + h + r$. At that point, we can prove (see Exercise 4.1) that if $f = Du$ for some $u \in k(t)$, then $h \in k[t]$, so we are left with integrating $h + r$ which is reduced. The algorithms of Sects. 5.7 to 5.11 can then be applied (with a minor modification in the nonlinear case, to prevent introducing a new logarithm), either proving that there is no such u, or reducing the problem to deciding whether an element $a \in k$ has an integral in $k(t)$.

If t is a primitive over k, then it follows from Theorem 4.4.2 and Lemma 5.1.2 that if a has an integral in $k(t)$, then $a = Dv + cDt$ where $v \in k$ and $c \in \text{Const}(k)$, and we are reduced to a limited integration problem in k. Otherwise, $\delta(t) \geq 1$, and it follows from Theorem 4.4.2 and Lemmas 3.4.2 and 5.1.2 that if a has an integral in $k(t)$, then $a = Dv$ where $v \in k$, and we are reduced to a similar problem in k.

When $f = Da/a$ for some $a \in k(t)^*$, then Corollary 9.3.1, 9.3.2 or 9.4.1 provide alternative algorithms: $f = Du$ for $u \in k(t)$ if and only if the linear equation (9.8), (9.12) or (9.21) has a solution in $\mathbb{Q}$. Corollary 9.3.2 also provides an alternative algorithm if $f = Db/(b^2 + 1)$ for some $b \in k(t)$, i.e. $f = \arctan(b)$.

It is obvious that the solution u is not unique, but that if $f = Du = Dv$ for $u, v \in k(t)$, then $u - v \in \text{Const}(k(t))$.

Recognizing Logarithmic Derivatives

The second problem is, given $f \in k(t)$, to determine whether there exists a nonzero $u \in k(t)$ such that $Du/u = f$, and to compute such an u if it exists.

We can prove (see Exercise 4.2) that if $f = Du/u$ for some nonzero $u \in k(t)$, then f is simple and that all the roots of the Rothstein–Trager resultant are integers. In that case, the residue reduction produces

$$g = \sum_{r_s(\alpha)=0} \alpha \frac{Dg_\alpha}{g_\alpha} = \frac{D\left(\prod_{r_s(\alpha)=0} g_\alpha^\alpha\right)}{\prod_{r_s(\alpha)=0} g_\alpha^\alpha} = \frac{Dv}{v}$$

where $v \in k(t)$ since the α's are all integers. Furthermore, Theorem 5.6.1 implies that if $f = Du/u$ for $u \in k(t)$, then $f - g \in k[t]$, so we are left with deciding whether an element p of $k[t]$ is the logarithmic derivative of an element of $k(t)$. If $p = Du/u$ for $u \in k(t)$, then it follows from Exercise 4.2 that $\deg(p) < \max(1, \delta(t))$ and from Corollary 4.4.2 that $u = p_1^{e_1} \dots p_n^{e_n}$ where $p_i \in S$ and $e_i \in \mathbb{Z}$.

If t is a primitive over k, then both p and u must be in k since $S = k$, so we are reduced to a similar problem in k.

If t is an hyperexponential over k, then $p \in k$ and $u = vt^e$ for $v \in k^*$ and $e \in \mathbb{Z}$, since $S^{\mathrm{irr}} = \{t\}$. We are thus reduced to deciding whether $p \in k$ can be written as

$$p = \frac{Dv}{v} + e\frac{Dt}{t}$$

for $v \in k^*$ and $e \in \mathbb{Z}$. This is a special case of the parametric logarithmic derivative problem, a variant of the limited integration problem, which is discussed in Chap. 7.

If t is a hypertangent over k and $\sqrt{-1} \notin k$, then $p = a + bt$ for $a, b \in k$, and $u = v(t^2 + 1)^e$ for $v \in k^*$ and $e \in \mathbb{Z}$, since $S^{\mathrm{irr}} = \{t^2 + 1\}$. We are thus reduced to deciding whether $a + bt$ can be written as

$$a + bt = \frac{Dv}{v} + e\frac{D(t^2 + 1)}{t^2 + 1} = \frac{Dv}{v} + 2e\frac{Dt}{t^2 + 1}t$$

which is equivalent to

$$a = \frac{Dv}{v} \quad \text{and} \quad \frac{b}{2}\frac{t^2 + 1}{Dt} \in \mathbb{Z}.$$

The second condition can be immediately verified, while the first is the problem of deciding whether an element of k is the logarithmic derivative of an element of k.

When $f = Db$ for some $b \in k(t)$, then Corollary 9.3.1, 9.3.2 or 9.4.1 provide alternative algorithms: f is the logarithmic derivative of a $k(t)$-radical if and only if the linear equation (9.9), (9.13) or (9.22) has a solution in $\mathbb{Q}$.

The solution u is not unique, but if $f = Du/u = Dv/v$ for $u, v \in k(t)\backslash\{0\}$, then $u/v \in \mathrm{Const}(k(t))$ (this is the case $n = m = 1$ of Lemma 5.12.1 below).

Recognizing Logarithmic Derivatives of $k(t)$-radicals

The third problem is, given $f \in k(t)$, to determine whether there exist a nonzero $n \in \mathbb{Z}$ and a nonzero $u \in k(t)$ such that $Du/u = nf$, and to compute such an n and u if they exist. We can prove (see Exercise 4.2) that if $nf = Du/u$ for some nonzero $n \in \mathbb{Z}$ and $u \in k(t)$, then f is simple and that all the roots of the Rothstein–Trager resultant are rational numbers. In that case, let m be a common denominator for the roots of the Rothstein–Trager resultant. Then, the residue reduction produces

$$g = \sum_{r_s(\alpha)=0} \alpha \frac{Dg_\alpha}{g_\alpha} = \frac{1}{m} \frac{D\left(\prod_{r_s(\alpha)=0} g_\alpha^{m\alpha}\right)}{\prod_{r_s(\alpha)=0} g_\alpha^{m\alpha}} = \frac{1}{m} \frac{Dv}{v}$$

where $v \in k(t)$ since the $m\alpha$ is an integer for each α. Furthermore, Theorem 5.6.1 implies that if $f = Du/(nu)$ for $n \in \mathbb{Z}$ and $u \in k(t)$, then $f - Dg \in k[t]$, so we are left with deciding whether an element p of $k[t]$ is the logarithmic derivative of a $k(t)$-radical. If $p = Du/(nu)$ for $n \in \mathbb{Z}$ and $u \in k(t)$, then it follows from Exercise 4.2 that $\deg(p) < \max(1, \delta(t))$ and from Corollary 4.4.2 that $u = p_1^{e_1} \dots p_s^{e_s}$ where $p_i \in S$ and $e_i \in \mathbb{Z}$.

If t is a primitive over k, then both p and u must be in k since $S = k$, so we are reduced to a similar problem in k.

If t is an hyperexponential over k, then $p \in k$ and $u = vt^e$ for $v \in k^*$ and $e \in \mathbb{Z}$, since $S^{\mathrm{irr}} = \{t\}$. We are thus reduced to deciding whether $p \in k$ can be written as

$$p = \frac{1}{n} \frac{Dv}{v} + \frac{e}{n} \frac{Dt}{t}$$

for $v \in k^*$ and $n, e \in \mathbb{Z}$. This is the parametric logarithmic derivative problem, a variant of the limited integration problem, which is discussed in Chap. 7.

If t is a hypertangent over k and $\sqrt{-1} \notin k$, then $p = a + bt$ for $a, b \in k$, and $u = v(t^2 + 1)^e$ for $v \in k^*$ and $e \in \mathbb{Z}$, since $S^{\mathrm{irr}} = \{t^2 + 1\}$. We are thus reduced to deciding whether $a + bt$ can be written as

$$a + bt = \frac{1}{n} \frac{Dv}{v} + \frac{e}{n} \frac{D(t^2 + 1)}{t^2 + 1} = \frac{1}{n} \frac{Dv}{v} + \frac{2e}{n} \frac{Dt}{t^2 + 1} t$$

which is equivalent to

$$na = \frac{Dv}{v} \quad \text{and} \quad \frac{b}{2} \frac{t^2 + 1}{Dt} \in \mathbb{Q}.$$

The second condition can be immediately verified, while the first is the problem of deciding whether an element of k is the logarithmic derivative of a k-radical.

When $f = Db$ for some $b \in k(t)$, then Corollary 9.3.1, 9.3.2 or 9.4.1 provide alternative algorithms: f is the logarithmic derivative of a $k(t)$-radical if and only if the linear equation (9.9), (9.13) or (9.22) has a solution in $\mathbb{Q}$.

The solution (n, u) is not unique, but any two solutions are related by the following lemma.

Lemma 5.12.1. *Let (K, D) be a differential field and $u, v \in K^*$. If*

$$\frac{1}{n}\frac{Du}{u} = \frac{1}{m}\frac{Dv}{v}$$

for nonzero $n, m \in \mathbb{Z}$, then

$$\frac{u^{\mathrm{lcm}(n,m)/n}}{v^{\mathrm{lcm}(n,m)/m}} \in \mathrm{Const}(K).$$

Proof. Let $c = u^{\mathrm{lcm}(n,m)/n}/v^{\mathrm{lcm}(n,m)/m}$. Then,

$$\frac{Dc}{c} = \frac{\mathrm{lcm}(n,m)}{n}\frac{Du}{u} - \frac{\mathrm{lcm}(n,m)}{m}\frac{Dv}{v} = \mathrm{lcm}(n,m)\left(\frac{1}{n}\frac{Du}{u} - \frac{1}{m}\frac{Dv}{v}\right) = 0$$

so $c \in \mathrm{Const}(K)$. $\square$

Exercises

Exercise 5.1. Let k be a differential field of characteristic 0, t a monomial over k, and $d \in k[t] \setminus \{0\}$. Let $d = d_1 d_2^2 \cdots d_n^n$ be a squarefree factorization of d. Show that $\mu(a/d) \leq n$ for any $a \in k[t]$, and that $\mu(a/d) = n$ if and only if $\gcd(a, d) = 1$.

Exercise 5.2. Rewrite the proof of Theorem 5.3.1 using Mack's linear version of the Hermite reduction instead of the quadratic version.

Exercise 5.3. Let k be a differential field of characteristic 0, t a monomial over k, and $f \in k(t) \setminus \{0\}$. Show that using only the extended Euclidean algorithm in $k[t]$, one can find $h_0, h_1, \ldots, h_q$ and $r \in k(t)$ such that $q \leq \mu(f)$, each h_i is simple, r is reduced, and $f = r + h_0 + Dh_1 + D^2h_2 + \ldots + D^qh_q$.

Exercise 5.4 (In-field integration). Let k be a differential field of characteristic 0 and t be a monomial over k. Write an algorithm that, given any $f \in k(t)$, returns either $g \in k(t)$ such that $Dg = f$, or "no solution" if f has no antiderivative in $k(t)$ (see Exercise 4.1).

Exercise 5.5 (Generalizations of Liouville's Theorem). Let k be a differential field of characteristic 0, $C = \mathrm{Const}(k)$, $f \in k$, t be a monomial over k and suppose that there exist an elementary extension E of $k(t)$ and $g \in E$ such that $Dg = f$.

a) Prove that

$$f = Dv + \sum_{i=1}^{n} c_i \frac{Du_i}{u_i} \tag{5.23}$$

has a solution $v \in k\langle t\rangle$, $c_1, \ldots, c_n \in \overline{C}$, and $u_1, \ldots, u_n \in S_{\overline{C}k[t]:\overline{C}k} \setminus \{0\}$.

b) Prove that if t is a nonlinear monomial over k, then (5.23) has a solution $v \in k\langle t \rangle$, $c_1, \ldots, c_n \in \overline{C}$, and $u_1, \ldots, u_n \in \overline{C}k^*$.

c) Prove that if $S_1^{irr} = S^{irr}$, then (5.23) has a solution $v \in k[t]$, $c_1, \ldots, c_n \in \overline{C}$, and $u_1, \ldots, u_n \in S_{\overline{C}k[t]:\overline{C}k} \setminus \{0\}$.

d) Prove that if t is a nonlinear monomial over k and $S_1^{irr} = S^{irr}$, then f has an elementary integral over k.

e) Prove that if t is an hyperexponential monomial over k, then f has an elementary integral over k.

f) Prove that if t is a primitive monomial over k, then (5.23) has a solution $v = at + b$, $c_1, \ldots, c_n \in \overline{C}$, and $u_1, \ldots, u_n \in \overline{C}k^*$, where $a \in C$ and $b \in k$.

Exercise 5.6. Decide which of the following integrals are elementary functions, and compute those that are elementary. Since the recursive problems involving the procedures **LimitedIntegrate**, **RischDE** and **CoupledDESystem** are trivial in these exercices, perform the portions allocated to those procedures by elementary methods.

a)
$$\int \tan(ax)^5 dx, \qquad a \in \mathbb{C}^*.$$

b)
$$\int x^n e^x dx, \qquad n \in \mathbb{Z}, n \neq 0.$$

c)
$$\int \frac{\log(x + a)}{x + b} dx, \qquad a, b \in \mathbb{C}, a \neq b.$$

d)
$$\int \frac{(x + 1)e^{x^2} + 1}{(e^{x^2})^2 - 1} dx$$

e)
$$\int \left(1 + \frac{x^{2-n}}{2 - n} + \frac{n - 1}{x^n} \right) e^x dx, \qquad n \in \mathbb{Z}, n \neq 2.$$

f)
$$\int \frac{2 + \tan(x)^2}{1 + (\tan(x) + x)^2} dx$$

g)
$$\int \frac{(3x - 2)\log(x)^3 + (x - 1)\log(x)^2 + 2x(x - 2)\log(x) + x^2}{x \log(x)^6 - 4x^2 \log(x)^5 + 6x^3 \log(x)^4 - 4x^4 \log(x)^3 + x^5 \log(x)^2} dx$$

6. The Risch Differential Equation

We describe in this chapter the solution to the Risch differential equation problem, *i.e.* given a differential field K of characteristic 0 and $f, g \in K$, to decide whether the equation

$$Dy + fy = g \tag{6.1}$$

has a solution in K, and to find one if there are some. We only study equation (6.1) in the transcendental case, *i.e.* when K is a simple monomial extension of a differential subfield k, so for the rest of this chapter, let k be a differential field of characteristic 0 and t be a monomial over k. We suppose that the coefficients f and g of our equation are in $k(t)$ and look for a solution $y \in k(t)$. The algorithm we present in this chapter proceeds as follows:

1. Compute the normal part of the denominator of any solution. This reduces the problem to finding a solution in $k\langle t \rangle$.
2. Compute the special part of the denominator of any solution. This reduces the problem to finding a solution in $k[t]$.
3. Bound the degree of any solution in $k[t]$.
4. Reduce equation (6.1) to one of a similar form but with $f, g \in k[t]$.
5. Find the solutions in $k[t]$ of bounded degree of the reduced equation.

6.1 The Normal Part of the Denominator

We show in this section that the normal part of the denominator of any solution of a Risch differential equation in a monomial extension is given by an explicit formula in terms of the coefficients of the equation, provided that the equation is adequately preprocessed. We describe first the required preprocessing.

Definition 6.1.1. *We say that $f \in k(t)$ is weakly normalized with respect to t if* $\mathrm{residue}_p(f)$ *is not a positive integer for any normal irreducible $p \in k[t]$ such that $f \in \mathcal{R}_p$.*

The motivation behind that definition is the following lemma, which gives a formula for the order of $Dy + fy$ at a normal polynomial whenever f is weakly normalized:

Lemma 6.1.1. *Let $f \in k(t) \setminus \{0\}$ be weakly normalized with respect to t, $y \in k(t) \setminus \{0\}$, and $p \in k[t]$ be normal irreducible. Then,*

$$\nu_p(y) < 0 \Longrightarrow \nu_p(Dy + fy) = \nu_p(y) + \min(\nu_p(f), -1).$$

Proof. Let $n = \nu_p(y)$, $m = \nu_p(f)$ and suppose that $n < 0$. Then, $\nu_p(Dy) = n - 1$ by Theorem 4.4.2, and $\nu_p(fy) = n + m$ by Theorem 4.1.1. If $m < -1$, then $\min(m, -1) = m$ and $\nu_p(Dy + fy) = \nu_p(fy)$ by Theorem 4.1.1, so $\nu_p(Dy + fy) = n + \min(m, -1)$. If $m > -1$, then $\min(m, -1) = -1$ and $\nu_p(Dy + fy) = \nu_p(Dy)$ by Theorem 4.1.1, so $\nu_p(Dy + fy) = n + \min(m, -1)$. Suppose now that $m = -1$. Then $\nu_p(pf) = 0$ so $f \in \mathcal{R}_p$, and $\nu_p(Dy + fy) \geq n-1$ by Theorem 4.1.1. By Corollary 4.4.2, $Dy/y \in \mathcal{R}_p$ and $\text{residue}_p(Dy/y) = n$. Since $\mathcal{R}_p$ is a vector space over k and residue_p is a linear map by Theorem 4.4.1, we get $Dy/y + f \in \mathcal{R}_p$ and $\text{residue}_p(Dy/y+f) = \text{residue}_p(Dy/y) + \text{residue}_p(f) = n + \text{residue}_p(f)$. Since f is weakly normalized, $\text{residue}_p(f)$ is not a positive integer, hence $\text{residue}_p(f) \neq -n$, so $\text{residue}_p(Dy/y+f) \neq 0$. By Theorem 4.4.1, this implies that $Dy/y+f \notin \mathcal{O}_p$, hence that $\nu_p(Dy/y+f) < 0$, so $\nu_p(Dy + fy) < n$. Therefore, $\nu_p(Dy + fy) = n - 1 = n + \min(m, -1)$. $\square$

Of course, the next step is, given $f \in k(t)$, testing whether f is weakly normalized with respect to t, and finding an adequate transformation otherwise. The following theorem shows that adding an appropriate logarithmic derivative to any $f \in k(t)$ makes it weakly normalized, and gives an explicit change of variable that transforms equation (6.1) into a similar one with a weakly normalized coefficient.

Theorem 6.1.1. *For any $f \in k(t)$, we can compute $q \in k[t]$ such that $\bar{f} = f - Dq/q$ is weakly normalized with respect to t. Furthermore, for any $g, y \in k(t)$,*

$$Dy + fy = g \iff Dz + \bar{f}z = qg$$

where $z = qy$.

Proof. Let $d = d_s d_n$ be a splitting factorization of the denominator of f, and $d_n = d_1 d_2^2 \cdots d_m^m$ be a squarefree factorization of d_n. Write $f = a/d_1 + b/c$ where $a, b, c \in k[t]$ and $\gcd(d_1, c) = 1$, and let z be a new indeterminate over k and $r = \text{resultant}_t(a - zDd_1, d_1) \in k[z]$. Let $n_1, \ldots, n_s$ be all the distinct positive integer roots of r, and

$$q = \prod_{i=1}^{s} \gcd(a - n_i Dd_1, d_1)^{n_i} \in k[t].$$

We now show that $\bar{f} = f - Dq/q$ is weakly normalized with respect to t. Let $p \in k[t]$ be normal irreducible and suppose that $\bar{f} \in \mathcal{R}_p$. By Corollary 4.4.2, $Dq/q \in \mathcal{R}_p$ and $\text{residue}_p(Dq/q) = \nu_p(q)$. Since $\mathcal{R}_p$ is a vector space over k and residue_p is a linear map by Theorem 4.4.1, we get $f \in \mathcal{R}_p$ and $\text{residue}_p(\bar{f}) = \text{residue}_p(f) - \nu_p(q)$. Let $\rho = \text{residue}_p(f)$. If ρ is not a positive integer, then residue$_p(\bar{f}) = \rho - \nu_p(q)$ is not a positive integer. Thus, suppose that ρ is a positive integer. Then $\rho \neq 0$, so $f \notin \mathcal{O}_p$ by Theorem 4.4.1, which implies that $p \mid d$. Since p is normal, this means that $p \mid d_n$, so $\nu_p(f) = -\nu_p(d_n) < 0$. Since $f \in \mathcal{R}_p$, we have $\nu_p(pf) \geq 0$, so $\nu_p(f) \geq -1$, hence $\nu_p(f) = -1$, so $\nu_p(d_n) = 1$. This implies that $p \mid d_1$ and $\gcd(p, d/d_1) = 1$, hence that $b/c \in \mathcal{O}_p$ and $a/d_1 \in \mathcal{R}_p$. Thus, residue$_p(b/c) = 0$, so $\rho = $ residue$_p(a/d_1)$. Since d_1 is normal, a/d_1 is simple. In addition $\rho \in k$ since ρ is an integer, hence residue$_p(a/d_1)$ is a root of r by Theorem 4.4.3. Thus, $\rho = n_j$ for some j, so $p \mid \gcd(a - n_j D d_1, d_1)$ by Lemma 4.4.3, which implies that $\nu_p(\gcd(a - n_j D d_1, d_1)) = 1$. For $i \neq j$, we have residue$_p(a/d_1) \neq n_i$, so $p \nmid \gcd(a - n_i D d_1, d_1)$ by Lemma 4.4.3, hence $\nu_p(\gcd(a - n_i D d_1, d_1)) = 0$. Therefore,

$$\nu_p(q) = \sum_{i=1}^{s} n_i \nu_p(\gcd(a - n_i D d_1, d_1)) = n_j .$$

Hence, residue$_p(\bar{f}) = \rho - \nu_p(q) = n_j - n_j = 0$, so f is weakly normalized with respect to t.
Let $g, y \in k(t)$ and $z = qy$. Then,

$$Dz + \bar{f}z = qDy + yDq + fqy - \frac{Dq}{q}qy = q(Dy + fy)$$

so $Dy + fy = g \iff Dz + \bar{f}z = qg$.   □

We note that in practice a full squarefree factorization of d_n is not necessary, since only d_1 is needed for computing q. The above proof gives an algorithm for weak-normalizing any element of $k(t)$.

WeakNormalizer(f, D) (* Weak normalization *)

(* Given a derivation D on $k[t]$ and $f \in k(t)$, return $q \in k[t]$ such that $f - Dq/q$ is weakly normalized with respect to t. *)

$(d_n, d_s) \leftarrow$ **SplitFactor**(denominator$(f), D$)
$g \leftarrow \gcd(d_n, dd_n/dt)$
$d^* \leftarrow d_n/g$
$d_1 \leftarrow d^*/\gcd(d^*, g)$
$(a, b) \leftarrow$ **ExtendedEuclidean**(denominator$(f)/d_1, d_1,$ numerator(f))
$r \leftarrow$ resultant$_t(a - zDd_1, d_1)$
$(n_1, \ldots, n_s) \leftarrow$ positive integer roots of r
return$(\prod_{i=1}^{s} \gcd(a - n_i D d_1, d_1)^{n_i})$

We can assume now that f is weakly normalized with respect to t in equation (6.1). Then, the following theorem gives an explicit formula for the normal part of the denominator of a solution.

Theorem 6.1.2. *Let $f \in k(t)$ be weakly normalized with respect to t and $g \in k(t)$. Let $y \in k(t)$ be such that $Dy + fy = g$. Let $d = d_s d_n$ be a splitting factorization of the denominator of f, and $e = e_s e_n$ be a splitting factorization of the denominator of g. Let $c = \gcd(d_n, e_n)$ and*

$$h = \frac{\gcd(e_n, de_n/dt)}{\gcd(c, dc/dt)} \in k[t] \,.$$

Then,

(i)

$$yh \in k\langle t \rangle \,.$$

(ii)

$$\frac{yh}{q} \notin k\langle t \rangle \quad \text{for any } q \in k[t] \setminus k \text{ such that } q \mid h \,.$$

Proof. (i) Let $q = yh \in k(t)$. In order to show that $q \in k\langle t \rangle$, we need to show that $\nu_p(q) \geq 0$ for any normal irreducible $p \in k[t]$. We have $\nu_p(q) = \nu_p(y) + \nu_p(h)$ by Theorem 4.1.1. If $\nu_p(y) \geq 0$, then $\nu_p(q) \geq \nu_p(h) \geq 0$ since $h \in k[t]$. So suppose now that $n = \nu_p(y) < 0$.

Case 1: $\nu_p(f) \geq 0$. Then, $\nu_p(Dy + fy) = \nu_p(y) - 1$ by Lemma 6.1.1. Since $g = Dy + fy$, this implies that $\nu_p(g) < 0$, hence that $p \mid e$. Since p is normal, $\gcd(p, e_s) = 1$, so $\nu_p(e_n) = -\nu_p(g) = 1 - n$.

Also, p does not divide d since $\nu_p(f) \geq 0$, so $\nu_p(c) = 0$, so $\nu_p(\gcd(c, dc/dt)) = 0$. Hence $\nu_p(h) = \nu_p(\gcd(e_n, de_n/dt)) = \nu_p(e_n) - 1 = -n$, so $\nu_p(q) = n - n = 0$.

Case 2: $\nu_p(f) < 0$. Then, $\nu_p(g) = \nu_p(Dy + fy) = \nu_p(f) + n$ by Lemma 6.1.1, so $n = \nu_p(g) - \nu_p(f)$. Since $n < 0$, this implies that $\nu_p(g) < \nu_p(f) < 0$, hence that $p \mid d$ and $p \mid e$. As above, since p is normal, $\gcd(p, d_s) = \gcd(p, e_s) = 1$, so $\nu_p(d_n) = -\nu_p(f) < -\nu_p(g) = \nu_p(e_n)$. Thus, $\nu_p(c) = \min(\nu_p(d_n), \nu_p(e_n)) = \nu_p(d_n) = -\nu_p(f) > 0$, so

$$\begin{aligned} \nu_p(h) &= \nu_p(\gcd(e_n, de_n/dt)) - \nu_p(\gcd(c, dc/dt)) \\ &= (\nu_p(e_n) - 1) - (\nu_p(c) - 1) = -\nu_p(g) + \nu_p(f) = -n \,, \end{aligned}$$

so $\nu_p(q) = n - n = 0$.

(ii) Let $q \in k[t] \setminus k$ and suppose that $q \mid h$. Let p be any irreducible factor of q in $k[t]$. Then $p \mid h$, so $p \mid e_n$, so p is normal with respect to D. In addition, $\min(\nu_p(Dy), \nu_p(fy)) \leq \nu_p(Dy + fy) = \nu_p(g) = -\nu_p(e_n) < 0$, so at least one of $\nu_p(Dy)$ or $\nu_p(fy)$ must be negative. If $\nu_p(f) \geq 0$, then $\nu_p(Dy) < 0$ or $\nu_p(y) < 0$, so $\nu_p(y) < 0$ in any case by Theorem 4.4.2. If $\nu_p(f) < 0$, then $p \mid d$, so $p \mid c$ and $\nu_p(h) = (\nu_p(e_n) - 1) - (\nu_p(c) - 1) > 0$, which implies that $\nu_p(e_n) > \nu_p(c) = \min(\nu_p(d_n), \nu_p(e_n))$, hence that $\nu_p(d_n) < \nu_p(e_n)$, so $\nu_p(f) > \nu_p(g)$. Since $\nu_p(g) = \nu_p(Dy + fy) = \nu_p(f) + \nu_p(y)$ by Lemma 6.1.1, we must have $\nu_p(y) < 0$ in this case also.

Thus, $\nu_p(y) < 0$. From the proof of part (i), this implies that $\nu_p(yh) = 0$, hence that $\nu_p(yh/q) = -\nu_p(q) < 0$, so $yh/q \notin k\langle t \rangle$. $\qquad \square$

Corollary 6.1.1. *Let $f \in k(t)$ be weakly normalized with respect to t, $g \in k(t)$, and d_n, e_n and h be as in Theorem 6.1.2. Then,*

(i) For any solution $y \in k(t)$ of $Dy + fy = g$, $q = yh \in k\langle t \rangle$ and q is a solution of

$$d_n h Dq + (d_n h f - d_n Dh)\, q = d_n h^2 g. \tag{6.2}$$

Conversely, for any solution $q \in k\langle t \rangle$ of (6.2), $y = q/h$ is a solution of $Dy + fy = g$.

(ii) If $Dy + fy = g$ has a solution in $k(t)$ then $e_n \mid d_n h^2$.

Proof. (i) Let $y \in k(t)$ be a solution of $Dy + fy = g$, and let $q = yh$. $q \in k\langle t \rangle$ by Theorem 6.1.2, and

$$Dq + \left(f - \frac{Dh}{h}\right) q = hDy + yDh + hfy - yDh = h(Dy + fy) = hg.$$

Multiplying through by $d_n h$ yields $d_n h Dq + (d_n h f - d_n Dh)q = d_n h^2 g$, so q is a solution of (6.2). Conversely, the same calculation shows that for any solution $q \in k\langle t \rangle$ of (6.2), $y = q/h$ is a solution of $Dy + fy = g$.

(ii) Suppose that $Dy + fy = g$ has a solution in $k(t)$. Then, (6.2) must have a solution $q \in k\langle t \rangle$. The denominator of $d_n f$ is d_s, which has no normal irreducible factor, so $d_n f \in k\langle t \rangle$. Since $k[t] \subseteq k\langle t \rangle$ and $k\langle t \rangle$ is a differential subring of $k(t)$ by Corollary 4.4.1, this implies that $d_n h Dq + (d_n h f - d_n Dh)q \in k\langle t \rangle$, hence that $d_n h^2 g \in k\langle t \rangle$. Let $p \in k[t]$ be any irreducible factor of e_n. Then p is normal, so we must have $\nu_p(d_n h^2 g) \geq 0$. Hence, $\nu_p(d_n h^2) \geq -\nu_p(g) = \nu_p(e_n)$. Since this holds for any irreducible factor of e_n, we have $e_n \mid d_n h^2$. $\square$

The above theorem and corollary give us an algorithm that either proves that a given Risch differential equation has no solution in a given monomial extension, or that reduces the equation to one over $k\langle t \rangle$.

RdeNormalDenominator(f, g, D)
(* Normal part of the denominator *)

(* Given a derivation D on $k[t]$ and $f, g \in k(t)$ with f weakly normalized with respect to t, return either "no solution", in which case the equation $Dy + fy = g$ has no solution in $k(t)$, or the quadruplet (a, b, c, h) such that $a, h \in k[t]$, $b, c \in k\langle t \rangle$, and for any solution $y \in k(t)$ of $Dy + fy = g$, $q = yh \in k\langle t \rangle$ satisfies $aDq + bq = c$. *)

$(d_n, d_s) \leftarrow$ **SplitFactor**(denominator$(f), D$)
$(e_n, e_s) \leftarrow$ **SplitFactor**(denominator$(g), D$)
$p \leftarrow \gcd(d_n, e_n)$
$h \leftarrow \gcd(e_n, de_n/dt)/\gcd(p, dp/dt)$
if $e_n \nmid d_n h^2$ **then return** "no solution"
return$(d_n h, d_n h f - d_n Dh, d_n h^2 g, h)$

Example 6.1.1. Let $k = \mathbb{Q}$ and let t be a monomial over k satisfying $Dt = 1$, *i.e.* $D = d/dt$, and consider the equation

$$Dy + y = \frac{1}{t} \qquad (6.3)$$

which arises from the integration of e^t/t. We have $f = 1$ and $g = 1/t$ so:

1. $(d_n, d_s) = \textbf{SplitFactor}(1, d/dt) = (1, 1)$
2. $(e_n, e_s) = \textbf{SplitFactor}(t, d/dt) = (t, 1)$
3. $p = \gcd(1, t) = 1$
4. $h = \gcd(t, 1)/\gcd(1, 1) = 1$

Since $t \nmid 1$, we conclude that (6.3) has no solution in $\mathbb{Q}(t)$, hence that $\int e^t/t \, dt$ is not an elementary function.

Example 6.1.2. Let $k = \mathbb{Q}(x)$ with $D = d/dx$, and let t be a monomial over k satisfying $Dt = 1 + t^2$, *i.e.* $t = \tan(x)$. Consider the equation

$$Dy + (t^2 + 1)y = \frac{1}{t^2} \qquad (6.4)$$

which arises from the integration of $e^{\tan(x)}/\tan(x)^2$. We have $f = t^2 + 1$ and $g = 1/t^2$ so:

1. $(d_n, d_s) = \textbf{SplitFactor}(t^2 + 1, D) = (1, t^2 + 1)$
2. $(e_n, e_s) = \textbf{SplitFactor}(t^2, D) = (t^2, 1)$
3. $p = \gcd(1, t^2) = 1$
4. $h = \gcd(t^2, 2t)/\gcd(1, 1) = t$
5. $d_n h^2 = t^2$ is divisible by t^2
6. $d_n h^2 g = 1$
7. $D_n h f - D_n D h = t(t^2 + 1) - (t^2 + 1) = (t - 1)(t^2 + 1)$

so any solution $y \in k(t)$ of (6.4) must be of the form $y = q/t$ where $q \in k\langle t \rangle$ is a solution of

$$tDq + (t - 1)(t^2 + 1)q = 1. \qquad (6.5)$$

6.2 The Special Part of the Denominator

As a result of Corollary 6.1.1, we are now reduced to finding solutions $q \in k\langle t \rangle$ of (6.2), which we rewrite as

$$aDq + bq = c \qquad (6.6)$$

where $a \in k[t]$ has no special factor, $b, c \in k\langle t \rangle$, $a \neq 0$, and t is a monomial over k. We give in this section algorithms that compute the denominator of any solution in $k\langle t \rangle$ of (6.6) for specific types of monomials, starting with a result valid for arbitrary monomial extensions.

Lemma 6.2.1. *Let t be a monomial over k, $p \in S^{\mathrm{irr}}$ and $a, b, y \in k(t)$ with $a \neq 0$ and $\nu_p(y) \neq 0$. Then,*

(i)   If $\nu_p(b) < \nu_p(a)$, then $\nu_p(aDy + by) = \nu_p(b) + \nu_p(y)$.
(ii) If $p \in S_1^{\mathrm{irr}}$ and $\nu_p(b) > \nu_p(a)$, then $\nu_p(aDy + by) = \nu_p(a) + \nu_p(y)$.
(iii) If $\nu_p(b) = \nu_p(a)$, then either $\nu_p(aDy + by) = \nu_p(a) + \nu_p(y)$, or

$$\pi_p\left(-\frac{b}{a}\right) = \nu_p(y)\, \pi_p\left(\frac{Dp}{p}\right) + \frac{D^*u}{u}$$

for some nonzero $u \in k[t]/(p)$, where D^* is the induced derivation (Lemma 3.1.2).

Proof. Since p is irreducible, we have $\nu_p(aDy) = \nu_p(a) + \nu_p(Dy)$ and $\nu_p(by) = \nu_p(b) + \nu_p(y)$ by Theorem 4.1.1. Furthermore, $\nu_p(Dy) \geq \nu_p(y)$ by Theorem 4.4.2, which implies that $Dy/y \in \mathcal{O}_p$.
(i) Suppose that $\nu_p(b) < \nu_p(a)$. Then,

$$\nu_p(by) = \nu_p(b) + \nu_p(y) < \nu_p(a) + \nu_p(y) \leq \nu_p(a) + \nu_p(Dy) = \nu_p(aDy)$$

which implies that $\nu_p(aDy + by) = \nu_p(by) = \nu_p(b) + \nu_p(y)$.
(ii) Suppose that $p \in S_1^{\mathrm{irr}}$ and that $\nu_p(b) > \nu_p(a)$. Then, $\nu_p(Dy) = \nu_p(y)$ by Theorem 4.4.2, so

$$\nu_p(aDy) = \nu_p(a) + \nu_p(Dy) = \nu_p(a) + \nu_p(y) < \nu_p(b) + \nu_p(y) = \nu_p(by)$$

which implies that $\nu_p(aDy + by) = \nu_p(aDy) = \nu_p(a) + \nu_p(y)$.
(iii) Suppose that $\nu_p(b) = \nu_p(a)$ and that $a \neq 0$. Then, $\nu_p(b/a) = \nu_p(b) - \nu_p(a) = 0$, so $b/a \in \mathcal{O}_p$. Furthermore,

$$\nu_p(aDy) = \nu_p(a) + \nu_p(Dy) \geq \nu_p(b) + \nu_p(y) = \nu_p(by)$$

so $\nu_p(aDy + by) \geq \nu_p(by) = \nu_p(ay)$. Suppose that $\nu_p(aDy + by) > \nu_p(ay)$. Then, $(aDy + by)/ay \in p\mathcal{O}_p$, so

$$0 = \pi_p\left(\frac{aDy + by}{ay}\right) = \pi_p\left(\frac{Dy}{y} + \frac{b}{a}\right) = \pi_p\left(\frac{Dy}{y}\right) + \pi_p\left(\frac{b}{a}\right)$$

since both Dy/y and b/a are in $\mathcal{O}_p$, and π_p is a field–homomorphism. Write now $y = p^{\nu_p(y)}z$ where $z \in \mathcal{O}_p$ and $\nu_p(z) = 0$. Then, $Dz \in \mathcal{O}_p$ by Lemma 4.2.1, and since $Dp/p \in \mathcal{O}_p$, we get

$$\begin{aligned}
-\pi_p\left(\frac{b}{a}\right) &= \pi_p\left(\frac{Dy}{y}\right) = \pi_p\left(\nu_p(y)\frac{Dp}{p} + \frac{Dz}{z}\right) \\
&= \pi_p\left(\nu_p(y)\frac{Dp}{p}\right) + \frac{\pi_p(Dz)}{\pi_p(z)} = \nu_p(y)\,\pi_p\left(\frac{Dp}{p}\right) + \frac{D^*\pi_p(z)}{\pi_p(z)}
\end{aligned}$$

since $D^* \circ \pi_p = \pi_p \circ D$ by Theorem 4.2.1.   □

Since $a \in k[t]$ and has no special factor in (6.6), this means that $\nu_p(a) = 0$ for any $p \in \mathcal{S}$, so Lemma 6.2.1 provides a lower bound for $\nu_p(q)$ where $q \in k\langle t \rangle$ is a solution of (6.6) in the following cases:

(i)   If $\nu_p(b) < 0$, then $\nu_p(q) \in \{0, \nu_p(c) - \nu_p(b)\}$.
(ii) If $\nu_p(b) > 0$ and $p \in \mathcal{S}_1^{\mathrm{irr}}$, then $\nu_p(q) \in \{0, \nu_p(c)\}$.

For $p \in \mathcal{S}^{\mathrm{irr}}$, once we have a lower bound $\nu_p(q) \geq n$ for some $n \leq 0$, replacing q by hp^n in (6.6) yields

$$a(p^n Dh + np^{n-1} h Dp) + bhp^n = c$$

hence

$$aDh + \left(b + na\frac{Dp}{p}\right) h = cp^{-n}. \tag{6.7}$$

Furthermore, $h \in k\langle t \rangle$ since $q \in k\langle t \rangle$, and $h \in \mathcal{O}_p$ since $\nu_p(q) \geq n$. Thus we are reduced to finding the solutions $h \in k\langle t \rangle \cap \mathcal{O}_p$ of (6.7). Note that $cp^{-n} \in k\langle t \rangle$ since $c \in k\langle t \rangle$, and $b + naDp/p \in k\langle t \rangle$ since $b \in k\langle t \rangle$, $a \in k[t]$ and $p \in \mathcal{S}$. The eventual power of p in the denominators of $b + naDp/p$ and cp^{-n} can be cleared by multiplying (6.7) by p^N where $N = \max(0, -\nu_p(b), n - \nu_p(c))$, ensuring that the coefficients of (6.7) are also in $k\langle t \rangle \cap \mathcal{O}_p$.

Since all the special polynomials are of the first kind in the monomial extensions we are considering in this section, we only have to find a lower bound for $\nu_p(q)$ in the potential cancellation case, $i.e.$ $\nu_p(b) = 0$. We consider this case separately for various kinds of monomial extensions.

The Primitive Case

If $Dt \in k$, then every squarefree polynomial is normal, so $k\langle t \rangle = k[t]$, which means that $a, b, c \in k[t]$ and any solution in $k\langle t \rangle$ of (6.6) must be in $k[t]$.

The Hyperexponential Case

If $Dt/t \in k$, then $k\langle t \rangle = k[t, t^{-1}]$, so we need to compute a lower bound on $\nu_t(q)$ where $q \in k\langle t \rangle$ is a solution of (6.6). In order to compute such a bound, we need to be able to decide whether an arbitrary element f of k can be written as

$$f = m\eta + \frac{Du}{u} \tag{6.8}$$

for some integer m and $u \in k^*$, where $\eta = Dt/t \in k$. As explained in Sect. 5.12, this is the parametric logarithmic derivative problem, a variant of the limited integration problem, which is discussed in Chap. 7. Since the integer m in a solution of (6.8) can appear in the lower bound computation, we first need to ensure that m is the same in all the solutions of (6.8).

Lemma 6.2.2. *Let K be a differential field of characteristic 0, and suppose that $\eta \in k^*$ is not the logarithmic derivative of a K-radical. Then, for $f \in K$, and any solutions (m, u) and (n, v) in $\mathbb{Z} \times K^*$ of (6.8), we have $n = m$ and $v/u \in \mathrm{Const}(K)$.*

Proof. Suppose that (m, u) and (n, v) are both solutions of (6.8). Then,

$$f = m\eta + \frac{Du}{u} = n\eta + \frac{Dv}{v}$$

which implies that

$$\frac{Dw}{w} = (m - n)\eta$$

where $w = v/u$. Since η is not the logarithmic derivative of a K-radical, the above implies that $m = n$ and that $Dw = 0$.   □

Lemma 6.2.3. *Suppose that t is an hyperexponential over k such that $\eta = Dt/t$ is not the logarithmic derivative of a k-radical. Let $a \in k[t], b, q \in k\langle t\rangle$ be such that $\gcd(a, t) = 1$, $\nu_t(b) = 0$, and $\nu_t(q) \neq 0$. Then, either*

$$\nu_t(aDq + bq) = \nu_t(q)$$

or

$$-\frac{b(0)}{a(0)} = \nu_t(q)\,\eta + \frac{Du}{u} \qquad \text{for some } u \in k^*.$$

Proof. Suppose that $\nu_t(aDq + bq) \neq \nu_t(q)$. Then, Lemma 6.2.1 implies that

$$\pi_t\left(-\frac{b}{a}\right) = \nu_t(q)\,\pi_t\left(\frac{Dt}{t}\right) + \frac{D^*u}{u}$$

for some nonzero $u \in k[t]/(t)$, where D^* is the induced derivation. But $k[t]/(t) \simeq k$ and D^* is an extension of D by Lemma 3.4.3, so $u \in k^*$ and $D^*u = Du$. Furthermore, $\pi_t(p) = p(0)$ for any $p \in k[t]$, so $\pi_t(Dt/t) = \pi_t(\eta) = \eta$, $\pi_t(a) = a(0)$ and $\pi_t(b) = b(0)$, which proves the lemma.   □

Since $t \in \mathcal{S}_1^{\mathrm{irr}}$ by Theorem 5.1.2, Lemmas 6.2.1 and 6.2.3 always provide a lower bound for $\nu_t(q)$ where $q \in k\langle t\rangle$ is a solution of (6.6): if $\nu_t(b) \neq 0$, then Lemma 6.2.1 provides the bound as explained earlier. Otherwise, $\nu_t(b) = 0$, so either $-b(0)/a(0) = m\eta + Du/u$ for some $m \in \mathbb{Z}$ and $u \in k^*$, in which case $\nu_t(q) \in \{0, m, \nu_t(c)\}$, or $\nu_t(q) \in \{0, \nu_t(c)\}$. Note that such an m is unique by Lemma 6.2.2 applied to k. Since $\mathcal{S}^{\mathrm{irr}} = \{t\}$, $k\langle t\rangle \cap \mathcal{O}_t = k[t]$, so having determined a lower bound for $\nu_t(q)$, we are left with finding solutions $h \in k[t]$ of (6.7).

RdeSpecialDenomExp(a, b, c, D)
(* Special part of the denominator – hyperexponential case *)

    (* Given a derivation D on $k[t]$ and $a \in k[t]$, $b, c \in k\langle t\rangle$ with $Dt/t \in k$,
    $a \neq 0$ and $\gcd(a, t) = 1$, return the quadruplet $(\bar{a}, \bar{b}, \bar{c}, h)$ such that
    $\bar{a}, \bar{b}, \bar{c}, h \in k[t]$ and for any solution $q \in k\langle t\rangle$ of $aDq + bq = c$, $r = qh \in k[t]$
    satisfies $\bar{a}Dr + \bar{b}r = \bar{c}$. *)

    $p \leftarrow t$ (* the monic irreducible special polynomial *)
    $n_b \leftarrow \nu_p(b)$, $n_c \leftarrow \nu_p(c)$
    $n \leftarrow \min(0, n_c - \min(0, n_b))$ (* $n \leq 0$ *)
    if $n_b = 0$ **then**   (* possible cancellation *)
        $\alpha \leftarrow$ **Remainder**$(-b/a, p)$ (* $\alpha = -b(0)/a(0) \in k$ *)
        if $\alpha = mDt/t + Dz/z$ for $z \in k^*$ and $m \in \mathbb{Z}$ **then** $n \leftarrow \min(n, m)$
    $N \leftarrow \max(0, -n_b, n - n_c)$ (* $N \geq 0$, for clearing denominators *)
    return$(ap^N, (b + naDp/p)p^N, cp^{N-n}, p^{-n})$

Example 6.2.1. Let $k = \mathbb{Q}(x)$ with $D = d/dx$, and let t be a monomial over k satisfying $Dt = t$, *i.e.* $t = e^x$, and consider the equation

$$\left(t^2 + 2xt + x^2\right) Dq + \left(\left(1 + \frac{1}{x^2}\right) t^2 + \left(2x - 1 - \frac{2}{x}\right) t + x^2\right) q = \frac{t}{x^2} - 1 + \frac{2}{x} \quad (6.9)$$

which arises from the integration of

$$\frac{e^x - x^2 + 2x}{\left(e^x + x\right)^2 x^2} e^{(x^2 - 1)/x + 1/(e^x + x)}.$$

We have $a = t^2 + 2xt + x^2$, $b = (1 + 1/x^2)t^2 + (2x - 1 - 2/x)t + x^2$, and $c = t/x^2 - 1 + 2/x$, hence

1. $n_b = \nu_t(b) = 0$, $n_c = \nu_t(c) = 0$
2. $n = \min(0, n_c - \min(0, n_b)) = 0$
3. $n_b = 0$, so $\alpha = -b(0)/a(0) = x^2/x^2 = -1$
4. $-1 = -Dt/t$, so $m = -1$ and $n = \min(n, m) = -1$
5. $N = \max(0, -n_b, n - n_c) = 0$

Hence, any solution $q \in k\langle t\rangle$ of (6.9) must be of the form $q = p/t$ for $p \in k[t]$ satisfying

$$\left(t^2 + 2xt + x^2\right) Dp + \left(\frac{t^2}{x^2} - \left(\frac{2}{x} - 1\right) t\right) p = \frac{t^2}{x^2} - \left(\frac{2}{x} - 1\right) t. \quad (6.10)$$

The Hypertangent Case

If $Dt/(t^2 + 1) \in k$ and $\sqrt{-1} \notin k$, then the only monic special irreducible is $t^2 + 1$, so we need to compute a lower bound on $\nu_{t^2+1}(q)$, where $q \in k\langle t\rangle$ is a solution of (6.6).

Lemma 6.2.4. *Suppose that $\sqrt{-1} \notin k$ and that t is an hypertangent over k such that $\eta = Dt/(t^2 + 1)$ is not the logarithmic derivative of a $k(\sqrt{-1})$-radical. Let $a \in k[t], b, q \in k\langle t\rangle$ be such that $\gcd(a, t^2 + 1) = 1$, $\nu_{t^2+1}(b) = 0$ and $\nu_{t^2+1}(q) \neq 0$. Then, either*

$$\nu_{t^2+1}(aDq + bq) = \nu_{t^2+1}(q)$$

or, writing $-b(\sqrt{-1})/a(\sqrt{-1}) = \alpha\sqrt{-1} + \beta$ for $\alpha, \beta \in k$, we have

$$-\frac{b(\sqrt{-1})}{a(\sqrt{-1})} = 2\nu_{t^2+1}(q)\,\eta\,\sqrt{-1} + \frac{Du}{u} \quad and \quad 2\beta = \frac{Dv}{v} \quad (6.11)$$

for some $u \in k(\sqrt{-1})^*$ and $v \in k^*$, and D is extended to $k(\sqrt{-1})$ via $D\sqrt{-1} = 0$.

Proof. Suppose that $\nu_{t^2+1}(aDq + bq) \neq \nu_t(q)$. Then, Lemma 6.2.1 implies that

$$\pi_{t^2+1}\left(-\frac{b}{a}\right) = \nu_{t^2+1}(q)\,\pi_{t^2+1}\left(\frac{D(t^2+1)}{t^2+1}\right) + \frac{D^*u}{u}$$

for some nonzero $u \in k[t]/(t^2 + 1)$, where D^* is the induced derivation. But $k[t]/(t^2 + 1) \simeq k(\sqrt{-1})$, and $(k(\sqrt{-1}), D^*)$ is an extension of (k, D) by Lemma 3.4.3, and D^* is the unique extension of D to $k(\sqrt{-1})$ by Theorem 3.2.3. Since $\sqrt{-1}$ is algebraic over $\mathrm{Const}(k)$, $D^*\sqrt{-1} = 0$ by Lemma 3.3.2, so $Du = D^*u$. Furthermore,

$$\frac{D(t^2+1)}{t^2+1} = 2t\frac{Dt}{t^2+1} = 2t\eta$$

so we get

$$-\frac{b(\gamma)}{a(\gamma)} = 2\nu_{t^2+1}(q)\,\eta\,\gamma + \frac{Du_\gamma}{u_\gamma}$$

for any $\gamma \in k(\sqrt{-1})$ such that $\gamma^2 + 1 = 0$. Taking $\gamma = \sqrt{-1}$ yields the first equality in (6.11). Let $\sigma : k(\sqrt{-1}) \to k(\sqrt{-1})$ be the automorphism that is the identity on k and that takes $\sqrt{-1}$ to $-\sqrt{-1}$. Applying $1 + \sigma$ to the first equality in (6.11) we get

$$
\begin{aligned}
2\beta &= (\alpha\sqrt{-1} + \beta) + (-\alpha\sqrt{-1} + \beta) \\
&= \left(2\nu_{t^2+1}(q)\,\eta\,\sqrt{-1} + \frac{Du}{u}\right) + \left(-2\nu_{t^2+1}(q)\,\eta\,\sqrt{-1} + \frac{D(u^\sigma)}{u^\sigma}\right) \\
&= \frac{Du}{u} + \frac{D(u^\sigma)}{u^\sigma} = \frac{D(uu^\sigma)}{uu^\sigma} = \frac{Dv}{v}
\end{aligned}
$$

where $v = uu^\sigma \in k^*$. $\qquad\square$

Since $t^2 + 1 \in \mathcal{S}_1^{\mathrm{irr}}$ by Theorem 5.10.1, Lemmas 6.2.1 and 6.2.4 always provide a lower bound for $\nu_{t^2+1}(q)$ where $q \in k\langle t \rangle$ is a solution of (6.6): if $\nu_{t^2+1}(b) \neq 0$, then Lemma 6.2.1 provides the bound as explained earlier. Otherwise, $\nu_{t^2+1}(b) = 0$, so either $-b(\sqrt{-1})/a(\sqrt{-1}) = m\eta\sqrt{-1} + Du/u$ for some $m \in \mathbb{Z}$ and $u \in k(\sqrt{-1})^*$, in which case $\nu_{t^2+1}(q) \in \{0, m, \nu_{t^2+1}(c)\}$, or $\nu_{t^2+1}(q) \in \{0, \nu_{t^2+1}(c)\}$. Note that such an m is unique by Lemma 6.2.2 applied to $k(\sqrt{-1})$. We also remark that the verification of (6.11) implies solving a parametric logarithmic derivative problem over $k(\sqrt{-1})$. This adjunction of $\sqrt{-1}$ is however temporary since only the integer $\nu_{t^2+1}(q)$ is used from the result, so the algorithm proceeds over k once this bound is determined. Since the necessary condition $2\beta = Dv/v$ is defined over k, it can be checked first, and $\sqrt{-1}$ needs to be introduced only if that condition is satisfied. Since $\mathcal{S}^{\mathrm{irr}} = \{t^2 + 1\}$, $k\langle t \rangle \cap \mathcal{O}_{t^2+1} = k[t]$, so having determined a lower bound for $\nu_{t^2+1}(q)$, we are left with finding solutions $h \in k[t]$ of (6.7).

There are analogues of Lemma 6.2.4 and the corresponding algorithm for fields containing $\sqrt{-1}$ (Exercise 6.1).

RdeSpecialDenomTan(a, b, c, D)
(* Special part of the denominator – hypertangent case *)

(* Given a derivation D on $k[t]$ and $a \in k[t]$, $b, c \in k\langle t \rangle$ with $Dt/(t^2+1) \in k$, $\sqrt{-1} \notin k$, $a \neq 0$ and $\gcd(a, t^2+1) = 1$, return the quadruplet $(\bar{a}, \bar{b}, \bar{c}, h)$ such that $\bar{a}, \bar{b}, \bar{c}, h \in k[t]$ and for any solution $q \in k\langle t \rangle$ of $aDq + bq = c$, $r = qh \in k[t]$ satisfies $\bar{a}Dr + \bar{b}r = \bar{c}$. *)

$p \leftarrow t^2 + 1$ (* the monic irreducible special polynomial *)
$n_b \leftarrow \nu_p(b)$, $n_c \leftarrow \nu_p(c)$
$n \leftarrow \min(0, n_c - \min(0, n_b))$ (* $n \leq 0$ *)
if $n_b = 0$ **then**   (* possible cancellation *)
    $\alpha\sqrt{-1} + \beta \leftarrow \mathbf{Remainder}(-b/a, p)$ (* $\alpha, \beta \in k$ *)
    $\eta \leftarrow Dt/(t^2 + 1)$ (* $\eta \in k$ *)
    if $2\beta = Dv/v$ for $v \in k^*$
    and $\alpha\sqrt{-1} + \beta = 2m\eta\sqrt{-1} + Dz/z$ for $z \in k(\sqrt{-1})^*$ and $m \in \mathbb{Z}$
    then $n \leftarrow \min(n, m)$
$N \leftarrow \max(0, -n_b, n - n_c)$ (* $N \geq 0$, for clearing denominators *)
return$(ap^N, (b + naDp/p)p^N, cp^{N-n}, p^{-n})$

Example 6.2.2. Continuing example 6.1.2, let $k = \mathbb{Q}(x)$ with $D = d/dx$, t be a monomial over k satisfying $Dt = 1 + t^2$, i.e. $t = \tan(x)$, and consider the solutions $q \in k\langle t \rangle$ of (6.5), which arises from the integration of $e^{\tan(x)}/\tan(x)^2$. We have $a = t$, $b = (t - 1)(t^2 + 1)$ and $c = 1$, hence

1. $n_b = \nu_{t^2+1}(b) = 1$, $n_c = \nu_{t^2+1}(b) = 0$
2. $n = \min(0, n_c - \min(0, n_b)) = 0$
3. $n_b \neq 0$, so $N = \max(0, -n_b, n - n_c) = 0$

Hence any solution of $q \in k\langle t \rangle$ of (6.5) must be in $k\langle t \rangle \cap \mathcal{O}_{t^2+1} = k[t]$.

6.3 Degree Bounds

As a result of the previous sections, we are now reduced to finding solutions $q \in k[t]$ of (6.7), which we rewrite as

$$aDq + bq = c \tag{6.12}$$

where $a, b, c \in k[t]$, $a \neq 0$, and t is a monomial over k. We give in this section algorithms that compute an upper bound on the degree in t of any solution in $k[t]$ of (6.12) for specific types of monomials, starting with a result valid for arbitrary monomial extensions.

Lemma 6.3.1. *Let t be a monomial over k and $a, b, q \in k[t]$ with $a \neq 0$ and $\deg(q) > 0$. Then,*

(i)   If $\deg(b) > \deg(a) + \max(0, \delta(t) - 1)$, then

$$\deg(aDq + bq) = \deg(b) + \deg(q).$$

(ii) If t is nonlinear and $\deg(b) < \deg(a) + \delta(t) - 1$, then

$$\deg(aDq + bq) = \deg(a) + \deg(p) + \delta(t) - 1.$$

(iii) If $\delta(t) \geq 1$ and $\deg(b) = \deg(a) + \delta(t) - 1$, then either

$$\deg(aDq + bq) = \deg(b) + \deg(q)$$

or

$$-\frac{\mathrm{lc}(b)}{\mathrm{lc}(a)} = \pi_\infty \left(\frac{Dq}{qt^{\delta(t)-1}} \right).$$

Proof.
(i) We have $\deg(Dq) \leq \deg(q) + \max(0, \delta(t) - 1)$ by Lemma 3.4.2, hence

$$
\begin{aligned}
\deg(aDq) = \deg(a) + \deg(Dq) &\leq \deg(q) + \deg(a) + \max(0, \delta(t) - 1) \\
&< \deg(q) + \deg(b) = \deg(bq)
\end{aligned}
$$

which implies that $\deg(aDq + bq) = \deg(bq) = \deg(b) + \deg(q)$.
(ii) If t is nonlinear, then $\deg(Dq) = \deg(q) + \delta(t) - 1$ by Lemma 3.4.2, hence

$$
\begin{aligned}
\deg(aDq) &= \deg(a) + \deg(Dq) \\
&= \deg(q) + \deg(a) + \delta(t) - 1 > \deg(q) + \deg(b) = \deg(bq)
\end{aligned}
$$

which implies that $\deg(aDq + bq) = \deg(aDq) = \deg(a) + \deg(q) + \delta(t) - 1$.
(iii) If $\delta(t) \geq 1$, then $\deg(Dq) \leq \deg(q) + \delta(t) - 1$ by Lemma 3.4.2, hence

$$
\begin{aligned}
\deg(aDq) &= \deg(a) + \deg(Dq) \\
&\leq \deg(q) + \deg(a) + \delta(t) - 1 = \deg(q) + \deg(b) = \deg(bq)
\end{aligned}
$$

which implies that $\deg(aDq + bq) \leq \deg(b) + \deg(q)$. Suppose that $\deg(aDq + bq) < \deg(b) + \deg(q)$. Then $\deg(aDq + bq) < \deg(a) + \deg(q) + \delta(t) - 1$, so $(aDq + bq)/(aqt^{\delta(t)-1}) \in t^{-1}\mathcal{O}_\infty$, which implies that

$$
\begin{aligned}
0 = \pi_\infty \left(\frac{aDq + bq}{aqt^{\delta(t)-1}} \right) &= \pi_\infty \left(\frac{b}{at^{\delta(t)-1}} + \frac{Dq}{qt^{\delta(t)-1}} \right) \\
&= \pi_\infty \left(\frac{b}{at^{\delta(t)-1}} \right) + \pi_\infty \left(\frac{Dq}{qt^{\delta(t)-1}} \right)
\end{aligned}
$$

since π_∞ is a ring–homomorphism and both $b/at^{\delta(t)-1}$ and $Dq/qt^{\delta(t)-1}$ are in $\mathcal{O}_\infty$. Since $\deg(b) = \deg(a) + \delta(t) - 1$, we have

$$
\pi_\infty \left(\frac{b}{at^{\delta(t)-1}} \right) = \frac{\mathrm{lc}(b)}{\mathrm{lc}(at^{\delta(t)-1})} = \frac{\mathrm{lc}(b)}{\mathrm{lc}(a)}
$$

and the lemma follows. $\square$

Lemma 6.3.1 provides an upper bound for $\deg(q)$ where $q \in k[t]$ is a solution of (6.12) in the following cases:

(i) If $\deg(b) > \deg(a) + \max(0, \delta(t) - 1)$, then $\deg(q) \in \{0, \deg(c) - \deg(b)\}$.
(ii) If $\deg(b) < \deg(a) + \delta(t) - 1$ and $\delta(t) \geq 2$, then

$$
\deg(q) \in \{0, \deg(c) - \deg(a) + 1 - \delta(t)\} .
$$

As a result, we only have to consider the cases $\deg(b) \leq \deg(a)$ for Louvillian monomials, and $\deg(b) = \deg(a) + \delta(t) - 1$ for nonlinear monomials. We consider those cases separately for various kinds of monomial extensions.

The Primitive Case

If $Dt \in k$, then, in order to compute an upper bound on $\deg(q)$, we need to decide whether an arbitrary element f of k can be written as

$$
f = m\eta + Du \tag{6.13}
$$

for some integer m and $u \in k$, where $\eta = Dt \in k$. Note that (6.13) is a limited integration problem in k, so it can be solved by applying the algorithm of Chap. 7 to f and η. Since the integer m in a solution of (6.13) can appear in the upper bound computation, we first need to ensure that m is the same in all the solutions of (6.13).

Lemma 6.3.2. *Suppose that t is a primitive over k such that $\eta = Dt$ is not the derivative of an element of k. Then, for $f \in k(t)$, and any solutions (m, u) and (n, v) in $\mathbb{Z} \times k$ of (6.13), we have $n = m$ and $v - u \in \mathrm{Const}(k)$.*

Proof. Suppose that (m, u) and (n, v) are both solutions of (6.13). Then,

$$f = m\eta + Du = n\eta + Dv$$

which implies that

$$Dw = (m - n)\eta$$

where $w = v - u$. Since η is not the derivative of an element of k, the above implies that $m = n$ and that $Dw = 0$. □

Lemma 6.3.3. *Suppose that t is a primitive over k such that $\eta = Dt$ is not the derivative of an element of k. Let $a, b, q \in k[t]$ be such that $a \neq 0$, $\deg(b) \leq \deg(a)$, and $\deg(q) > 0$. Then,*

(i)   If $\deg(b) < \deg(a) - 1$, then

$$\deg(aDq + bq) \in \{\deg(a) + \deg(q), \deg(a) + \deg(q) - 1\}.$$

(ii) If $\deg(b) = \deg(a) - 1$, then either

$$\deg(aDq + bq) \in \{\deg(a) + \deg(q), \deg(a) + \deg(q) - 1\}.$$

or

$$-\frac{\mathrm{lc}(b)}{\mathrm{lc}(a)} = \deg(q)\, \eta + Du \qquad \text{for some } u \in k.$$

(iii) If $\deg(b) = \deg(a)$, then either

$$\deg(aDq + bq) \in \{\deg(a) + \deg(q), \deg(a) + \deg(q) - 1\}$$

or

$$-\frac{\mathrm{lc}(b)}{\mathrm{lc}(a)} = \frac{D(\mathrm{lc}(q))}{\mathrm{lc}(q)} \quad \text{and} \quad -\frac{\mathrm{lc}(a\, D(\mathrm{lc}(q)) + b\,\mathrm{lc}(q))}{\mathrm{lc}(a)\mathrm{lc}(q)} = \deg(q)\, \eta + Du$$

for some $u \in k$.

Proof. Since $\deg(q) > 0$, we have $\deg(Dq) \in \{\deg(q), \deg(q) - 1\}$ by Lemma 5.1.2.

(i) If $\deg(b) < \deg(a) - 1$, then

$$
\begin{aligned}
\deg(aDq) &= \deg(a) + \deg(Dq) \\
&> (\deg(b) + 1) + (\deg(q) - 1) = \deg(b) + \deg(q) = \deg(bq)
\end{aligned}
$$

which implies that

$$\deg(aDq + bq) = \deg(aDq) \in \{\deg(a) + \deg(q), \deg(a) + \deg(q) - 1\}.$$

(ii) Suppose that $\deg(b) = \deg(a) - 1$. If $\deg(Dq) = \deg(q)$, then

$$
\begin{aligned}
\deg(aDq) = \deg(a) + \deg(Dq) &= \deg(b) + 1 + \deg(q) \\
&> \deg(b) + \deg(q) = \deg(bq)
\end{aligned}
$$

which implies that $\deg(aDq + bq) = \deg(aDq) = \deg(a) + \deg(q)$. Otherwise, $\deg(Dq) = \deg(q) - 1$, so $D(\mathrm{lc}(q)) = 0$ by Lemma 5.1.2, which implies that

$$\mathrm{lc}(Dq) = \deg(q)\, \eta \, \mathrm{lc}(q) + Dv$$

where $v \in k$ is the coefficient of $t^{\deg(q)-1}$ in q. In addition, we have

$$
\begin{aligned}
\deg(aDq) = \deg(a) + \deg(Dq) &= (\deg(b) + 1) + (\deg(q) - 1) \\
&= \deg(b) + \deg(q) = \deg(bq)
\end{aligned}
$$

which implies that $\deg(aDq + bq) \le \deg(a) + \deg(q) - 1$. Suppose that $\deg(aDq + bq) < \deg(a) + \deg(q) - 1$. Then, $(aDq + bq)t/(aq) \in t^{-1}\mathcal{O}_\infty$, which implies that

$$
0 = \pi_\infty\left(\frac{(aDq + bq)t}{aq}\right) = \pi_\infty\left(\frac{tb}{a} + \frac{tDq}{q}\right) = \pi_\infty\left(\frac{tb}{a}\right) + \pi_\infty\left(\frac{tDq}{q}\right)
$$

since π_∞ is a ring–homomorphism, and both tb/a and tDq/q are in $\mathcal{O}_\infty$. Since $\deg(b) = \deg(a) - 1$ and $\deg(Dq) = \deg(q) - 1$, we have

$$
\pi_\infty\left(\frac{tb}{a}\right) = \frac{\mathrm{lc}(tb)}{\mathrm{lc}(a)} = \frac{\mathrm{lc}(b)}{\mathrm{lc}(a)}
$$

and

$$
\pi_\infty\left(\frac{tDq}{q}\right) = \frac{\mathrm{lc}(tDq)}{\mathrm{lc}(q)} = \frac{\mathrm{lc}(Dq)}{\mathrm{lc}(q)} = \frac{\deg(q)\, \eta\, \mathrm{lc}(q) + Dv}{\mathrm{lc}(q)} = \deg(q)\,\eta + Du
$$

where $u = v/\mathrm{lc}(q) \in k$.

(iii) Suppose that $\deg(b) = \deg(a)$. If $\deg(Dq) = \deg(q) - 1$, then

$$
\begin{aligned}
\deg(aDq) = \deg(a) + \deg(Dq) &= \deg(b) + \deg(q) - 1 \\
&< \deg(b) + \deg(q) = \deg(bq)
\end{aligned}
$$

which implies that $\deg(aDq + bq) = \deg(bq) = \deg(a) + \deg(q)$. Otherwise, $\deg(Dq) = \deg(q)$, which implies that

$$\deg(aDq) = \deg(a) + \deg(Dq) = \deg(b) + \deg(q) = \deg(bq)$$

hence that $\deg(aDq + bq) \le \deg(a) + \deg(q)$. Suppose that $\deg(aDq + bq) < \deg(a) + \deg(q)$. Then, $(aDq + bq)/(aq) \in t^{-1}\mathcal{O}_\infty$, which implies that

$$
0 = \pi_\infty\left(\frac{aDq + bq}{aq}\right) = \pi_\infty\left(\frac{b}{a} + \frac{Dq}{q}\right) = \pi_\infty\left(\frac{b}{a}\right) + \pi_\infty\left(\frac{Dq}{q}\right)
$$

since π_∞ is a ring–homomorphism, and both b/a and Dq/q are in $\mathcal{O}_\infty$. Since $\deg(b) = \deg(a)$ and $\deg(Dq) = \deg(q)$, we have

$$
\pi_\infty\left(\frac{b}{a}\right) = \frac{\mathrm{lc}(b)}{\mathrm{lc}(a)} \quad \text{and} \quad \pi_\infty\left(\frac{Dq}{q}\right) = \frac{\mathrm{lc}(Dq)}{\mathrm{lc}(q)} = \frac{D(\mathrm{lc}(q))}{\mathrm{lc}(q)}
$$

which implies that

$$-\frac{\mathrm{lc}(b)}{\mathrm{lc}(a)} = \frac{Du}{u} \tag{6.14}$$

where $u = \mathrm{lc}(q)$. Write $p = u^{-1}q \in k[t]$. Then, $\deg(p) = \deg(q)$ and

$$aDq + bq = aD(up) + bup = ADp + Bp$$

where $A = ua$ and $B = aDu + bu$. Note that $\deg(A) = \deg(a)$, and $\deg(B) < \deg(A)$, since (6.14) implies that the coefficient of $t^{\deg(a)}$ in B is 0. Suppose first that $\deg(B) < \deg(A) - 1$. Then, (i) implies that

$$\deg(ADp + Bp) \in \{\deg(A) + \deg(p), \deg(A) + \deg(p) - 1\}$$

hence that

$$\deg(aDq + bq) \in \{\deg(a) + \deg(q), \deg(a) + \deg(q) - 1\}.$$

Suppose finally that $\deg(B) = \deg(A) - 1$. Then, (ii) implies that either

$$\deg(ADp + Bp) \in \{\deg(A) + \deg(p), \deg(A) + \deg(p) - 1\}$$

or

$$-\frac{\mathrm{lc}(B)}{\mathrm{lc}(A)} = \deg(p)\,\eta + Dv$$

for some $v \in k$. Noting that $\deg(p) = \deg(q)$, $\deg(A) = \deg(a)$, and $\mathrm{lc}(A) = \mathrm{lc}(a)\mathrm{lc}(q)$ completes the proof. $\square$

Lemmas 6.3.1 and 6.3.3 always provide an upper bound for $\deg(q)$ where $q \in k[t]$ is a solution of (6.12): if $\deg(b) > \deg(a)$, then Lemma 6.3.1 implies that $\deg(q) \in \{0, \deg(c) - \deg(b)\}$. If $\deg(b) < \deg(a) - 1$, then Lemma 6.3.3 implies that $\deg(q) \in \{0, \deg(c) - \deg(a), \deg(c) - \deg(a) + 1\}$. If $\deg(b) = \deg(a) - 1$, then either $-\mathrm{lc}(b)/\mathrm{lc}(a) = m\eta + Du$ for some $m \in \mathbb{Z}$ and $u \in k$, in which case $\deg(q) \in \{0, m, \deg(c) - \deg(a), \deg(c) - \deg(a) + 1\}$, or $\deg(q) \in \{0, \deg(c) - \deg(a), \deg(c) - \deg(a) + 1\}$. Note that such an m is unique by Lemma 6.3.2.

Finally, if $\deg(b) = \deg(a)$, then either $-\mathrm{lc}(b)/\mathrm{lc}(a) = Du/u$ for some $u \in k^*$ and $-\mathrm{lc}(aDu + bu)/(u\,\mathrm{lc}(a)) = m\eta + Dv$ for some $m \in \mathbb{Z}$ and $v \in k$, in which case $\deg(q) \in \{0, m, \deg(c) - \deg(a), \deg(c) - \deg(a) + 1\}$, or $\deg(q) \in \{0, \deg(c) - \deg(a), \deg(c) - \deg(a) + 1\}$. We can compute such an u by a variant of the integration algorithm (Sect. 5.12). Although it is not unique, if $-\mathrm{lc}(b)/\mathrm{lc}(a) = Du/u = Dv/v$ for $u, v \in k^*$, then $u = cv$ for some $c \in \mathrm{Const}(k)$ by Lemma 5.12.1, which implies that

$$\frac{\mathrm{lc}(aDu + bu)}{\mathrm{lc}(a)u} = \frac{\mathrm{lc}(acDv + bcv)}{\mathrm{lc}(a)cv} = \frac{c\,(\mathrm{lc}(aDv + bv))}{c\mathrm{lc}(a)v} = \frac{\mathrm{lc}(aDv + bv)}{\mathrm{lc}(a)v}$$

so the solution we use does not affect the bound m, which is unique by Lemma 6.3.2.

RdeBoundDegreePrim(a, b, c, D)
(* Bound on polynomial solutions – primitive case *)

(* Given a derivation D on $k[t]$ and $a, b, c \in k[t]$ with $Dt \in k$ and $a \neq 0$, return $n \in \mathbb{Z}$ such that $\deg(q) \leq n$ for any solution $q \in k[t]$ of $aDq + bq = c$. *)

$d_a \leftarrow \deg(a)$, $d_b \leftarrow \deg(b)$, $d_c \leftarrow \deg(c)$
if $d_b > d_a$ **then** $n \leftarrow \max(0, d_c - d_b)$ **else** $n \leftarrow \max(0, d_c - d_a + 1)$
if $d_b = d_a - 1$ **then**   (* possible cancellation *)
  $\alpha \leftarrow -\mathrm{lc}(b)/\mathrm{lc}(a)$
  if $\alpha = mDt + Dz$ for $z \in k$ and $m \in \mathbb{Z}$ **then** $n \leftarrow \max(n, m)$
if $d_b = d_a$ **then**   (* possible cancellation *)
  $\alpha \leftarrow -\mathrm{lc}(b)/\mathrm{lc}(a)$
  if $\alpha = Dz/z$ for $z \in k^*$ **then**
    $\beta \leftarrow -\mathrm{lc}(aDz + bz)/(z\,\mathrm{lc}(a))$
    if $\beta = mDt + Dw$ for $w \in k$ and $m \in \mathbb{Z}$ **then** $n \leftarrow \max(n, m)$
return n

Example 6.3.1. Let $k = \mathbb{Q}(x, t_0)$ with $D = d/dx$, where t_0 is a monomial over $\mathbb{Q}(x)$ satisfying $Dt_0/t_0 = 1/x^2$, *i.e.* $t_0 = \exp(-1/x)$, and let t be a monomial over k satisfying $Dt = 1/x$, *i.e.* $t = \log(x)$. Consider

$$t^2 Dy - \left(\frac{1}{x^2}t^2 + \frac{1}{x}\right) y = (2x - 1)t^4 + \frac{t_0 + x}{x}t^3 - \frac{t_0 + 4x^2}{2x}t^2 + xt \quad (6.15)$$

which arises from the integration of

$$\left((2x - 1)\log(x)^2 + \frac{e^{-1/x} + x}{x}\log(x) - \frac{e^{-1/x} + 4x^2}{2x} + \frac{x}{\log(x)}\right) e^{1/\log(x) + 1/x}.$$

Theorem 6.1.2 gives $h = 1$, so any solution in $k(t)$ must be in $k\langle t \rangle = k[t]$. We have $a = t^2$, $b = -t^2/x^2 - 1/x$ and

$$c = (2x - 1)t^4 + \frac{t_0 + x}{x}t^3 - \frac{t_0 + 4x^2}{2x}t^2 + xt$$

hence

1. $d_a = \deg(a) = 2$, $d_b = \deg(b) = 2$, $d_c = \deg(c) = 4$
2. $n = \max(0, d_c - d_a + 1) = 3$
3. d_a is equal to d_b, so
   a) $\alpha = -\mathrm{lc}(b)/\mathrm{lc}(a) = 1/x^2$
   b) α is equal to $D(t_0)/t_0$, so
      i. $\beta = -\mathrm{lc}(aDt_0 + bt_0)/(t_0\mathrm{lc}(a)) = -1/x$
      ii. β is equal to $-Dt$, so $n = \max(n, -1) = 3$

So any solution in $k[t]$ of (6.15) must have degree at most 3.

In the specific case where $D = d/dt$, then $Du = 0$ for any $u \in k$, so in particular, $-\mathrm{lc}(b)/\mathrm{lc}(a)$ is not of the form Du/u for $u \in k$. This yields a simpler form of Lemma 6.3.3 for that case, together with a simpler algorithm.

Corollary 6.3.1. *Suppose that t is transcendental over k and that $D = d/dt$. Let $a, b, q \in k[t]$ be such that $a \neq 0$ and $\deg(q) > 0$. Then,*

(i)   If $\deg(b) > \deg(a) - 1$ then, $\deg(aDq + bq) = \deg(b) + \deg(q)$.
(ii) If $\deg(b) < \deg(a) - 1$, then $\deg(aDq + bq) = \deg(a) + \deg(q) - 1$.
(iii) If $\deg(b) = \deg(a) - 1$, then either $\deg(aDq + bq) = \deg(b) + \deg(q)$, or

$$-\frac{\mathrm{lc}(b)}{\mathrm{lc}(a)} = \deg(q).$$

RdeBoundDegreeBase(a, b, c)
(* Bound on polynomial solutions – base case *)

    (* Given $a, b, c \in k[t]$ with $a \neq 0$, return $n \in \mathbb{Z}$ such that $\deg(q) \leq n$ for any solution $q \in k[t]$ of $adq/dt + bq = c$. *)

    $d_a \leftarrow \deg(a), \ d_b \leftarrow \deg(b), \ d_c \leftarrow \deg(c)$
    $n \leftarrow \max(0, d_c - \max(d_b, d_a - 1))$
    if $d_b = d_a - 1$ **then**   (* possible cancellation *)
        $m \leftarrow -\mathrm{lc}(b)/\mathrm{lc}(a)$
        if $m \in \mathbb{Z}$ **then** $n \leftarrow \max(0, m, d_c - d_b)$
    return n

Example 6.3.2. Let $k = \mathbb{Q}$ and let t be a monomial over k satisfying $Dt = 1$, *i.e.* $D = d/dt$, and consider the equation

$$Dy - 2ty = 1$$

which arises from the integration of $e^{-t^2} dt$. Theorem 6.1.2 gives $h = 1$, so any solution in $k(t)$ must be in $k\langle t \rangle$. Lemma 6.2.1 shows that $\nu_t(y) \geq 0$ for any solution, hence any solution in $k(t)$ must be in $k[t]$. We have $a = c = 1$ and $b = -2t$, hence

1. $d_a = \deg(a) = 0, \ d_b = \deg(b) = 1, \ d_c = \deg(c) = 0$
2. $n = \max(0, d_c - \max(d_b, d_a - 1)) = 0.$

So any solution in $k(t)$ must be in $k = \mathbb{Q}$. Since t is transcendental over $\mathbb{Q}$, $-2ty \neq 1$ for any $y \in \mathbb{Q}$, which implies that

$$\int e^{-t^2} dt$$

is not an elementary function.

The Hyperexponential Case

Lemma 6.3.4. *Suppose that t is an hyperexponential over k such that $\eta = Dt/t$ is not the logarithmic derivative of a k-radical. Let $a, b, q \in k[t]$ be such that $a \neq 0$, $\deg(b) \leq \deg(a)$, and $\deg(q) > 0$. Then,*

(i) If $\deg(b) < \deg(a)$, then $\deg(aDq + bq) = \deg(a) + \deg(q)$.
(ii) If $\deg(b) = \deg(a)$, then either $\deg(aDq + bq) = \deg(b) + \deg(q)$, or

$$-\frac{\mathrm{lc}(b)}{\mathrm{lc}(a)} = \deg(q)\,\eta + \frac{D(\mathrm{lc}(q))}{\mathrm{lc}(q)}.$$

Proof. Since $\deg(q) > 0$, we have $\deg(Dq) = \deg(q)$ by Lemma 5.1.2.
(i) If $\deg(b) < \deg(a)$, then

$$\deg(aDq) = \deg(a) + \deg(Dq) > \deg(b) + \deg(q) = \deg(bq)$$

which implies that $\deg(aDq + bq) = \deg(aDq) = \deg(a) + \deg(q)$.
(ii) Suppose that $\deg(b) = \deg(a)$ and $\deg(aDq + bq) \neq \deg(b) + \deg(q)$. Since $\delta(t) = 1$, Lemma 6.3.1 implies that

$$
\begin{aligned}
-\frac{\mathrm{lc}(b)}{\mathrm{lc}(a)} &= \pi_\infty\left(\frac{Dq}{q}\right) \\
&= \frac{\mathrm{lc}(Dq)}{\mathrm{lc}(q)} = \frac{D(\mathrm{lc}(q)) + \deg(q)\,\eta\,\mathrm{lc}(q)}{\mathrm{lc}(q)} = \deg(q)\,\eta + \frac{D(\mathrm{lc}(q))}{\mathrm{lc}(q)}.
\end{aligned}
$$

$\square$

Lemmas 6.3.1 and 6.3.4 always provide an upper bound for $\deg(q)$ where $q \in k[t]$ is a solution of (6.12): if $\deg(b) > \deg(a)$, then Lemma 6.3.1 implies that $\deg(q) \in \{0, \deg(c) - \deg(b)\}$. If $\deg(b) < \deg(a)$, then Lemma 6.3.4 implies that $\deg(q) \in \{0, \deg(c) - \deg(a)\}$. Finally, if $\deg(b) = \deg(a)$, then either $-\mathrm{lc}(b)/\mathrm{lc}(a) = m\,\eta + Du/u$ for some $m \in \mathbb{Z}$ and $u \in k^*$, in which case $\deg(q) \in \{0, m, \deg(c) - \deg(b)\}$, or $\deg(q) \in \{0, \deg(c) - \deg(b)\}$. Note that such an m is unique by Lemma 6.2.2.

RdeBoundDegreeExp(a, b, c, D)
(* Bound on polynomial solutions – hyperexponential case *)

(* Given a derivation D on $k[t]$ and $a, b, c \in k[t]$ with $Dt/t \in k$ and $a \neq 0$, return $n \in \mathbb{Z}$ such that $\deg(q) \leq n$ for any solution $q \in k[t]$ of $aDq + bq = c$. *)

$d_a \leftarrow \deg(a)$, $d_b \leftarrow \deg(b)$, $d_c \leftarrow \deg(c)$, $n \leftarrow \max(0, d_c - \max(d_b, d_a))$
if $d_a = d_b$ **then**   (* possible cancellation *)
$\quad \alpha \leftarrow -\mathrm{lc}(b)/\mathrm{lc}(a)$
$\quad$ **if** $\alpha = mDt/t + Dz/z$ for $z \in k^*$ and $m \in \mathbb{Z}$ **then** $n \leftarrow \max(n, m)$
return n

Example 6.3.3. Continuing example 6.2.1, let $k = \mathbb{Q}(x)$ with $D = d/dx$, t be a monomial over k satisfying $Dt = t$, *i.e.* $t = e^x$, and consider the solutions in $k[t]$ of (6.10). We have $a = t^2 + 2xt + x^2$, $b = c = t^2/x^2 - (2/x - 1)t$, hence

1. $d_a = d_b = d_c = 2$
2. $d_a = d_b$, so $\alpha = -\mathrm{lc}(b)/\mathrm{lc}(a) = -1/x^2$
3. $n = \max(0, d_c - \max(d_b, d_a)) = 0$
4. $-1/x^2$ cannot be written in the form $m + Dz/z$ for $m \in \mathbb{Z}$ and $z \in \mathbb{Q}(x)$

Hence any solution $p \in k[t]$ of (6.10) must be of degree 0, *i.e.* in $\mathbb{Q}(x)$.

The Nonlinear Case

Lemma 6.3.5. *Suppose that t is a nonlinear monomial over k, and let $a, b, q \in k[t]$ be such that $a \neq 0$, $\deg(b) = \deg(a) + \delta(t) - 1$, and $\deg(q) > 0$. Then, either $\deg(aDq + bq) = \deg(b) + \deg(q)$, or*

$$-\frac{\mathrm{lc}(b)}{\mathrm{lc}(a)} = \deg(q)\,\lambda(t)\,.$$

Proof. Suppose that $\deg(aDq + bq) \neq \deg(b) + \deg(q)$. Then, Lemma 6.3.1 implies that

$$-\frac{\mathrm{lc}(b)}{\mathrm{lc}(a)} = \pi_\infty\left(\frac{Dq}{qt^{\delta(t)-1}}\right) = \frac{\mathrm{lc}(Dq)}{\mathrm{lc}(qt^{\delta(t)-1})} = \frac{\mathrm{lc}(Dq)}{\mathrm{lc}(q)}\,.$$

Furthermore, $\mathrm{lc}(Dq) = \deg(q)\mathrm{lc}(q)\lambda(t)$ by Lemma 3.4.2, so $-\mathrm{lc}(b)/\mathrm{lc}(a) = \deg(q)\,\lambda(t)$. $\qquad\square$

Lemmas 6.3.1 and 6.3.5 always provide an upper bound for $\deg(q)$ where $q \in k[t]$ is a solution of (6.12): if $\deg(b) \neq \deg(a) + \delta(t) - 1$, then Lemma 6.3.1 provides the bound as explained earlier. Otherwise, either $-\mathrm{lc}(b)/\mathrm{lc}(a) = m\lambda(t)$ for some $m \in \mathbb{Z}$, in which case $\deg(q) \in \{0, m, \deg(c) - \deg(b)\}$, or $\deg(q) \in \{0, \deg(c) - \deg(b)\}$.

RdeBoundDegreeNonLinear(a, b, c, D)
(* Bound on polynomial solutions – nonlinear case *)

(* Given a derivation D on $k[t]$ and $a, b, c \in k[t]$ with $\deg(Dt) \geq 2$ and $a \neq 0$, return $n \in \mathbb{Z}$ such that $\deg(q) \leq n$ for any solution $q \in k[t]$ of $aDq + bq = c$. *)

$d_a \leftarrow \deg(a),\ d_b \leftarrow \deg(b),\ d_c \leftarrow \deg(c),\ \delta \leftarrow \deg(Dt),\ \lambda \leftarrow \mathrm{lc}(Dt)$
$n \leftarrow \max(0, d_c - \max(d_a + \delta - 1, d_b))$
if $d_b = d_a + \delta - 1$ **then**   (* possible cancellation *)
    $m \leftarrow -\mathrm{lc}(b)/(\lambda\,\mathrm{lc}(a))$
    if $m \in \mathbb{Z}$ **then** $n \leftarrow \max(0, m, d_c - d_b)$
return n

Example 6.3.4. Continuing examples 6.1.2 and 6.2.2, let $k = \mathbb{Q}(x)$ with $D = d/dx$, t be a monomial over k satisfying $Dt = 1 + t^2$, *i.e.* $t = \tan(x)$, and consider the the solutions $q \in k[t]$ of (6.5). We have $a = t$, $b = (t-1)(t^2+1)$ and $c = 1$, hence

1. $d_a = \deg(a) = 1$, $d_b = \deg(b) = 3$, $d_c = \deg(c) = 0$
2. $\delta = \delta(t) = 2$, $\lambda = \mathrm{lc}(1 + t^2) = 1$
3. $n = \max(0, d_c - \max(d_a + \delta - 1, d_b)) = 0$
4. $d_b \neq d_a + \delta - 1$

Hence any solution $q \in k[t]$ of (6.5) must be of degree 0, *i.e.* in $\mathbb{Q}(x)$.

6.4 The SPDE Algorithm

We are now reduced to finding solutions $q \in k[t]$ of (6.12) and we have an upper bound n on $\deg(q)$. We present here an algorithm of Rothstein [68] that either reduces equation (6.12) to one with $a = 1$, or proves that it has no solutions of degree at most n in $k[t]$. This algorithm is based on the following theorem.

Theorem 6.4.1. *Let $a, b, c \in k[t]$ with $a \neq 0$ and $\gcd(a, b) = 1$. Let $z, r \in k[t]$ be such that $c = az + br$ and either $r = 0$ or $\deg(r) < \deg(a)$. Then, for any solution $q \in k[t]$ of $aDq + bq = c$, $h = (q - r)/a \in k[t]$, and h is a solution of*

$$aDh + (b + Da)h = z - Dr. \tag{6.16}$$

Conversely, for any solution $h \in k[t]$ of (6.16), $q = ah + r$ is a solution of $aDq + bq = c$.

Proof. Let $q \in k[t]$ be a solution of $aDq + bq = c$. Then, $aDq + bq = az + br$, so $b(q - r) = a(z - Dq)$, so $a \mid b(q - r)$. Since $\gcd(a, b) = 1$, this implies that $a \mid q - r$, hence that $h = (q - r)/a \in k[t]$. We then have:

$$
\begin{aligned}
aDh + (b + Da)h &= a\left(\frac{Dq - Dr}{a} - \frac{(q-r)Da}{a^2}\right) + \frac{b(q-r) + (q-r)Da}{a} \\
&= Dq - Dr + \frac{b(q-r)}{a} \\
&= \frac{(az + br) - bq}{a} - Dr + \frac{bq - br}{a} = z - Dr.
\end{aligned}
$$

Conversely, let $h \in k[t]$ be a solution of (6.16), and let $q = ah + r$. Then,

$$
\begin{aligned}
aDq + bq &= a^2 Dh + ahDa + aDr + abh + br \\
&= a(aDh + (b + Da)h) + aDr + br \\
&= a(z - Dr) + aDr + br = az + br = c.
\end{aligned}
$$

□

Theorem 6.4.1 reduces (6.12) to (6.16), which is an equation of the same type. However, if (6.12) has a solution q of degree n, then the corresponding solution h of (6.16) must have degree at most $n - \deg(a)$ since $q = ah + r$ and $\deg(r) < \deg(a)$. Thus, if $\deg(a) > 0$ and $\gcd(a, b) = 1$, we can use Theorem 6.4.1 to reduce the degree of the unknown polynomial. The hypothesis that $\gcd(a, b) = 1$ is not a restriction: if (6.12) has a solution in $k[t]$, then $c \in (a, b)$, so $g = \gcd(a, b)$ must divide c, in which case we can divide a, b and c by g in order to get an equivalent equation with $\gcd(a, b) = 1$. Note that this step reduces the degree of a. If $\gcd(a, b) \nmid c$, we can conclude that (6.12) has no solution in $k[t]$. We can repeat this until either we have proven that (6.12) has no solution of degree at most n in $k[t]$, or until $\deg(a) = 0$ i.e. $a \in k^*$, at which point we divide the equation by a and we get an equation of the type (6.12) with $a = 1$. This is the SPDE[1] algorithm of Rothstein [67, 68].

SPDE(a, b, c, D, n) (* Rothstein's SPDE algorithm *)

(* Given a derivation D on $k[t]$, an integer n and $a, b, c \in k[t]$ with $a \neq 0$, return either "no solution", in which case the equation $aDq + bq = c$ has no solution of degree at most n in $k[t]$, or the tuple $(\bar{b}, \bar{c}, m, \alpha, \beta)$ such that $\bar{b}, \bar{c}, \alpha, \beta \in k[t]$, $m \in \mathbb{Z}$, and any solution $q \in k[t]$ of degree at most n of $aDq + bq = c$ must be of the form $q = \alpha h + \beta$, where $h \in k[t]$, $\deg(h) \leq m$ and $Dh + \bar{b}h = \bar{c}$. *)

if $n < 0$ **then**
      if $c = 0$ **then return**$(0, 0, 0, 0, 0)$ **else return** "no solution"
$g \leftarrow \gcd(a, b)$
if $g \nmid c$ **then return** "no solution"
$a \leftarrow a/g, b \leftarrow b/g, c \leftarrow c/g$
if $\deg(a) = 0$ **then return**$(b/a, c/a, n, 1, 0)$
$(r, z) \leftarrow$ **ExtendedEuclidean**(b, a, c)(* $br + az = c$, $\deg(r) < \deg(a)$ *)
$u \leftarrow$ **SPDE**$(a, b + Da, z - Dr, D, n - \deg(a))$
if $u =$ "no solution" **then return** "no solution"
$(\bar{b}, \bar{c}, m, \alpha, \beta) \leftarrow u$
(* The solutions of (6.16) are $h = \alpha s + \beta$ where $Ds + \bar{b}s = \bar{c}$ *)
return$(\bar{b}, \bar{c}, m, a\alpha, a\beta + r)$ (* $ah + r = a\alpha s + a\beta + r$ *)

Example 6.4.1. Continuing examples 6.1.2, 6.2.2 and 6.3.4, let $k = \mathbb{Q}(x)$ with $D = d/dx$, t be a monomial over k satisfying $Dt = 1 + t^2$, i.e. $t = \tan(x)$, and consider the the solutions in $k[t]$ of (6.5). We have $a = t$, $b = (t-1)(t^2+1) = t^3 - t^2 + t - 1$, $c = 1$ and $n = 0$ from example 6.3.4, hence

1. $g = \gcd(a, b) = 1$
2. $(r, z) =$ **ExtendedEuclidean**$(t^3 - t^2 + t - 1, t, 1) = (-1, t^2 - t + 1)$
3. $b + Da = t^3 - t^2 + t - 1 + Dt = t^3 + t$

---

[1] Special **P**olynomial **D**ifferential **E**quation

4. $z - Dr = t^2 - t + 1$
5. recursive call, $\mathbf{SPDE}(t, t^3 + t, t^2 - t + 1, D, -1)$:
    a) $-1 < 0$ and $t^2 - t + 1 \neq 0$, so return "no solution"

Thus (6.5) has no solution in $k[t]$, hence it has no solution in $k\langle t \rangle$. This implies that (6.4) has no solution in $k(t)$, hence that

$$\int \frac{e^{\tan(x)}}{\tan(x)^2} \, dx$$

is not an elementary function.

Example 6.4.2. Continuing examples 6.2.1 and 6.3.3, let $k = \mathbb{Q}(x)$ with $D = d/dx$, t be a monomial over k satisfying $Dt = t$, i.e. $t = e^x$, and consider the solutions in $k[t]$ of (6.10). We have $a = t^2 + 2xt + x^2$, $b = c = t^2/x^2 - (2/x - 1)t$ and $n = 0$ from example 6.3.3, hence

1. $g = \gcd(a, b) = 1$
2. $(r, z) = \mathbf{ExtendedEuclidean}(b, a, c) = (1, 0)$
3. $b + Da = (2x^2 + 1)t^2/x^2 + (2x^3 + 3x^2 - 2x)t/x^2 + 2x$
4. $z - Dr = 0$
5. recursive call, $\mathbf{SPDE}(t, b + Da, 0, D, -2)$:
    a) $-2 < 0$ so return $(0, 0, 0, 0, 0)$
6. $\bar{b} = \bar{c} = m = \alpha = \beta = 0$
7. return $(0, 0, 0, 0, 1)$

Thus any solution in $k[t]$ of degree at most 0 of (6.10) must be of the form $0h + 1 = 1$ where $Dh = 0$. It follows that $p = 1$ is a solution of (6.10). Going back to example 6.2.1, this implies that $q = 1/t$ is a solution of (6.9), hence that

$$\int \frac{e^x - x^2 + 2x}{(e^x + x)^2 x^2} e^{(x^2 - 1)/x + 1/(e^x + x)} dx = \frac{1}{e^x} e^{(x^2 - 1)/x + 1/(e^x + x)}.$$

Example 6.4.3. Continuing example 6.3.1, let $k = \mathbb{Q}(x, t_0)$ with $D = d/dx$, where t_0 is a monomial over $\mathbb{Q}(x)$ satisfying $Dt_0/t_0 = 1/x^2$, i.e. $t_0 = \exp(-1/x)$, t be a monomial over k satisfying $Dt = 1/x$, i.e. $t = \log(x)$, and consider the solutions in $k[t]$ of (6.15). We have $a = t^2$, $b = -t^2/x^2 - 1/x$,

$$c = (2x - 1)t^4 + \frac{t_0 + x}{x} t^3 - \frac{t_0 + 4x^2}{2x} t^2 + xt$$

and $n = 3$ from example 6.3.1, hence

1. $g = \gcd(a, b) = 1$
2.

$$
\begin{aligned}
(r, z) &= \mathbf{ExtendedEuclidean}(b, a, c) \\
&= \left(-x^2 t, (2x - 1)t^2 + \frac{t_0}{x} t - \frac{t_0 + 4x^2}{2x} \right)
\end{aligned}
$$

3.
$$b + Da = -\frac{t^2}{x^2} + \frac{2}{x}t - \frac{1}{x}$$

4.
$$z - Dr = (2x - 1)t^2 + \frac{t_0 + 2x^2}{x}t - \frac{t_0 + 2x^2}{2x}$$

5. recursive call, $\mathbf{SPDE}(a_1 = t^2, b_1 = b + Da, c_1 = z - Dr, D, 1)$:
   a) $g = \gcd(a_1, b_1) = 1$
   b)

$$(r_1, z_1) = \mathbf{ExtendedEuclidean}(b_1, a_1, c_1) = \left(x^2 + \frac{t_0}{2}, \frac{4x^3 + t_0}{2x^2} \right)$$

   c)
$$b_1 + Da_1 = -\frac{t^2}{x^2} + \frac{4}{x}t - \frac{1}{x}, z_1 - Dr_1 = 0$$

   d) recursive call, $\mathbf{SPDE}(t^2, -t^2/x^2 + 4t/x - 1/x, 0, D, -1)$:
      i. $-1 < 0$ so return $(0, 0, 0, 0, 0)$
   e) $\bar{b}_1 = \bar{c}_1 = m_1 = \alpha_1 = \beta_1 = 0$
   f) return $(0, 0, 0, 0, x^2 + t_0/2)$
6. $\bar{b} = \bar{c} = m = \alpha = 0, \beta = x^2 + t_0/2$
7. return $(0, 0, 0, 0, (x^2 + t_0/2)t^2 - x^2 t)$

Thus any solution in $k[t]$ of degree at most 3 of (6.15) must be of the form $0h + (x^2 + t_0/2)t^2 - x^2 t$ where $Dh = 0$. It follows that

$$y = \left(\frac{t_0}{2} + x^2 \right) t^2 - x^2 t$$

is a solution of (6.15), hence that

$$\int \phi(x) e^{1/\log(x) + 1/x} dx = \left(\left(\frac{e^{-1/x}}{2} + x^2 \right) \log(x)^2 - x^2 \log(x) \right) e^{1/\log(x) + 1/x}$$

where

$$\phi(x) = (2x - 1) \log(x)^2 + \frac{e^{-1/x} + x}{x} \log(x) - \frac{e^{-1/x} + 4x^2}{2x} + \frac{x}{\log(x)}.$$

Example 6.4.4. Let $k = \mathbb{Q}$ and t be a monomial over k satisfying $Dt = 1$, i.e. $D = d/dt$, and consider the solutions in $k[t]$ of arbitrary degree n of

$$(t^2 + t + 1) Dq - (2t + 1)q = \frac{1}{2}t^5 + \frac{3}{4}t^4 + t^3 - t^2 + 1. \tag{6.17}$$

We have $a = t^2 + t + 1$, $b = -2t - 1$, and $c = t^5/2 + 3t^4/4 + t^3 - t^2 + 1$ so:

1. $g = \gcd(a, b) = 1$
2. $(r, z) = \mathbf{ExtendedEuclidean}(b, a, c) = (5t/4, t^3/2 + t^2/4 + t/4 + 1)$

3. $b + Da = -2t - 1 + D(t^2 + t + 1) = 0$
4. recursive call, $\mathbf{SPDE}(t^2 + t + 1, 0, t^3/2 + t^2/4 + t/4 - 1/4, D, n - 2)$:
   a) $g = \gcd(t^2 + t + 1, 0) = t^2 + t + 1$
   b) $g \mid c$ so $a = 1$, $b = 0$, $c = t/2 - 1/4$
   c) $\deg(a) = 0$, so return $(\bar{b}, \bar{c}, m, \alpha, \beta) = (0, t/2 - 1/4, n - 2, 1, 0)$
5. return $(0, t/2 - 1/4, n - 2, t^2 + t + 1, 5t/4)$

so any solution $q \in k[t]$ of degree at most n of (6.17) must be of the form $q = (t^2 + t + 1)h + 5t/4$ where $h \in k[t]$ has degree at most $n - 2$ and satisfies

$$Dh = \frac{1}{2}t - \frac{1}{4}. \tag{6.18}$$

6.5 The Non-Cancellation Cases

We are now reduced to finding solutions in $k[t]$ of the following equation:

$$Dq + bq = c \tag{6.19}$$

where $b, c \in k[t]$ and t is a monomial over k. Furthermore, we have an upper bound n on $\deg(q)$. We describe in this section an algorithm that can be used in any monomial extension whenever the leading terms of Dq and bq do not sum to 0. Sufficient conditions for this are either $D = d/dt$, or $\deg(b) > \max(0, \delta(t) - 1)$, or t is nonlinear and either $\deg(b) \neq \delta(t) - 1$ or $\deg(b)/\lambda(t)$ is not a negative integer. Since there is no cancellation between the leading terms of Dq and bq in those cases, we call them the *non-cancellation* cases.

Lemma 6.5.1. *Let $b, q \in k[t]$ with $q \neq 0$.*

(i)   *If $b \neq 0$ and either $D = d/dt$ or $\deg(b) > \max(0, \delta(t) - 1)$, then the leading monomial of $Dq + bq$ is*

$$\mathrm{lc}(b)\mathrm{lc}(q)t^{\deg(q) + \deg(b)}.$$

(ii) *If $\deg(q) > 0$, $\deg(b) < \delta(t) - 1$, and either $\delta(t) \geq 2$ or $D = d/dt$, then the leading monomial of $Dq + bq$ is*

$$\deg(q)\mathrm{lc}(q)\lambda(t)t^{\deg(q) + \delta(t) - 1}.$$

(iii) *If $\delta(t) \geq 2$, $\deg(b) = \delta(t) - 1$, $\deg(q) > 0$ and $\deg(q) \neq -\mathrm{lc}(b)/\lambda(t)$, then the leading monomial of $Dq + bq$ is*

$$(\deg(q)\lambda(t) + \mathrm{lc}(b))\,\mathrm{lc}(q)t^{\deg(q) + \delta(t) - 1}.$$

Proof. (i) Suppose that $b \neq 0$. If $D = d/dt$, then

$$\deg(Dq) < \deg(q) \leq \deg(q) + \deg(b) = \deg(bq)$$

so $\deg(Dq + bq) = \deg(b) + \deg(q)$ and the leading coefficient of $Dq + bq$ is the leading coefficient of bq, which is the product of the leading coefficients of b and q. If t is an arbitrary monomial and $\deg(b) > m = \max(0, \delta(t) - 1)$, then, by Lemma 3.4.2, $\deg(Dq) \leq \deg(q) + m$, so so $\deg(Dq) < \deg(q) + \deg(b) = \deg(bq)$. Hence, $\deg(Dq + bq) = \deg(bq) = \deg(b) + \deg(q)$ as previously, and the leading coefficient of $Dq + bq$ is the leading coefficient of bq, which is the product of the leading coefficients of b and q.

(ii) Suppose that $\deg(q) > 0$, $\deg(b) < \delta(t) - 1$, and either $\delta(t) \geq 2$ or $D = d/dt$. If $\delta(t) \geq 2$, then $\deg(Dq) = \deg(q) + \delta(t) - 1$ by Lemma 3.4.2, so $\deg(Dq) > \deg(q) + \deg(b) = \deg(bq)$. Hence, $\deg(Dq + bq) = \deg(Dq) = \deg(q) + \delta(t) - 1$, and the leading coefficient of $Dq + bq$ is the leading coefficient of Dq, which is $\deg(q)\mathrm{lc}(q)\lambda(t)$ by Lemma 3.4.2. If $D = d/dt$, then $\delta(t) = 0$, so $\deg(b) < 0$ which implies that $b = 0$, so $\deg(Dq + bq) = \deg(Dq) = \deg(q) - 1$, and the leading coefficient of $Dq + bq$ is the leading coefficient of Dq which is $\deg(q)\mathrm{lc}(q)\lambda(t)$ since $\lambda(t) = 1$.

(iii) Suppose that $\delta(t) \geq 2$, $\deg(b) = \delta(t) - 1$, $\deg(q) > 0$ and $\deg(q) \neq -\mathrm{lc}(b)/\lambda(t)$. Then, $\deg(Dq) = \deg(q) + \delta(t) - 1$ by Lemma 3.4.2, so $\deg(Dq) = \deg(bq)$. The leading coefficient of Dq is $\deg(q)\mathrm{lc}(q)\lambda(t)$ by Lemma 3.4.2, and the leading coefficient of bq is $\mathrm{lc}(b)\mathrm{lc}(q)$. Since $\deg(q)\lambda(t) + \mathrm{lc}(b) \neq 0$ by hypothesis, we get that the leading coefficient of $Dq + bq$ is $(\deg(q)\lambda(t) + \mathrm{lc}(b))\mathrm{lc}(q)$ and the degree of $Dq + bq$ is $\deg(q) + \delta(t) - 1$.   □

Lemma 6.5.1 yields the following algorithms for finding the solutions of equation (6.19) whenever one of its hypotheses is satisfied.

When deg(b) is Large Enough

Suppose that $b \neq 0$, and that either $D = d/dt$ or $\deg(b) > \max(0, \delta(t) - 1)$. Then, for any solution $q \in k[t] \setminus \{0\}$ of $Dq + bq = c$, we must have $\deg(q) + \deg(b) = \deg(c)$, so $\deg(q) = \deg(c) - \deg(b)$ and $\mathrm{lc}(b)\mathrm{lc}(q) = \mathrm{lc}(c)$. This gives the leading monomial ut^n of any such q, and replacing q by $ut^n + h$ in (6.19), we get

$$D(ut^n) + Dh + but^n + bh = c$$

so

$$Dh + bh = c - D(ut^n) - but^n$$

which is an equation of the same type as (6.19) with the same b as before. Hence the hypotheses of part (i) of Lemma 6.5.1 are satisfied again, so we can repeat this process, but with a bound of $n - 1$ on $\deg(h)$. This bound will decrease at every pass through this process, guaranteeing termination.

PolyRischDENoCancel1(b, c, D, n) (* Poly Risch d.e. – no cancellation *)

(* Given a derivation D on $k[t]$, n either an integer or $+\infty$, and $b, c \in k[t]$ with $b \neq 0$ and either $D = d/dt$ or $\deg(b) > \max(0, \delta(t)-1)$, return either "no solution", in which case the equation $Dq + bq = c$ has no solution of degree at most n in $k[t]$, or a solution $q \in k[t]$ of this equation with $\deg(q) \leq n$. *)

$q \leftarrow 0$
while $c \neq 0$ **do**
    $m \leftarrow \deg(c) - \deg(b)$
    if $n < 0$ or $m < 0$ or $m > n$ **then return** "no solution"
    $p \leftarrow (\mathrm{lc}(c)/\mathrm{lc}(b))\, t^m$
    $q \leftarrow q + p$
    $n \leftarrow m - 1$
    $c \leftarrow c - Dp - bp$
return q

Example 6.5.1. Let $k = \mathbb{Q}(x)$ with $D = d/dx$, and let t be a monomial over k satisfying $Dt = 1 + t^2$, *i.e.* $t = \tan(x)$, and consider the equation

$$Dy + (t^2 + 1)y = t^3 + (x + 1)t^2 + t + x + 2 \qquad (6.20)$$

which arises fom the integration of

$$\left(\tan(x)^3 + (x + 1)\tan(x)^2 + \tan(x) + x + 2\right) e^{\tan(x)}.$$

Theorem 6.1.2 gives $h = 1$, so any solution in $k(t)$ must be in $k\langle t \rangle$. Lemma 6.2.1 shows that $\nu_{t^2+1}(y) \geq 0$ for any solution, hence any solution in $k(t)$ must be in $k[t]$, so looking for solutions in $k[t]$ of arbitrary degree we get: $b = t^2 + 1$, $c = t^3 + (x + 1)t^2 + t + x + 2$, $n = +\infty$ and

m	p	q	n	c
1	t	t	0	$xt^2 + x + 1$
0	x	$t + x$	-1	0

so $y = t + x$ is a solution of (6.20), hence

$$\int (\tan(x)^3 + (x + 1)\tan(x)^2 + \tan(x) + x + 2)e^{\tan(x)}dx = (\tan(x) + x)e^{\tan(x)}.$$

When $\deg(b)$ is Small Enough

Suppose that $\deg(b) < \delta(t) - 1$ and either $D = d/dt$, which implies that $b = 0$, or $\delta(t) \geq 2$. Let $q \in k[t]$ be a solution of $Dq + bq = c$.
If $\deg(q) > 0$, then $\deg(q) + \delta(t) - 1 = \deg(c)$, so $\deg(q) = \deg(c) + 1 - \delta(t)$ and $\deg(q)\mathrm{lc}(q)\lambda(t) = \mathrm{lc}(c)$. This yields the leading monomial ut^n of q, and replacing q by $ut^n + h$ in the equation yields a similar equation with a lower degree bound on its solution.

If $q \in k$, then: if $b \in k^*$, then $Dq + bq \in k$, so either $c \in k$, in which case we are reduced to solving a Risch differential equation of type (6.1) over k, or $\deg(c) > 0$ and (6.19) has no solution in k, hence in $k[t]$. If $\deg(b) > 0$, then the leading term of $Dq + bq$ is $q \, \mathrm{lc}(b) t^{\deg(b)}$, so either $\deg(c) = \deg(b)$, in which case $q = \mathrm{lc}(c)/\mathrm{lc}(b)$ is the only potential solution, or $\deg(c) \neq \deg(b)$ and (6.19) has no solution in k, hence in $k[t]$.

PolyRischDENoCancel2(b, c, D, n) (* Poly Risch d.e. – no cancellation *)

(* Given a derivation D on $k[t]$, n either an integer or $+\infty$, and $b, c \in k[t]$ with $\deg(b) < \delta(t) - 1$ and either $D = d/dt$ or $\delta(t) \geq 2$, return either "no solution", in which case the equation $Dq + bq = c$ has no solution of degree at most n in $k[t]$, or a solution $q \in k[t]$ of this equation with $\deg(q) \leq n$, or the tuple (h, b_0, c_0) such that $h \in k[t]$, $b_0, c_0 \in k$, and for any solution $q \in k[t]$ of degree at most n of $Dq + bq = c$, $y = q - h$ is a solution in k of $Dy + b_0 y = c_0$. *)

$q \leftarrow 0$
while $c \neq 0$ **do**
$\quad$ **if** $n = 0$ **then** $m \leftarrow 0$ **else** $m \leftarrow \deg(c) - \delta(t) + 1$
$\quad$ **if** $n < 0$ or $m < 0$ or $m > n$ **then return** "no solution"
$\quad$ **if** $m > 0$ **then** $p \leftarrow (\mathrm{lc}(c)/(m\,\lambda(t)))\, t^m$
$\quad$ **else** $\qquad\qquad\qquad\qquad\qquad\qquad\qquad\qquad\qquad$ (* $m = 0$ *)
$\quad\qquad$ **if** $\deg(b) \neq \deg(c)$ **then return** "no solution"
$\quad\qquad$ **if** $\deg(b) = 0$ **then return** (q, b, c)
$\quad\qquad$ $p \leftarrow \mathrm{lc}(c)/\mathrm{lc}(b)$
$\quad$ $q \leftarrow q + p$
$\quad$ $n \leftarrow m - 1$
$\quad$ $c \leftarrow c - Dp - bp$
return q

Example 6.5.2. Continuing example 6.4.4, let $k = \mathbb{Q}$, t be a monomial over k satisfying $Dt = 1$, *i.e.* $D = d/dt$, and consider the solutions $h \in k[t]$ of arbitrary degree n of (6.18). We get $c = t/2 - 1/4$, $n = +\infty$ and

m	p	q	n	c
2	$t^2/4$	$t^2/4$	1	$-1/4$
1	$-t/4$	$t^2/4 - t/4$	0	0

so $h = t^2/4 - t/4$ is a solution of (6.18). Going back to example 6.4.4, this implies that

$$q = (t^2 + t + 1)\left(\frac{1}{4}t^2 - \frac{1}{4}t\right) + \frac{5}{4}t = \frac{1}{4}t^4 + t$$

is a solution of (6.17). This example illustrates that in the case $b = 0$, the algorithm **PolyRischDENoCancel2** is computing exactly an integral of c, taking into account the degree constraints. Using the integration algorithm for that purpose would not be more efficient.

When $\delta(t) \geq 2$ and $\deg(b) = \delta(t) - 1$

In that case, we have cancellation only when $\deg(q) = -\mathrm{lc}(b)/\lambda(t)$, which implies in particular that $-\mathrm{lc}(b)/\lambda(t)$ is a positive integer. Let $q \in k[t]$ be a solution of $Dq + bq = c$.

If $\deg(q) > 0$ and $\deg(q) \neq -\mathrm{lc}(b)/\lambda(t)$, then $\deg(q) + \delta(t) - 1 = \deg(c)$, so $deg(q) = \deg(c) + 1 - \delta(t)$ and $(\deg(q)\lambda(t) + \mathrm{lc}(b))\mathrm{lc}(q) = \mathrm{lc}(c)$. This yields the leading monomial ut^n of q, and replacing q by $ut^n + h$ in the equation yields a similar equation with a lower degree bound on its solution. We can repeat this as long as the new degree bound is greater than $-\mathrm{lc}(b)/\lambda(t)$, or until we have a complete solution if $-\mathrm{lc}(b)/\lambda(t)$ is not a positive integer.

If $q \in k$, then the leading term of $Dq + bq$ is $q \, \mathrm{lc}(b)t^{\delta(t)-1}$, so either $\deg(c) = \delta(t) - 1$, in which case $q = \mathrm{lc}(c)/\mathrm{lc}(b)$ is the only potential solution, or $\deg(c) \neq \delta(t) - 1$ and (6.19) has no solution in k, hence in $k[t]$.

PolyRischDENoCancel3(b, c, D, n) (* Poly Risch d.e. – no cancellation *)

(* Given a derivation D on $k[t]$ with $\delta(t) \geq 2$, n either an integer or $+\infty$, and $b, c \in k[t]$ with $\deg(b) = \delta(t) - 1$, return either "no solution", in which case the equation $Dq + bq = c$ has no solution of degree at most n in $k[t]$, or a solution $q \in k[t]$ of this equation with $\deg(q) \leq n$, or the tuple $(h, m, \bar{c})$ such that $h \in k[t]$, $m \in \mathbb{Z}$, $\bar{c} \in k[t]$, and for any solution $q \in k[t]$ of degree at most n of $Dq + bq = c$, $y = q - h$ is a solution in $k[t]$ of degree at most m of $Dy + by = \bar{c}$. *)

$q \leftarrow 0$
if $-\mathrm{lc}(b)/\lambda(t) \in \mathbb{N}$ then $M \leftarrow -\mathrm{lc}(b)/\lambda(t)$ else $M \leftarrow -1$
while $c \neq 0$ do
    $m \leftarrow \max(M, \deg(c) - \delta(t) + 1)$
    if $n < 0$ or $m < 0$ or $m > n$ then return "no solution"
    $u \leftarrow m\lambda(t) + \mathrm{lc}(b)$
    if $u = 0$ then return(q, m, c)
    if $m > 0$ then $p \leftarrow (\mathrm{lc}(c)/u) \, t^m$
    else   (* $m = 0$ *)
        if $\deg(c) \neq \delta(t) - 1$ then return "no solution"
        $p \leftarrow \mathrm{lc}(c)/\mathrm{lc}(b)$
    $q \leftarrow q + p$
    $n \leftarrow m - 1$
    $c \leftarrow c - Dp - bp$
return q

Example 6.5.3. Let $k = \mathbb{Q}(x)$ with $D = d/dx$, and let t be a monomial over k satisfying $Dt = 1 + t^2$, *i.e.* $t = \tan(x)$, and consider the equation

$$Dy + (1 - t)y = t^3 + t^2 - 2xt - 2x \tag{6.21}$$

which arises from the integration of

$$\left(\tan(x)^3 + \tan(x)^2 - 2x\tan(x) - 2x\right) e^{x - \log(1+\tan(x)^2)/2}.$$

Theorem 6.1.2 gives $h = 1$, so any solution in $k(t)$ must be in $k\langle t\rangle$. Since $-b(\sqrt{-1}) = \sqrt{-1} - 1$ and -2 is not the logarithmic derivative of an element of k, Lemma 6.2.4 implies that $\nu_{t^2+1}(y) \geq 0$ for any solution, hence any solution in $k(t)$ must be in $k[t]$, so looking for solutions in $k[t]$ of arbitrary degree we get: $b = 1 - t$, $c = t^3 + t^2 - 2xt - 2x$, $n = +\infty$, $M = 1$ and

m	u	p	q	n	c
2	1	t^2	t^2	1	$-2(x+1)t - 2x$
1	0				

so any solution of (6.21) must be of the form $y = t^2 + q$ where $q \in k[t]$ is a solution of degree at most 1 of

$$Dq + (1 - t)q = -2(x+1)t - 2x. \tag{6.22}$$

6.6 The Cancellation Cases

We finally study equation (6.19) whenever the non-cancellation cases do not hold, *i.e.* in one of the following cases:

1. $\delta(t) \leq 1$, $b \in k$ and $D \neq d/dt$,
2. $\delta(t) \geq 2$, $\deg(b) = \delta(t) - 1$, and $\deg(q) = -\mathrm{lc}(b)/\lambda(t)$.

We present in this sections algorithms for the above cases for specific types of monomials.

The Primitive Case

If $Dt \in k$, then $\delta(t) = 0$, so the only cases not handled by Lemma 6.5.1 are $b = 0$ or $b \in k^*$. If $b = 0$, then (6.19) becomes $Dq = c$ for $c \in k[t]$, which is an integration problem in $k[t]$, and deciding whether it has a solution in $k[t]$ can be done by the in-field integration algorithm (Sect. 5.12), so suppose now that $b \in k^*$.

If $b = Du/u$ for some $u \in k^*$, which can also be checked by a variant of the integration algorithm (Sect. 5.12), then (6.19) becomes $Dq + qDu/u = c$, *i.e.* $D(uq) = uc$ which is as earlier an integration problem in $k[t]$.

If b is not of the form Du/u for some $u \in k^*$, then $D(\mathrm{lc}(q)) + b\,\mathrm{lc}(q) \neq 0$, so the leading monomial of $Dq + bq$ is $(D(\mathrm{lc}(q)) + b\,\mathrm{lc}(q))t^{\deg(q)}$. This implies that $\deg(q) = \deg(c)$, and that $\mathrm{lc}(q)$ is a solution in k^* of

$$Dy + by = \mathrm{lc}(c) \tag{6.23}$$

which is a Risch differential equation in k. If it has no solution in k, then (6.19) has no solution in $k[t]$. Otherwise, Lemma 5.9.1 implies that it has a unique solution which must then be $\mathrm{lc}(q)$. This gives the leading monomial $yt^{\deg(c)}$ of any solution q, and as earlier, replacing q by $yt^{\deg(c)} + h$ in (6.19) yields an equation of the same type with a lower degree bound on its solution, and a lower degree right hand side.

PolyRischDECancelPrim(b, c, D, n)
(* Poly Risch d.e., cancellation – primitive case *)

   (* Given a derivation D on $k[t]$, n either an integer or $+\infty$, $b \in k$ and
   $c \in k[t]$ with $Dt \in k$ and $b \neq 0$, return either "no solution", in which
   case the equation $Dq + bq = c$ has no solution of degree at most n in
   $k[t]$, or a solution $q \in k[t]$ of this equation with $\deg(q) \leq n$. *)

   if $b = Dz/z$ for $z \in k^*$ **then**
       if $zc = Dp$ for $p \in k[t]$ and $\deg(p) \leq n$ **then** **return**(p/z)
       else return "no solution"
   if $c = 0$ **then return** 0
   if $n < \deg(c)$ **then return** "no solution"
   $q \leftarrow 0$
   while $c \neq 0$ **do**
       $m \leftarrow \deg(c)$
       if $n < m$ **then return** "no solution"
       $s \leftarrow$ **RischDE**$(b, \mathrm{lc}(c))$ (* $Ds + bs = \mathrm{lc}(c)$ *)
       if $s = $ "no solution" **then return** "no solution"
       $q \leftarrow q + st^m$
       $n \leftarrow m - 1$
       $c \leftarrow c - bst^m - D(st^m)$ (* $\deg(c)$ becomes smaller *)
   return q

The Hyperexponential Case

If $Dt/t = \eta \in k$, then $\delta(t) = 1$, so the only cases not handled by Lemma 6.5.1 are $b = 0$ or $b \in k^*$. If $b = 0$, then (6.19) becomes $Dq = c$ for $c \in k[t]$, which is an integration problem in $k[t]$, and deciding whether it has a solution in $k[t]$ can be done by a variant of the integration algorithm (Sect. 5.12), so suppose now that $b \in k^*$.

   If $b = Du/u + m\eta$ for some $u \in k^*$ and $m \in \mathbb{Z}$, then (6.19) becomes $Dq + (Du/u + m\eta)q = c$, i.e. $D(uqt^m) = uct^m$ which is an integration problem in $k\langle t\rangle$, and deciding whether it has a solution in $k\langle t\rangle$ can be done by a variant of the integration algorithm (Sect. 5.12).

   Suppose finally that b is not of the form $Du/u + m\eta$ for some $u \in k^*$ and $m \in \mathbb{Z}$. Then $D(\mathrm{lc}(q)) + \deg(q)\, \eta \, \mathrm{lc}(q) + b\, \mathrm{lc}(q) \neq 0$, so the leading monomial of $Dq + bq$ is $(D(\mathrm{lc}(q)) + \deg(q)\,\eta\,\mathrm{lc}(q) + b\,\mathrm{lc}(q))t^{\deg(q)}$. This implies that $\deg(q) = \deg(c)$, and that $\mathrm{lc}(q)$ is a solution in k^* of

$$Dy + (b + \deg(q)\,\eta)\,y = \mathrm{lc}(c) \qquad (6.24)$$

which is a Risch differential equation in k. If it has no solution in k, then (6.19) has no solution in $k[t]$. Otherwise, Lemma 5.9.1 implies that it has a unique solution which must then be $\mathrm{lc}(q)$. This gives the leading monomial $yt^{\deg(c)}$ of any solution q, and as earlier, replacing q by $yt^{\deg(c)} + h$ in (6.19) yields an equation of the same type with a lower degree bound on its solution, and a lower degree right hand side.

PolyRischDECancelExp(b, c, D, n)
(* Poly Risch d.e., cancellation – hyperexponential case *)

    (* Given a derivation D on $k[t]$, n either an integer or $+\infty$, $b \in k$ and $c \in k[t]$ with $Dt/t \in k$ and $b \neq 0$, return either "no solution", in which case the equation $Dq + bq = c$ has no solution of degree at most n in $k[t]$, or a solution $q \in k[t]$ of this equation with $\deg(q) \leq n$. *)

    if $b = Dz/z + mDt/t$ for $z \in k^*$ and $m \in \mathbb{Z}$ **then**
        if $czt^m = Dp$ for $p \in k\langle t\rangle$ and $q = p/(zt^m) \in k[t]$ and $\deg(q) \leq n$
        then return(q)
      else return "no solution"
    if $c = 0$ **then return** 0
    if $n < \deg(c)$ **then return** "no solution"
    $q \leftarrow 0$
    while $c \neq 0$ **do**
        $m \leftarrow \deg(c)$
        if $n < m$ **then return** "no solution"
        $s \leftarrow$ **RischDE**$(b + mDt/t, \mathrm{lc}(c))$ (* $Ds + (b + mDt/t)s = \mathrm{lc}(c)$ *)
        if $s =$ "no solution" **then return** "no solution"
        $q \leftarrow q + st^m$
        $n \leftarrow m - 1$
        $c \leftarrow c - bst^m - D(st^m)$ (* $\deg(c)$ becomes smaller *)
    return q

The Nonlinear Case

If $\delta(t) \geq 2$, then we must have $\deg(b) = \delta(t) - 1$ and $\mathrm{lc}(b) = -n\lambda(t)$ where $n > 0$ is the bound on $\deg(q)$. There is no general algorithm for solving equation (6.19) in this case. If however $\mathcal{S}^{\mathrm{irr}} \neq \emptyset$, then the following can be done: for $p \in \mathcal{S}^{\mathrm{irr}}$, applying π_p to (6.19) and using the fact that $D^* \circ \pi_p = \pi_p \circ D$ where D^* is the induced derivation on $k[t]/(p)$ (Theorem 4.2.1), we get

$$D^*q^* + \pi_p(b)q^* = \pi_p(c) \qquad (6.25)$$

where $q^* = \pi_p(q)$. Assuming that we have an algorithm for solving (6.25) in $k[t]/(p)$, we can then solve (6.19) as follows: if (6.25) has no solution in $k[t]/(p)$, then (6.19) has no solution in $k[t]$. Otherwise, let $q^* \in k[t]/(p)$ be a solution of (6.25), and let $r \in k[t]$ be such that $\deg(r) < \deg(p)$ and $\pi_p(r) = q^*$. Note that $\pi_p(Dr + br) = \pi_p(c)$, so $p \mid c - Dr - br$. In addition, $\pi_p(q) = \pi_p(r)$, so $h = (q - r)/p \in k[t]$ and we have

$$c = Dq + bq = p\left(Dh + \left(b + \frac{Dp}{p}\right)h\right) + Dr + br$$

so h is a solution in $k[t]$ of degree at most $\deg(q) - \deg(p)$ of

$$Dh + \left(b + \frac{Dp}{p}\right)h = \frac{c - Dr - br}{p} \tag{6.26}$$

which is an equation of type (6.19), but with a lower bound on the degree of its solution.

There are cases when (6.25) can be solved, for example if there exists $p \in \mathcal{S}^{\mathrm{irr}}$ with $\deg(p) = 1$. Then, $k[t]/(p) \simeq k$, so (6.25) is a Risch differential equation in k. Another possibility is if $\mathcal{S}^{\mathrm{irr}} \cap \mathrm{Const}(k)[t] \neq \emptyset$, in which case taking $p = t - \alpha$ where α is a constant root of an irreducible special, we get $k[t]/(p) \simeq k(\alpha)$, so (6.25) is a Risch differential equation in $k(\alpha)$. This is the case when t is an hypertangent monomial with $\alpha = \pm\sqrt{-1}$. Taking $p = t - \alpha$ can also be done with α not constant, but (6.25) is then a Risch differential equation in a nonconstant algebraic extension of $k(t)$, and no algorithms are known for such curves when t is a nonlinear monomial. Although the techniques of [13, 58] are probably generalizable to such curves, they would not yield a practical algorithm in their current form.

The Hypertangent Case

If $Dt/(t^2 + 1) = \eta \in k$, then $\delta(t) = 2$, so the only case not handled by Lemma 6.5.1 is $b = b_0 - n\eta t$ where $b_0 \in k$ and $n > 0$ is the bound on $\deg(q)$. In such extensions, the method outlined above provides a complete algorithm: if $\sqrt{-1} \in k$, then $\mathcal{S}^{\mathrm{irr}} = \{t - \sqrt{-1}, t + \sqrt{-1}\}$, and (6.25) is simply a Risch differential equation over k.

If $\sqrt{-1} \notin k$, then taking $p = t^2 + 1 \in \mathcal{S}^{\mathrm{irr}}$, (6.25) becomes

$$Dq^* + (b_0 - n\eta\sqrt{-1})q^* = c(\sqrt{-1}) \tag{6.27}$$

where D is extended to $k[t]/(p) \simeq k(\sqrt{-1})$ by $D\sqrt{-1} = 0$. One possibility is to view (6.27) as a Risch differential equation in $k(\sqrt{-1})$ and to solve it recursively. If it has no solution in $k(\sqrt{-1})$, then (6.19) has no solution in $k[t]$. Otherwise, if $u + v\sqrt{-1}$ is a solution of (6.27) with $u, v \in k$, then letting $r = u + vt$, $h = (q - r)/p$ is a solution in $k[t]$ of degree at most $n - 2$ of (6.26). It is also possible however to avoid introducing $\sqrt{-1}$ by considering the real and imaginary parts of (6.27): writing $q^* = u + v\sqrt{-1}$, we get

$$\begin{pmatrix} Du \\ Dv \end{pmatrix} + \begin{pmatrix} b_0 & n\eta \\ -n\eta & b_0 \end{pmatrix} = \begin{pmatrix} c_0 \\ c_1 \end{pmatrix} \tag{6.28}$$

where $c_0 + c_1 t$ is the remainder of c by $t^2 + 1$. This is the coupled differential system introduced in Sect. 5.10. If it has no solution in k, then (6.19) has no solution in $k[t]$. Otherwise, if $(u, v) \in k^2$ is a solution of (6.28), then letting $r = u + vt$, $h = (q - r)/p$ is a solution in $k[t]$ of degree at most $n - 2$ of (6.26).

PolyRischDECancelTan(b_0, c, D, n)
(* Poly Risch d.e., degenerate cancellation – tangent case *)

(* Given a derivation D on $k[t]$, $n \in \mathbb{Z}$, $b_0 \in k$ and $c \in k[t]$ with $Dt/(t^2 + 1) \in k$, $\sqrt{-1} \notin k$ and $n \geq 0$, return either "no solution", in which case the equation $Dq + (b_0 - ntDt/(t^2 + 1))q = c$ has no solution of degree at most n in $k[t]$, or a solution $q \in k[t]$ of this equation with $\deg(q) \leq n$. *)

if $n = 0$ **then**
  if $c \in k$ **then**
    if $b_0 \neq 0$ **then return** **RischDE**(b_0, c)
    else if $\int c = q \in k$ **then return**(q) **else return** "no solution"
  else return "no solution"
$p \leftarrow t^2 + 1$ (* the monic irreducible special polynomial *)
$\eta \leftarrow Dt/p$ (* $t = \tan(\int \eta)$ *)
$(\bar{c}, c_1 t + c_0) \leftarrow$ **PolyDivide**(c, p) (* $c(\sqrt{-1}) = c_1\sqrt{-1} + c_0$ *)
(* CoupledDESystem will be given in Chap. 8 *)
$(u, v) \leftarrow$ **CoupledDESystem**$(b_0, -n\eta, c_0, c_1)$
(* $Du + b_0 u + n\eta v = c_0$, $Dv - n\eta u + b_0 v = c_1$ *)
if $(u, v) =$ "no solution" **then return** "no solution"
if $n = 1$ **then return**$(ut + v)$
$r \leftarrow u + vt$
$c \leftarrow (c - Dr - (b_0 - n\eta)r)/p$ (* this division is always exact *)
$h \leftarrow$ **PolyRischDECancelTan**$(b_0, c, D, n - 2)$
if $h =$ "no solution" **then return** "no solution"
return$(ph + r)$

Example 6.6.1. Continuing example 6.5.3, let $k = \mathbb{Q}(x)$ with $D = d/dx$, t be a monomial over k satisfying $Dt = 1 + t^2$, *i.e.* $t = \tan(x)$, and consider the solutions $q \in k[t]$ of degree at most 1 of (6.22). We have $b = 1 - t$, $b_0 = 1$, $c = -2(x + 1)t - 2x$, $n = 1$ and:

1. $p = t^2 + 1$
2. $\eta = Dt/p = 1$
3. $(\bar{c}, c_1 t + c_0) =$ **PolyDivide**$(-2(x + 1)t - 2x, p) = (0, -2(x + 1)t - 2x)$
4. Since

$$\begin{pmatrix} Du \\ Dv \end{pmatrix} + \begin{pmatrix} 1 & 1 \\ -1 & 1 \end{pmatrix} \begin{pmatrix} u \\ v \end{pmatrix} = \begin{pmatrix} -2x \\ -2(x + 1) \end{pmatrix}$$

has the solution $u = 0$ and $v = -2x$,
$$(u, v) = \textbf{CoupledDESystem}(1, -1, -2x, -2(x+1)) = (0, -2x).$$

Thus, $q = -2xt$ is a solution of degree at most 1 of (6.22). Going back to example 6.5.3, this implies that $y = t^2 - 2xt$ is a solution of (6.21), hence that

$$\int \left(\tan(x)^3 + \tan(x)^2 - 2x\tan(x) - 2x\right) e^{x - \log(1+\tan(x)^2)/2}\, dx =$$

$$\left(\tan(x)^2 - 2x\tan(x)\right) e^{x - \log(1+\tan(x)^2)/2}.$$

Note that the above can also be written as

$$\int \left(\tan(x)^3 + \tan(x)^2 - 2x\tan(x) - 2x\right) \cos(x)e^x\, dx =$$

$$\left(\tan(x)^2 - 2x\tan(x)\right) \cos(x)e^x.$$

Exercises

Exercise 6.1. Prove the following analogue of Lemma 6.2.4 for fields containing $\sqrt{-1}$: let k be a differential field of characteristic 0 containing $\sqrt{-1}$, t be a hypertangent over k such that $\eta = Dt/(t^2 + 1)$ is not the logarithmic derivative of a k-radical. Let $a \in k[t], b, q \in k\langle t\rangle$ be such that $\gcd(a, t^2 + 1) = 1$, $\nu_{t-\sqrt{-1}}(b) = \nu_{t+\sqrt{-1}}(b) = 0$. Let $\epsilon = \pm 1$ and suppose that $\nu_{t-\epsilon\sqrt{-1}}(q) \neq 0$. Then, either

$$\nu_{t-\epsilon\sqrt{-1}}(aDq + bq) = \nu_{t-\epsilon\sqrt{-1}}(q)$$

or

$$-\frac{b(\epsilon\sqrt{-1})}{a(\epsilon\sqrt{-1})} = 2\nu_{t-\epsilon\sqrt{-1}}(q)\,\eta\,\epsilon\,\sqrt{-1} + \frac{Du}{u}$$

for some $u \in k^*$.

7. Parametric Problems

We describe in this chapter solutions to several integration-related problems involving parameters. Those problems arise as subproblems in the integration algorithm: the limited integration problem, which arises from integrating polynomials in a primitive extension (Sect. 5.8) and the parametric logarithmic derivative problem, which arises from recognizing logarithmic derivatives (Sect. 5.12) and from bounding orders and degrees of solutions of the Risch differential equation (Sects. 6.2 and 6.3). The common thread between those problems is that they ask whether there exists constants for which a given parametric differential equation has a solution in a given differential field.

7.1 The Parametric Risch Differential Equation

We present first the classical parametric problem, namely the *parametric Risch differential equation*, which is a Risch differential equation where we replace the right hand side $g \in K$ by the linear combination $\sum_{i=1}^{m} c_i g_i$ with $g_i \in K$. The problem is then to determine all the constants $c_i \in \mathrm{Const}(K)$ for which the equation

$$Dy + fy = \sum_{i=1}^{m} c_i g_i \qquad (7.1)$$

has a solution in K, and of course to compute such solutions. This problem, which does not arise when we integrate only transcendental elementary functions, shows up in the integration of nonelementary functions, or in integration in terms of nonelementary functions [5, 20, 21, 38, 39]. In addition, the problem of limited integration can be seen as a special case of this problem. Note that the set of constants $(c_1, \ldots, c_m)$ for which (7.1) has a solution in K forms a linear subspace of $\mathrm{Const}(K)^m$. This motivates the following formal definition of the parametric Risch differential equation problem: given a differential field K of characteristic 0 and $f, g_1, \ldots, g_m \in K$, to compute $h_1, \ldots, h_r \in K$, a matrix A with $m + r$ columns and entries in $\mathrm{Const}(K)$ such that (7.1) has a solution $c_1, \ldots, c_m \in \mathrm{Const}(K)$ and $y \in K$ if and only if

$$y = \sum_{j=1}^{r} d_j h_j$$

where $d_1, \ldots, d_r \in \text{Const}(K)$ and

$$A \left(c_1, \ldots, c_m, d_1, \ldots, d_r\right)^T = 0 \,.$$

As in Chap. 6, we only study equation (7.1) in the transcendental case, *i.e.* when K is a simple monomial extension of a differential subfield k, so for the rest of this section, let k be a differential field of characteristic 0 and t be a monomial over k. We assume in addition that $\text{Const}(k(t)) = \text{Const}(k)$. We suppose that the coefficients f and $g_1, \ldots, g_m$ of our equation are in $k(t)$ and look for solutions $c_1, \ldots, c_m \in \text{Const}(k)$ and $y \in k(t)$. It turns out that the algorithms of Chap. 6 can be easily generalized to the parametric problem.

The Normal Part of the Denominator

Since $\nu_p(\sum_{i=1}^m c_i g_i) \geq \min_{1 \leq i \leq m}(\nu_p(g_i))$ for any irreducible $p \in k[t]$, part (i) of Theorem 6.1.2 generalizes to parametric equations. Of course, part (ii) does not generalize since the above inequality can be strict.

Theorem 7.1.1. *Let $f \in k(t)$ be weakly normalized with respect to t and $g_1, \ldots, g_m \in k(t)$. Let $c_1, \ldots, c_m \in \text{Const}(k)$ and $y \in k(t)$ be such that $Dy + fy = \sum_{i=1}^m c_i g_i$. Let $d = d_s d_n$ be a splitting factorization of the denominator of f, e be a least common multiple of the denominators of the g_i's, and $e = e_s e_n$ be a splitting factorization of e. Let $c = \gcd(d_n, e_n)$ and*

$$h = \frac{\gcd(e_n, de_n/dt)}{\gcd(c, dc/dt)} \in k[t] \,.$$

Then, $yh \in k\langle t \rangle$.

Proof. Let $q = yh \in k(t)$. In order to show that $q \in k\langle t \rangle$, we need to show that $\nu_p(q) \geq 0$ for any normal irreducible $p \in k[t]$. We have $\nu_p(q) = \nu_p(y) + \nu_p(h)$ by Theorem 4.1.1. If $\nu_p(y) \geq 0$, then $\nu_p(q) \geq \nu_p(h) \geq 0$ since $h \in k[t]$. So suppose now that $n = \nu_p(y) < 0$. Let $g = \sum_{i=1}^m c_i g_i \in k(t)$ and $\mu = \min_{1 \leq i \leq m}(\nu_p(g_i))$. We then have $\nu_p(g) \geq \mu$. In addition, $eg \in k[t]$ since e is a least common multiple of the denominators of the g_i's, so $\nu_p(e) + \nu_p(g) = \nu_p(eg) \geq 0$, which implies that $\nu_p(e) \geq -\nu_p(g)$.
Case 1: $\nu_p(f) \geq 0$. Then, $\nu_p(Dy + fy) = \nu_p(y) - 1$ by Lemma 6.1.1. Since $g = Dy + fy$, this implies that $\nu_p(g) < 0$, hence that $p \mid e$. Since p is normal, $\gcd(p, e_s) = 1$, so $\nu_p(e_n) = \nu_p(e) \geq -\nu_p(g) = 1 - n$.
Also, p does not divide d since $\nu_p(f) \geq 0$, so $\nu_p(c) = 0$, so $\nu_p(\gcd(c, dc/dt)) = 0$. Hence $\nu_p(h) = \nu_p(\gcd(e_n, de_n/dt)) = \nu_p(e_n) - 1 \geq -n$, so $\nu_p(q) = n + \nu_p(h) \geq n - n = 0$.
Case 2: $\nu_p(f) < 0$. Then, $\nu_p(g) = \nu_p(Dy + fy) = \nu_p(f) + n$ by Lemma 6.1.1, so $n = \nu_p(g) - \nu_p(f)$. Since $n < 0$, this implies that $\nu_p(g) < \nu_p(f) < 0$, hence that $p \mid d$ and $p \mid e$. As above, since p is normal, $\gcd(p, d_s) = \gcd(p, e_s) = 1$, so $\nu_p(d_n) = -\nu_p(f) < -\nu_p(g) \geq \nu_p(e_n)$. Thus, $\nu_p(c) = \min(\nu_p(d_n), \nu_p(e_n)) = \nu_p(d_n) = -\nu_p(f) > 0$, so

$$\nu_p(h) = \nu_p(\gcd(e_n, de_n/dt)) - \nu_p(\gcd(c, dc/dt))$$
$$= (\nu_p(e_n) - 1) - (\nu_p(c) - 1)$$
$$= \nu_p(e) + \nu_p(f) \geq -\nu_p(g) + \nu_p(f) = -n ,$$

so $\nu_p(q) = n + \nu_p(h) \geq n - n = 0$. $\square$

Corollary 7.1.1. *Let* $f \in k(t)$ *be weakly normalized with respect to* t, $g_1, \ldots, g_m \in k(t)$, *and* d_n, e_n *and* h *be as in Theorem 7.1.1. Then, for any solution* $c_1, \ldots, c_m \in \mathrm{Const}(k)$ *and* $y \in k(t)$ *of* $Dy + fy = \sum_{i=1}^m c_i g_i$, $q = yh \in k\langle t \rangle$ *and* q *is a solution of*

$$d_n h Dq + (d_n h f - d_n Dh) q = \sum_{i=1}^m c_i (d_n h^2 g_i) . \tag{7.2}$$

Conversely, for any solution $c_1, \ldots, c_m \in \mathrm{Const}(k)$ *and* $q \in k\langle t \rangle$ *of (7.2),* $y = q/h$ *is a solution of* $Dy + fy = \sum_{i=1}^m c_i g_i$.

Proof. Let $c_1, \ldots, c_m \in \mathrm{Const}(k)$ and $y \in k(t)$ be a solution of $Dy + fy = g$, and let $q = yh$. $q \in k\langle t \rangle$ by Theorem 7.1.1, and

$$Dq + \left(f - \frac{Dh}{h} \right) q = hDy + yDh + hfy - yDh = h(Dy + fy) = h \sum_{i=1}^m c_i g_i .$$

Multiplying through by $d_n h$ yields $d_n h Dq + (d_n h f - d_n Dh)q = d_n h^2 \sum_{i=1}^m c_i g_i$, so q is a solution of (7.2). Conversely, the same calculation shows that for any solution $c_1, \ldots, c_m \in \mathrm{Const}(k)$ and $q \in k\langle t \rangle$ of (7.2), $y = q/h$ is a solution of $Dy + fy = \sum_{i=1}^m c_i g_i$. $\square$

The above theorem and corollary give us an algorithm that reduces a given parametric Risch differential equation to one over $k\langle t \rangle$.

ParamRdeNormalDenominator$(f, g_1, \ldots, g_m, D)$
(* Normal part of the denominator *)

> (* Given a derivation D on $k[t]$ and $f, g_1, \ldots, g_m \in k(t)$ with f weakly normalized with respect to t, return the tuple $(a, b, G_1, \ldots, G_m, h)$ such that $a, h \in k[t]$, $b \in k\langle t \rangle$, $G_1, \ldots, G_m \in k(t)$, and for any solution $c_1, \ldots, c_m \in \mathrm{Const}(k)$ and $y \in k(t)$ of $Dy + fy = \sum_{i=1}^m c_i g_i$, $q = yh \in k\langle t \rangle$ satisfies $aDq + bq = \sum_{i=1}^m c_i G_i$. *)
>
> $(d_n, d_s) \leftarrow \mathbf{SplitFactor}(\mathrm{denominator}(f), D)$
> $(e_n, e_s) \leftarrow \mathbf{SplitFactor}(\mathrm{lcm}(\mathrm{denominator}(g_1), \ldots, \mathrm{denominator}(g_m)), D)$
> $p \leftarrow \gcd(d_n, e_n)$
> $h \leftarrow \gcd(e_n, de_n/dt)/\gcd(p, dp/dt)$
> $\mathbf{return}(d_n h, d_n h f - d_n Dh, d_n h^2 g_1, \ldots, d_n h^2 g_m, h)$

The Special Part of the Denominator

As a result of Corollary 7.1.1, we are now reduced to finding solutions $c_1, \ldots, c_m \in \mathrm{Const}(k)$ and $q \in k\langle t \rangle$ of (7.2), which we rewrite as

$$aDq + bq = \sum_{i=1}^{m} c_i g_i \qquad (7.3)$$

where $a \in k[t]$ has no special factor, $b \in k\langle t \rangle$, $g_1, \ldots, g_m \in k(t)$, $a \neq 0$, and t is a monomial over k. Since $\nu_p(\sum_{i=1}^{m} c_i g_i) \geq \min_{1 \leq i \leq m}(\nu_p(g_i))$ for any irreducible $p \in k[t]$, and since $a \in k[t]$ and has no special factor in (7.3), Lemma 6.2.1 provides a lower bound for $\nu_p(q)$ as in the nonparametric case:

(i) If $\nu_p(b) < 0$, then $\nu_p(q) \geq \min(0, \min_{1 \leq i \leq m}(\nu_p(g_i)) - \nu_p(b))$.
(ii) If $\nu_p(b) > 0$ and $p \in \mathcal{S}_1^{\mathrm{irr}}$, then $\nu_p(q) \geq \min(0, \min_{1 \leq i \leq m}(\nu_p(g_i)))$.

For $p \in \mathcal{S}^{\mathrm{irr}}$, once we have a lower bound $\nu_p(q) \geq n$ for some $n \leq 0$, replacing q by hp^n in (7.3) yields

$$a(p^n Dh + np^{n-1} h Dp) + bhp^n = \sum_{i=1}^{m} c_i g_i$$

hence

$$aDh + \left(b + na\frac{Dp}{p} \right) h = \sum_{i=1}^{m} c_i \left(g_i p^{-n} \right) . \qquad (7.4)$$

Furthermore, $h \in k\langle t \rangle$ since $q \in k\langle t \rangle$, and $h \in \mathcal{O}_p$ since $\nu_p(q) \geq n$. Thus we are reduced to finding the solutions $c_1, \ldots, c_m \in \mathrm{Const}(k)$ and $h \in k\langle t \rangle \cap \mathcal{O}_p$ of (7.4). Note that $b + naDp/p \in k\langle t \rangle$ since $b \in k\langle t \rangle$, $a \in k[t]$ and $p \in \mathcal{S}$. The eventual power of p in the denominator of $b + naDp/p$ can be cleared by multiplying (6.7) by p^N where $N = \max(0, -\nu_p(b))$, ensuring that the coefficients of the left hand side of (7.4) are also in $k\langle t \rangle \cap \mathcal{O}_p$.

Since all the special polynomials are of the first kind in the monomial extensions we are considering in this section, we only have to find a lower bound for $\nu_p(q)$ in the potential cancellation case, i.e. $\nu_p(b) = 0$. We consider this case separately for various kinds of monomial extensions.

The Primitive Case. If $Dt \in k$, then every squarefree polynomial is normal, so $k\langle t \rangle = k[t]$, which means that $a, b \in k[t]$ and any solution in $q \in k\langle t \rangle$ of (7.3) must be in $k[t]$.

The Hyperexponential Case. If $Dt/t = \eta \in k$, then $k\langle t \rangle = k[t, t^{-1}]$, so we need to compute a lower bound on $\nu_t(q)$ where $c_1, \ldots, c_m \in \mathrm{Const}(k)$ and $q \in k\langle t \rangle$ is a solution of (7.3). Since $t \in \mathcal{S}_1^{\mathrm{irr}}$ by Theorem 5.1.2, Lemmas 6.2.1 and 6.2.3 always provide a lower bound for $\nu_t(q)$: if $\nu_t(b) \neq 0$, then Lemma 6.2.1 provides the bound as explained earlier. Otherwise, $\nu_t(b) = 0$, so either $-b(0)/a(0) = s\eta + Du/u$ for some $s \in \mathbb{Z}$ and $u \in k^*$, in which case

$$\nu_t(q) \geq \min\left(0, s, \min_{1 \leq i \leq m}(\nu_t(g_i))\right).$$

Otherwise,

$$\nu_t(q) \geq \min\left(0, \min_{1 \leq i \leq m}(\nu_t(g_i))\right).$$

Note that such an s is unique by Lemma 6.2.2 applied to k. Since $\mathcal{S}^{\mathrm{irr}} = \{t\}$, $k\langle t \rangle \cap \mathcal{O}_t = k[t]$, so having determined a lower bound for $\nu_t(q)$, we are left with finding solutions $c_1, \ldots, c_m \in \mathrm{Const}(k)$ and $h \in k[t]$ of (7.4).

ParamRdeSpecialDenomExp$(a, b, g_1, \ldots, g_m, D)$
(* Special part of the denominator – hyperexponential case *)

(* Given a derivation D on $k[t]$ and $a \in k[t]$, $b \in k\langle t \rangle$ and $g_1, \ldots, g_m \in k\langle t \rangle$ with $Dt/t \in k$, $a \neq 0$ and $\gcd(a, t) = 1$, return the tuple $(\bar{a}, \bar{b}, \overline{g_1}, \ldots, \overline{g_m}, h)$ such that $\bar{a}, \bar{b}, h \in k[t]$, $\overline{g_1}, \ldots, \overline{g_m} \in k\langle t \rangle$, and for any solution $c_1, \ldots, c_m \in \mathrm{Const}(k)$ and $q \in k\langle t \rangle$ of $aDq + bq = \sum_{i=1}^{m} c_i g_i$, $r = qh \in k[t]$ satisfies $\bar{a}Dr + \bar{b}r = \sum_{i=1}^{m} c_i \overline{g_i}$. *)

$p \leftarrow t$ (* the monic irreducible special polynomial *)
$n_b \leftarrow \nu_p(b)$, $n_c \leftarrow \min_{1 \leq i \leq m}(\nu_p(g_i))$
$n \leftarrow \min(0, n_c - \min(0, n_b))$ (* $n \leq 0$ *)
if $n_b = 0$ **then**   (* possible cancellation *)
    $\alpha \leftarrow$ **Remainder**$(-b/a, p)$ (* $\alpha = -b(0)/a(0) \in k$ *)
    if $\alpha = sDt/t + Dz/z$ for $z \in k^*$ and $s \in \mathbb{Z}$ **then** $n \leftarrow \min(n, s)$
$N \leftarrow \max(0, -n_b)$ (* $N \geq 0$, for clearing denominators *)
return$(ap^N, (b + naDp/p)p^N, g_1 p^{N-n}, \ldots, g_m p^{N-n}, p^{-n})$

The Hypertangent Case. If $Dt/(t^2 + 1) = \eta \in k$ and $\sqrt{-1} \notin k$, then the only monic special irreducible is $t^2 + 1$, so we need to compute a lower bound on $\nu_{t^2+1}(q)$, where $c_1, \ldots, c_m \in \mathrm{Const}(k)$ and $q \in k\langle t \rangle$ is a solution of (7.3). Since $t^2 + 1 \in \mathcal{S}_1^{\mathrm{irr}}$ by Theorem 5.10.1, Lemmas 6.2.1 and 6.2.4 always provide a lower bound for $\nu_{t^2+1}(q)$: if $\nu_{t^2+1}(b) \neq 0$, then Lemma 6.2.1 provides the bound as explained earlier. Otherwise, $\nu_{t^2+1}(b) = 0$, so either $-b(\sqrt{-1})/a(\sqrt{-1}) = s\,\eta\,\sqrt{-1} + Du/u$ for some $s \in \mathbb{Z}$ and $u \in k(\sqrt{-1})^*$, in which case

$$\nu_{t^2+1}(q) \geq \min\left(0, s, \min_{1 \leq i \leq m}(\nu_{t^2+1}(g_i))\right).$$

Otherwise,

$$\nu_{t^2+1}(q) \geq \min\left(0, \min_{1 \leq i \leq m}(\nu_{t^2+1}(g_i))\right).$$

Note that such an s is unique by Lemma 6.2.2 applied to $k(\sqrt{-1})$. The remarks made in the nonparametric case about adjoining $\sqrt{-1}$ temporarily remain valid in this case. Since $\mathcal{S}^{\mathrm{irr}} = \{t^2 + 1\}$, $k\langle t \rangle \cap \mathcal{O}_{t^2+1} = k[t]$, so having determined a lower bound for $\nu_{t^2+1}(q)$, we are left with finding solutions $c_1, \ldots, c_m \in \mathrm{Const}(k)$ and $h \in k[t]$ of (7.4).

ParamRdeSpecialDenomTan$(a, b, g_1, \ldots, g_m, D)$
(* Special part of the denominator – hypertangent case *)

(* Given a derivation D on $k[t]$ and $a \in k[t]$, $b \in k\langle t \rangle$ and $g_1, \ldots, g_m \in$
$k(t)$ with $Dt/(t^2 + 1) \in k$, $\sqrt{-1} \notin k$, $a \neq 0$ and $\gcd(a, t^2 + 1) = 1$, return
the tuple $(\bar{a}, \bar{b}, \overline{g_1}, \ldots, \overline{g_m}, h)$ such that $\bar{a}, \bar{b}, h \in k[t]$, $\overline{g_1}, \ldots, \overline{g_m} \in k(t)$,
and for any solution $c_1, \ldots, c_m \in \text{Const}(k)$ and $q \in k\langle t \rangle$ of $aDq + bq =$
$\sum_{i=1}^{m} c_i g_i$, $r = qh \in k[t]$ satisfies $\bar{a} Dr + \bar{b} r = \sum_{i=1}^{m} c_i \overline{g_i}$. *)

$p \leftarrow t^2 + 1$ (* the monic irreducible special polynomial *)
$n_b \leftarrow \nu_p(b)$, $n_c \leftarrow \min_{1 \leq i \leq m}(\nu_p(g_i))$
$n \leftarrow \min(0, n_c - \min(0, n_b))$ (* $n \leq 0$ *)
if $n_b = 0$ **then**   (* possible cancellation *)
    $\alpha\sqrt{-1} + \beta \leftarrow$ **Remainder**$(-b/a, p)$ (* $\alpha, \beta \in k$ *)
    $\eta \leftarrow Dt/(t^2 + 1)$ (* $\eta \in k$ *)
    if $2\beta = Dv/v$ for $v \in k^*$
        and $\alpha\sqrt{-1} + \beta = 2s\eta\sqrt{-1} + Dz/z$ for $z \in k(\sqrt{-1})^*$ and $s \in \mathbb{Z}$
        then $n \leftarrow \min(n, s)$
$N \leftarrow \max(0, -n_b)$ (* $N \geq 0$, for clearing denominators *)
return$(ap^N, (b + naDp/p)p^N, g_1 p^{N-n}, \ldots, g_m p^{N-n}, p^{-n})$

The Linear Constraints on the Constants

As a result of the previous paragraphs, we are now reduced to finding solutions $c_1, \ldots, c_m \in \text{Const}(k)$ and $q \in k[t]$ of (7.4), which we rewrite as:

$$aDq + bq = c_1 g_1 + \ldots + c_m g_m \tag{7.5}$$

where $a, b \in k[t]$, $g_1 \ldots, g_m \in k(t)$, $a \neq 0$, and t is a monomial over k. In addition, dividing (7.5) by $\gcd(a, b)$ if needed, we can assume without loss of generality that $\gcd(a, b) = 1$. We show that if any v_i is not in $k[t]$, then we can obtain linear constraints on the c_i's, and reduce (7.5) to a similar equation with the right hand side in $k[t]$.

Lemma 7.1.1. *Let $a, b, q \in k[t]$, $g_1, \ldots, g_m \in k(t)$ and $c_1, \ldots, c_m \in \text{Const}(k)$
be such that $aDq + bq = c_1 g_1 + \ldots + c_m g_m$. Let d_i be the denominator of g_i
for $1 \leq i \leq m$, $d = \text{lcm}(d_1, \ldots, d_m)$, and $q_1, \ldots, q_m, r_1, \ldots, r_m$ be such that
$dg_i = dq_i + r_i$ and either $r_i = 0$ or $\deg(r_i) < \deg(d)$ for each i. Then,*

$$\sum_{i=1}^{m} c_i r_i = 0 \tag{7.6}$$

and

$$aDq + bq = c_1 q_1 + \ldots + c_m q_m . \tag{7.7}$$

Proof. Since $g_i = q_i + r_i/d$ for each i, we obtain from (7.5) that

$$\frac{\sum_{i=1}^{m} c_i r_i}{d} = aDq + bq - \sum_{i-1}^{m} c_i q_i \in k[t] .$$

Since $\deg\left(\sum_{i=1}^{m} c_i r_i\right) < \deg(d)$, it follows that $\sum_{i=1}^{m} c_i r_i$ must be equal to 0, hence that $aDq + bq = \sum_{i=1}^{m} c_i q_i$.   □

Equating the coefficients of the powers of t on both sides of (7.6) yields a homogeneous system of linear equations for the c_i's, *i.e.* a matrix M with coefficients in $k(t)$ such that

$$M \begin{pmatrix} c_1 \\ c_2 \\ \vdots \\ c_m \end{pmatrix} = 0. \tag{7.8}$$

LinearConstraints$(a, b, g_1, \ldots, g_m, D)$
(* Generate linear constraints on the constants *)

(* Given a derivation D on $k(t)$, $a, b \in k[t]$ and $g_1, \ldots, g_m \in k(t)$, return $q_1, \ldots, q_m \in k[t]$ and a matrix M with entries in $k(t)$ such that for any solution $c_1, \ldots, c_m \in \mathrm{Const}(k)$ and $p \in k[t]$ of $aDp + bp = c_1 g_1 + \ldots + c_m g_m$, $(c_1, \ldots, c_m)$ is a solution of $Mx = 0$, and p and the c_i satisfy $aDp + bp = c_1 q_1 + \ldots + c_m q_m$. *)

$d \leftarrow \mathrm{lcm}(\mathrm{denominator}(g_1), \ldots, \mathrm{denominator}(g_m))$
for $i \leftarrow 1$ **to** m **do** $(q_i, r_i) \leftarrow$ **PolyDivide**(dv_i, d) (* $dv_i = q_i d + r_i$ *)
if $r_1 = \ldots = r_m = 0$ **then** $n = -1$ **else** $n \leftarrow \max(\deg(r_1), \ldots, \deg(r_m))$
for $i \leftarrow 0$ **to** n **do for** $j \leftarrow 1$ **to** m **do** $M_{ij} \leftarrow$ **coefficient**(r_j, t^i)
return$(q_1, \ldots, q_m, M)$

Example 7.1.1. Let $k = \mathbb{Q}$, t be a monomial over k satisfying $Dt = 1$, *i.e.* $D = d/dt$, and consider the equation

$$Dp = c_1 \frac{2t^3 + 3t + 1}{t^2 - 1} + c_2 \frac{1}{t - 1} + c_3 \frac{1}{t + 1}. \tag{7.9}$$

We have $a = 1$, $b = 0$, $g_1 = (2t^3 + 3t + 1)/(t^2 - 1)$, $g_2 = 1/(t-1)$, $g_3 = 1/(t+1)$ and:

1. $d = \mathrm{lcm}(t^2 - 1, t - 1, t + 1) = t^2 - 1$
2. $dg_1 = 2t^3 + 3t + 1 = 2td + 5t + 1$, $dg_2 = t + 1$, $dg_3 = t - 1$, so $q_1 = 2t$, $q_2 = q_3 = 0$, $r_1 = 5t + 1$, $r_2 = t + 1$ and $r_3 = t - 1$.
3. Equation (7.6) becomes $c_1(5t + 1) + c_2(t + 1) + c_3(t - 1) = 0$, which yields the linear system

$$\begin{pmatrix} 5 & 1 & 1 \\ 1 & 1 & -1 \end{pmatrix} \begin{pmatrix} c_1 \\ c_2 \\ c_3 \end{pmatrix} = 0 \tag{7.10}$$

which has the solution space $(c_1, c_2, c_3) = (\lambda, -3\lambda, -2\lambda)$ for any $\lambda \in \mathbb{Q}$.

4. Replacing c_1, c_2 and c_3 by the above solution in (7.9) yields

$$Dp = \lambda \left(\frac{2t^3 + 3t + 1}{t^2 - 1} - \frac{3}{t-1} - \frac{2}{t+1} \right) = 2\lambda t \qquad (7.11)$$

which is now a parametric Risch differential equation with polynomial right-hand side.

Since we are interested only in the constant solutions of (7.8), we need to reduce it to an equivalent system with coefficients in Const(k). An algorithm for this reduction is provided in the following lemma.

Lemma 7.1.2. *Let (K, D) be a differential field, A be a matrix with coefficients in K, and $\mathbf{u}$ be a vector with coefficients in K. Then, using only elementary row operations on A and $\mathbf{u}$, we can either prove that $Ax = \mathbf{u}$ has no constant solution, or we can compute a matrix B and a vector $\mathbf{v}$, both with coefficients in $\text{Const}(K)$, such that the constant solutions of $Ax = \mathbf{u}$ are exactly all the solutions of $Bx = \mathbf{v}$. Furthermore, if $\mathbf{u} = 0$, then $\mathbf{v} = 0$.*

Proof. Let $C = \text{Const}(K)$, and write R_i for the ith row of A, and a_{ij} for the jth entry of R_i. By applying the usual Gaussian elimination, we can compute an equivalent system in row-reduced echelon form, so suppose that A is in that form. If all the entries of A are in C, let $B = A$ and $\mathbf{v} = \mathbf{u}$. Otherwise, let j be the smallest index such that the jth column of A has a non-constant entry, and let i be such that $a_{ij} \notin C$. Then, $Da_{ij} \neq 0$, so we add the row

$$R_{m+1} = \frac{DR_i}{Da_{ij}} = \left(\frac{Da_{i1}}{Da_{ij}}, \ldots, \frac{Da_{ir}}{Da_{ij}} \right)$$

at the bottom of A, and the entry $u_{m+1} = Du_i/Da_{ij}$ at the bottom of $\mathbf{u}$. By our choice of j, the first nonzero entry in R_{m+1} is a 1 in column j, so we add adequate multiples of R_{m+1} to all the other rows to ensure that $a_{ij} = 0$ for $i = 1 \ldots m$. We now have a new matrix $\tilde{A}$ and a new vector $\tilde{\mathbf{u}}$ with one more row, but with only constant entries in columns 1 through j. Repeating this, we eventually obtain a matrix B and a vector $\mathbf{v}$ such that all the entries of B are in C. By construction, $\mathbf{v} = 0$ if $\mathbf{u} = 0$. Since we have only added extra rows to A and performed elementary row operations to A, any solution of $Ax = \mathbf{u}$ must be a solution of $Bx = \mathbf{v}$.

Case 1, $\mathbf{v}$ *has a nonconstant entry:* let x be a constant solution of $Ax = \mathbf{u}$. Then all the entries of Bx are constant, in contradiction with $Bx = \mathbf{v}$. Hence $Ax = \mathbf{u}$ has no constant solution if $\mathbf{v}$ has a nonconstant entry.

Case 2, all the entries of $\mathbf{v}$ *are in C:* we have already seen that any constant solution of $Ax = \mathbf{u}$ must be a solution of $Bx = \mathbf{v}$. Conversely, let x be a solution of $Bx = \mathbf{v}$. Then all the entries of x are in C, since B and $\mathbf{v}$ are both constant. In order for x to satisfy $Ax = \mathbf{u}$, it only has to satisfy $R_{m+1}x = u_{m+1}$, where R_{m+1} is the extra row added in the reduction step. But,

$$R_{m+1}x = \frac{(Da_{i1})x_1 + \ldots + (Da_{ir})x_r}{Da_{ij}} = \frac{D\left(a_{i1}x_1 + \ldots + a_{ir}x_r\right)}{Da_{ij}}$$

$$= \frac{D\left(R_i x\right)}{Da_{ij}} = \frac{Du_i}{Da_{ij}} = u_{m+1}$$

so x is a constant solution of $Ax = \mathbf{u}$.   □

ConstantSystem$(M, \mathbf{u}, D)$
(* Generate a system for the constant solutions *)

(* Given a differential field (K, D) with constant field C, a matrix A and a vector $\mathbf{u}$ with coefficients in K, returns a matrix B with coefficients in C and a vector $\mathbf{v}$ such that either $\mathbf{v}$ has coefficients in C, in which case the solutions in C of $Ax = \mathbf{u}$ are exactly all the solutions of $Bx = \mathbf{v}$, or $\mathbf{v}$ has a nonconstant coefficient, in which case $Ax = \mathbf{u}$ has no constant solution. *)

$(A, \mathbf{u}) \leftarrow \mathbf{RowEchelon}(A, \mathbf{u})$
$m \leftarrow$ number of rows of A
while A is not constant **do**
    $j \leftarrow$ minimal index such that that the jth column of A is not constant
    $i \leftarrow$ any index such that $a_{ij} \notin C$,
    $R_i \leftarrow i^{\text{th}}$ row of A
    $R_{m+1} = D(R_i)/D(a_{ij})$, $u_{m+1} \leftarrow D(u_i)/D(a_{ij})$
    for $s \leftarrow 1$ **to** m **do**
        $R_s \leftarrow R_s - a_{sj} R_{m+1}$
        $u_s \leftarrow u_s - a_{sj} u_{m+1}$
    $A \leftarrow A \cup R_{m+1}, \mathbf{u} \leftarrow \mathbf{u} \cup u_{m+1}$ (* vertical concatenation *)
$\text{return}(A, \mathbf{u})$

Example 7.1.2. Let $k = \mathbb{Q}(x)$ with $D = d/dx$, and consider the system

$$Ax = \mathbf{u} \text{ where } A = \begin{pmatrix} -\frac{x+3}{x-1} & \frac{x+1}{x-1} & 1 \\ -x-3 & x+1 & x-1 \\ 2\frac{x+3}{x-1} & 0 & 0 \end{pmatrix} \text{ and } \mathbf{u} = \begin{pmatrix} \frac{x+1}{x-1} \\ x+1 \\ 0 \end{pmatrix} \quad (7.12)$$

1. RowEchelon$(A, \mathbf{u})$ yields

$$A = \begin{pmatrix} 1 & 0 & 0 \\ 0 & 1 & \frac{x-1}{x+1} \\ 0 & 0 & 0 \end{pmatrix} \quad \text{and} \quad \mathbf{u} = \begin{pmatrix} 0 \\ 1 \\ 0 \end{pmatrix}$$

2. $j = 3$, $i = 2$, $R_2 = \begin{pmatrix} 0 & 1 & \frac{x-1}{x+1} \end{pmatrix}$,
3.

$$R_4 = \frac{DR_2}{D((x-1)/(x+1))} = \begin{pmatrix} 0 & 0 & 1 \end{pmatrix}, u_4 = \frac{Du_2}{D((x-1)/(x+1))} = 0.$$

4. Adding (R_4, u_4) to $(A, \mathbf{u})$ yields

$$A = \begin{pmatrix} 1 & 0 & 0 \\ 0 & 1 & \frac{x-1}{x+1} \\ 0 & 0 & 0 \\ 0 & 0 & 1 \end{pmatrix} \quad \text{and} \quad \mathbf{u} = \begin{pmatrix} 0 \\ 1 \\ 0 \\ 0 \end{pmatrix}$$

5. Finally, adding $-(x-1)/(x+1)R_4$ to R_2 yields

$$A = \begin{pmatrix} 1 & 0 & 0 \\ 0 & 1 & 0 \\ 0 & 0 & 0 \\ 0 & 0 & 1 \end{pmatrix} \quad \text{and} \quad \mathbf{u} = \begin{pmatrix} 0 \\ 1 \\ 0 \\ 0 \end{pmatrix}$$

which both have constant entries. The above constant system has the unique solution

$$x = \begin{pmatrix} 0 \\ 1 \\ 0 \end{pmatrix}$$

which is thus the unique constant solution of (7.12). Note that (7.12) has a one-dimensional affine space of solutions over $\mathbb{Q}(x)$, namely

$$x = \begin{pmatrix} 0 \\ 1 \\ 0 \end{pmatrix} + w \begin{pmatrix} 0 \\ \frac{1-x}{1+x} \\ 1 \end{pmatrix} \quad \text{for any } w \in \mathbb{Q}(x).$$

Using Lemmas 7.1.1 and 7.1.2, we can produce a constant homogeneous linear system for the c_i's. If its kernel has dimension 0, then the only solution of 7.5 is $q = c_1 = \ldots = c_m = 0$. Otherwise, a basis of its kernel allows us to express some of the c_i's in terms of others, thereby decreasing m and reducing the problem to solving equation (7.7).

Degree Bounds

As a result of Lemma 7.1.1, we are now reduced to finding solutions $c_1, \ldots, c_m$ in $\mathrm{Const}(k)$ and $q \in k[t]$ of (7.7) where $a, b, q_1, \ldots, q_m \in k[t]$, $a \neq 0$, and t is a monomial over k. Since $\deg(\sum_{i=1}^{m} c_i q_i) \leq \max_{1 \leq i \leq m}(\deg(q_i))$, Lemma 6.3.1 provides an upper bound for $\deg(q)$ as in the nonparametric case:

(i) If $\deg(b) > \deg(a) + \max(0, \delta(t) - 1)$, then

$$\deg(q) \leq \max(0, \max_{1 \leq i \leq m} (\deg(q_i)) - \deg(b)).$$

(ii) If $\deg(b) < \deg(a) + \delta(t) - 1$ and $\delta(t) \geq 2$, then

$$\deg(q) \leq \max(0, \max_{1 \leq i \leq m} (\deg(q_i)) - \deg(a) + 1 - \delta(t)).$$

As a result, we only have to consider the cases $\deg(b) \leq \deg(a)$ for Louvillian monomials, and $\deg(b) = \deg(a) + \delta(t) - 1$ for nonlinear monomials. We consider those cases separately for various kinds of monomial extensions.

The Primitive Case. If $Dt = \eta \in k$, then Lemmas 6.3.1 and 6.3.3 always provide an upper bound for $\deg(q)$ as in the nonparametric case: if $\deg(b) > \deg(a)$, then Lemma 6.3.1 implies that

$$\deg(q) \geq \max(0, \max_{1 \leq i \leq m} (\deg(q_i)) - \deg(b)).$$

If $\deg(b) < \deg(a) - 1$, then Lemma 6.3.3 implies that

$$\deg(q) \geq \max(0, \max_{1 \leq i \leq m} (\deg(q_i)) - \deg(a) + 1).$$

If $\deg(b) = \deg(a) - 1$, then either $-\mathrm{lc}(b)/\mathrm{lc}(a) = s\eta + Du$ for some $s \in \mathbb{Z}$ and $u \in k$, in which case

$$\deg(q) \geq \max(0, s, \max_{1 \leq i \leq m} (\deg(q_i)) - \deg(a) + 1).$$

Otherwise, $\deg(q) \geq \max(0, \max_{1 \leq i \leq m}(\deg(q_i)) - \deg(a) + 1)$. Note that such an s is unique by Lemma 6.3.2. Finally, if $\deg(b) = \deg(a)$, then either $-\mathrm{lc}(b)/\mathrm{lc}(a) = Du/u$ for some $u \in k^*$ and $-\mathrm{lc}(aDu + bu)/(u\,\mathrm{lc}(a)) = s\eta + Dv$ for some $s \in \mathbb{Z}$ and $v \in k$, in which case

$$\deg(q) \geq \max(0, s, \max_{1 \leq i \leq m} (\deg(q_i)) - \deg(a) + 1).$$

Otherwise, $\deg(q) \geq \max(0, \max_{1 \leq i \leq m}(\deg(q_i)) - \deg(a) + 1)$. We can compute such an u by a variant of the integration algorithm (Sect. 5.12). Lemma 6.3.2 implies that the choice of u does not affect s.

ParamRdeBoundDegreePrim$(a, b, q_1, \ldots, q_m, D)$
(* Bound on polynomial solutions – primitive case *)

(* Given a derivation D on $k[t]$ and $a, b, q_1, \ldots, q_m \in k[t]$ with $Dt \in k$ and $a \neq 0$, return $n \in \mathbb{Z}$ such that $\deg(q) \leq n$ for any solution $c_1, \ldots, c_m \in \mathrm{Const}(k)$ and $q \in k[t]$ of $aDq + bq = \sum_{i=1}^{m} c_i q_i$. *)

$d_a \leftarrow \deg(a)$, $d_b \leftarrow \deg(b)$, $d_c \leftarrow \max_{1 \leq i \leq m}(\deg(q_i))$
if $d_b > d_a$ then $n \leftarrow \max(0, d_c - d_b)$ else $n \leftarrow \max(0, d_c - d_a + 1)$
if $d_b = d_a - 1$ then   (* possible cancellation *)
  $\alpha \leftarrow -\mathrm{lc}(b)/\mathrm{lc}(a)$
  if $\alpha = sDt + Dz$ for $z \in k$ and $s \in \mathbb{Z}$ then $n \leftarrow \max(n, s)$
if $d_b = d_a$ then   (* possible cancellation *)
  $\alpha \leftarrow -\mathrm{lc}(b)/\mathrm{lc}(a)$
  if $\alpha = Dz/z$ for $z \in k^*$ then
    $\beta \leftarrow -\mathrm{lc}(aDz + bz)/(z\,\mathrm{lc}(a))$
    if $\beta = sDt + Dw$ for $w \in k$ and $s \in \mathbb{Z}$ then $n \leftarrow \max(n, s)$
return n

In the specific case where $D = d/dt$, Corollary 6.3.1 yields a simpler algorithm, as in the nonparametric case.

ParamRdeBoundDegreeBase$(a, b, q_1, \ldots, q_m)$
(* Bound on polynomial solutions – base case *)

    (* Given $a, b, q_1, \ldots, q_m \in k[t]$ with $a \neq 0$, return $n \in \mathbb{Z}$ such that $\deg(q) \leq n$ for any solution $c_1, \ldots, c_m \in k$ and $q \in k[t]$ of

$$a\frac{dq}{dt} + bq = \sum_{i=1}^{m} c_i q_i \,.$$

    *)

$d_a \leftarrow \deg(a)$, $d_b \leftarrow \deg(b)$, $d_c \leftarrow \max_{1 \leq i \leq m}(\deg(q_i))$
$n \leftarrow \max(0, d_c - \max(d_b, d_a - 1))$
if $d_b = d_a - 1$ **then**   (* possible cancellation *)
    $s \leftarrow -\mathrm{lc}(b)/\mathrm{lc}(a)$
    if $s \in \mathbb{Z}$ **then** $n \leftarrow \max(0, s, d_c - d_b)$
return n

Example 7.1.3. Let $k = \mathbb{Q}$, t be a monomial over k satisfying $Dt = 1$, *i.e.* $D = d/dt$, N be a positive integer, and consider the parametric Risch differential equation

$$Dy + y = c_1 t^N + c_2 \,. \tag{7.13}$$

We have $f = 1$, $g_1 = t^N$ and $g_2 = 1$, so by Theorem 7.1.1, any solution $y \in k(t)$ must be in $k\langle t \rangle = k[t]$. We then have $a = b = 1$, so $d_a = d_b = 0$, $d_c = \max(N, 0) = N$ and $n = \max(0, N - \max(0, -1)) = N$, which implies that any solution $y \in k[t]$ of (7.13) has degree at most N.

The Hyperexponential Case. If $Dt/t = \eta \in k$, then Lemmas 6.3.1 and 6.3.4 always provide an upper bound for $\deg(q)$ as in the nonparametric case: if $\deg(b) > \deg(a)$, then Lemma 6.3.1 implies that

$$\deg(q) \geq \max(0, \max_{1 \leq i \leq m}(\deg(q_i)) - \deg(b)) \,.$$

If $\deg(b) < \deg(a)$, then Lemma 6.3.4 implies that

$$\deg(q) \geq \max(0, \max_{1 \leq i \leq m}(\deg(q_i)) - \deg(a)) \,.$$

At last, if $\deg(a) = \deg(b)$, then either $-\mathrm{lc}(b)/\mathrm{lc}(a) = s\eta + Du/u$ for some $s \in \mathbb{Z}$ and $u \in k^*$, in which case

$$\deg(q) \geq \max(0, s, \max_{1 \leq i \leq m}(\deg(q_i)) - \deg(b)) \,,$$

or

$$\deg(q) \geq \max(0, \max_{1 \leq i \leq m}(\deg(q_i)) - \deg(b)) \,.$$

Note that such an s is unique by Lemma 6.2.2.

ParamRdeBoundDegreeExp$(a, b, q_1, \ldots, q_m, D)$
(* Bound on polynomial solutions – hyperexponential case *)

(* Given a derivation D on $k[t]$ and $a, b, q_1, \ldots, q_m \in k[t]$ with $Dt/t \in k$ and $a \neq 0$, return $n \in \mathbb{Z}$ such that $\deg(q) \leq n$ for any solution $c_1, \ldots, c_m \in \text{Const}(k)$ and $q \in k[t]$ of

$$aDq + bq = \sum_{i=1}^{m} c_i q_i \,.$$

*)

$d_a \leftarrow \deg(a), \; d_b \leftarrow \deg(b), \; d_c \leftarrow \max_{1 \leq i \leq m}(\deg(q_i))$
$n \leftarrow \max(0, d_c - \max(d_b, d_a))$ (* $n \geq 0$ *)
if $d_a = d_b$ **then**   (* possible cancellation *)
     $\alpha \leftarrow -\text{lc}(b)/\text{lc}(a)$
     if $\alpha = sDt/t + Dz/z$ for $z \in k^*$ and $s \in \mathbb{Z}$ **then** $n \leftarrow \max(n, s)$
return n

The Nonlinear Case. If $\delta(t) \geq 2$, then Lemmas 6.3.1 and 6.3.5 always provide an upper bound for $\deg(q)$ as in the nonparametric case: if $\deg(b) \neq \deg(a) + \delta(t) - 1$, then Lemma 6.3.1 provides the bound as explained earlier. Otherwise, either $-\text{lc}(b)/\text{lc}(a) = s\lambda(t)$ for some $s \in \mathbb{Z}$, in which case

$$\deg(q) \geq \max(0, s, \max_{1 \leq i \leq m}(\deg(q_i)) - \deg(b)) \,,$$

or

$$\deg(q) \geq \max(0, \max_{1 \leq i \leq m}(\deg(q_i)) - \deg(b)) \,.$$

ParamRdeBoundDegreeNonLinear$(a, b, q_1, \ldots, q_m, D)$
(* Bound on polynomial solutions – nonlinear case *)

(* Given a derivation D on $k[t]$ and $a, b, q_1, \ldots, q_m \in k[t]$ with $\deg(Dt) \geq 2$ and $a \neq 0$, return $n \in \mathbb{Z}$ such that $\deg(q) \leq n$ for any solution $c_1, \ldots, c_m \in \text{Const}(k)$ and $q \in k[t]$ of $aDq + bq = \sum_{i=1}^{m} c_i q_i$. *)

$d_a \leftarrow \deg(a), \; d_b \leftarrow \deg(b), \; d_c \leftarrow \max_{1 \leq i \leq m}(\deg(q_i))$
$\delta \leftarrow \deg(Dt), \; \lambda \leftarrow \text{lc}(Dt)$
$n \leftarrow \max(0, d_c - \max(d_a + \delta - 1, d_b))$
if $d_b = d_a + \delta - 1$ **then**   (* possible cancellation *)
     $s \leftarrow -\text{lc}(b)/(\lambda \, \text{lc}(a))$
     if $s \in \mathbb{Z}$ **then** $n \leftarrow \max(0, s, d_c - d_b)$
return n

The Parametric SPDE Algorithm

We are now reduced to finding solutions $c_1, \ldots, c_m$ in $\mathrm{Const}(k)$ and $q \in k[t]$ of (7.7) and we have an upper bound n on $\deg(q)$. Theorem 6.4.1 and the SPDE algorithm of Sect. 6.4 generalize to the parametric case.

Theorem 7.1.2. *Let $a, b, q_1, \ldots, q_m \in k[t]$ with $a \neq 0$ and $\gcd(a, b) = 1$. Let $z_1, \ldots, z_m, r_1, \ldots, r_m \in k[t]$ be such that for each i, $q_i = az_i + br_i$ and either $r_i = 0$ or $\deg(r_i) < \deg(a)$, and let $r = \sum_{i=1}^{m} c_i r_i$. Then, for any solution $c_1, \ldots, c_m \in \mathrm{Const}(k)$ and $q \in k[t]$ of $aDq + bq = \sum_{i=1}^{m} c_i q_i$, $p = (q-r)/a \in k[t]$, and p is a solution of*

$$aDp + (b + Da)p = c_1(z_1 - Dr_1) + \ldots + c_m(z_m - Dr_m). \qquad (7.14)$$

Conversely, for any solution $c_1, \ldots, c_m \in \mathrm{Const}(k)$ and $p \in k[t]$ of (7.14), $q = ap + r$ is a solution of (7.7).

Proof. Let $c_1, \ldots, c_m \in \mathrm{Const}(k)$ and $q \in k[t]$ be a solution of (7.7). Then,

$$aDq + bq = a \sum_{i=1}^{m} c_i z_i + br$$

so $b(q - r) = a\left(\sum_{i=1}^{m} c_i z_i - Dq\right)$, so $a \mid b(q-r)$. Since $\gcd(a, b) = 1$, this implies that $a \mid q - r$, hence that $p = (q-r)/a \in k[t]$. We then have:

$$
\begin{aligned}
aDp + (b + Da)p &= a\left(\frac{Dq - Dr}{a} - \frac{(q-r)Da}{a^2}\right) + \frac{b(q-r) + (q-r)Da}{a} \\
&= Dq - Dr + \frac{b(q-r)}{a} \\
&= \frac{\left(a\sum_{i=1}^{m} c_i z_i + br\right) - bq}{a} - Dr + \frac{bq - br}{a} \\
&= \sum_{i=1}^{m} c_i z_i - Dr = \sum_{i=1}^{m} c_i(z_i - Dr_i).
\end{aligned}
$$

Conversely, let $c_1, \ldots, c_m \in \mathrm{Const}(k)$ and $p \in k[t]$ be a solution of (7.14), and let $q = ap + r$. Then,

$$
\begin{aligned}
aDq + bq &= a^2 Dp + apDa + aDr + abp + br \\
&= a\left(aDp + (b + Da)p\right) + aDr + br \\
&= a \sum_{i=1}^{m} c_i(z_i - Dr_i) + aDr + br \\
&= a \sum_{i=1}^{m} c_i z_i + br = \sum_{i=1}^{m} c_i(az_i + br_i) = \sum_{i=1}^{m} c_i q_i.
\end{aligned}
$$

$\square$

Theorem 7.1.2 reduces (7.7) to (7.14), which is an equation of the same type. If the coefficients a and b of the new equation have a nontrivial gcd, we divide it by that gcd, obtaining an equation of type (7.5) and reapply the linear constraints algorithm, obtaining a new equation of type (7.7). However, in all cases if (7.7) has a solution q of degree n, then the corresponding solution of the new equation must have degree at most $n - \deg(a)$ since $q = ap + r$ and $\deg(r) < \deg(a)$. Thus, if $\deg(a) > 0$, we can use Theorem 7.1.2 and Lemma 7.1.1 to reduce the degree of the unknown polynomial. We can repeat this until $\deg(a) = 0$ *i.e.* $a \in k^*$, at which point we divide the equation by a and we get an equation of type (7.7) with $a = 1$.

ParSPDE$(a, b, q_1, \ldots, q_m, D, n)$ (* Parametric SPDE algorithm *)

(* Given a derivation D on $k[t]$, an integer n and $a, b, q_1, \ldots, q_m \in k[t]$ with $\deg(a) > 0$ and $\gcd(a, b) = 1$, return $(\bar{a}, \bar{b}, \overline{q_1}, \ldots, \overline{q_m}, r_1, \ldots, r_m, \bar{n})$ such that for any solution $c_1, \ldots, c_m \in \mathrm{Const}(k)$ and $q \in k[t]$ of degree at most n of $aDq + bq = c_1 q_1 + \ldots + c_m q_m$, $p = (q - c_1 r_1 - \ldots - c_m r_m)/a$ has degree at most $\bar{n}$ and satisfies

$$\bar{a} Dp + \bar{b} p = c_1 \overline{q_1} + \ldots + c_m \overline{q_m} \, .$$

*)

for $i \leftarrow 1$ **to** m **do**   (* $br_i + az_i = q_i, \deg(r_i) < \deg(a)$ *)
  $(r_i, z_i) \leftarrow$ **ExtendedEuclidean**(b, a, q_i)
return$(a, b + Da, z_1 - Dr_1, \ldots, z_m - Dr_m, r_1, \ldots, r_m, n - \deg(a))$

Example 7.1.4. Let $k = \mathbb{Q}(x)$ with $D = d/dx$, t be a monomial over k satisfying $Dt = 1/x$, *i.e.* $t = \log(x)$, and let us search for a polynomial solution of arbitrary degree n of

$$tDq - \frac{1}{x} q = c_1 x - c_2 x t \, . \tag{7.15}$$

We have $a = t$, $b = -1/x$, $m = 2$, $q_1 = x$ and $q_2 = -xt$ so:

1. $(r_1, z_1) =$ **ExtendedEuclidean**$(-1/x, t, x) = (-x^2, 0)$,
   $(r_2, z_2) =$ **ExtendedEuclidean**$(-1/x, t, -xt) = (0, -x)$
2.

$$b + Da = -\frac{1}{x} + Dt = -\frac{1}{x} + \frac{1}{x} = 0, z_1 - Dr_1 = 2x, z_2 - Dr_2 = -x \, .$$

So (7.15) is reduced to

$$tDp = 2c_1 x - c_2 x$$

where $p = (q + x^2)/t \in k[t]$ and $\deg(p) \leq 0$. We have $\gcd(a, b) = t$ in the above equation, so it becomes

$$Dp = c_1 \frac{2x}{t} - c_2 \frac{x}{t}$$

which is of type (7.5). Calling **LinearConstraints**$(1, 0, 2x/t, -x/t)$ gives:

1. $d = \text{lcm}(t, t) = t$
2. $(q_1, r_1) = $ **PolyDivide**$(2x, t) = (0, 2x)$,
   $(q_2, r_2) = $ **PolyDivide**$(-x, t) = (0, -x)$
3. $n = \max(\deg(r_1), \deg(r_2)) = 1$
4.

$$M = \begin{pmatrix} 0 & 0 \\ 2x & -x \end{pmatrix}$$

so we are reduced to the equation $Dp = 0$ and the linear constraints $M(c_1, c_2)^T = 0$. Calling **ConstantSystem**$(M, 0)$ yields the constant system

$$\begin{pmatrix} 0 & 0 \\ 0 & 0 \\ 2 & -1 \end{pmatrix} \begin{pmatrix} c_1 \\ c_2 \end{pmatrix} = 0$$

which has the 1-dimensional solution space $(c_1, c_2) = (\lambda, 2\lambda)$. Since $Dp = 0$ has the 1-dimensional solution space $p = c$ for any $c \in \mathbb{Q}$, we get the 2-dimensional solution space

$$(q = \mu t - \lambda x^2, c_1 = \lambda, c_2 = 2\lambda)$$

of (7.34). Given the formal definition of the parametric Risch differential equation problem, this solution space is represented by $q = d_1 h_1 + d_2 h_2$ where $h_1 = t$, $h_2 = x^2$ and

$$\begin{pmatrix} 2 & -1 & 0 & 0 \\ 1 & 0 & 0 & 1 \end{pmatrix} \begin{pmatrix} c_1 \\ c_2 \\ d_1 \\ d_2 \end{pmatrix} = 0 \,.$$

The Non-Cancellation Cases

We are now reduced to finding solutions $c_1, \ldots, c_m \in \text{Const}(k)$ and $q \in k[t]$ of the following equation:

$$Dq + bq = \sum_{i=1}^{m} c_i q_i \tag{7.16}$$

where $b, q_1, \ldots, q_m \in k[t]$ and t is a monomial over k. Furthermore, we have an upper bound n on $\deg(q)$. As in the nonparametric case, Lemma 6.5.1 provides algorithms for all the non-cancellation cases.

When deg(b) is Large Enough. Suppose that $b \neq 0$, and that either $D = d/dt$ or $\deg(b) > \max(0, \delta(t) - 1)$. Then, for any solution $q = y_n t^n + \ldots + y_0 \in k[t]$ of (7.16), Lemma 6.5.1 implies that $\deg(Dq + bq) \leq n + \deg(b)$ and equating the coefficients of $t^{n+\deg(b)}$ on both sides yields

$$\mathrm{lc}(b) \, y_n = \sum_{i=1}^{m} c_i \, \mathrm{coefficient}(q_i, t^{n+\deg(b)}) \,.$$

Replacing q by $h + \sum_{i=1}^{m} c_i s_{in} t^n$ in (7.16), where

$$s_{in} = \frac{\mathrm{coefficient}(q_i, t^{n+\deg(b)})}{\mathrm{lc}(b)} \quad \in k \,, \tag{7.17}$$

we get

$$Dh + \sum_{i=1}^{m} c_i D s_{in} + \sum_{i=1}^{m} c_i b s_{in} + bh = \sum_{i=1}^{m} c_i q_i$$

which is equivalent to

$$Dh + bh = \sum_{i=1}^{m} c_i \left(q_i - D(s_{in} t^n) - b s_{in} t^n \right)$$

which is an equation of the same type as (7.16) with the same b as before. Hence the hypotheses of part (i) of Lemma 6.5.1 are satisfied again, so we can repeat this process, but with a bound of $n - 1$ on $\deg(h)$. Note that although b remains the same, the right side of (7.16) changes at every pass, so we must recompute the q_i's that appear in (7.17). The bound on $\deg(q)$ will decrease at every pass through this process, guaranteeing termination. After finishing the case $n = 0$, we get that any solution $q \in k[t]$ of the initial equation with $\deg(q) \leq n$ must be of the form $q = \sum_{i=1}^{m} c_i h_i$ where

$$h_i = \sum_{j=0}^{n} s_{ij} t^j \quad \in k[t] \,.$$

Replacing q by that form in (7.16) with the original q_i's yields

$$\sum_{i=1}^{m} c_i (q_i - Dh_i - bh_i) = 0 \,.$$

The left side is an element of $k[t]$, so setting all its coefficients to 0 yields a homogeneous linear system of the form $M(c_1, \ldots, c_m)^T = 0$, where M has entries in k. The same system can also be obtained from the last q_i's when $n = 0$ and the equation $\sum_{i=1}^{m} c_i (q_i - Ds_{i0} - bs_{i0}) = 0$, and this is how it is obtained in the algorithm below. By Lemma 7.1.2, we can compute an equivalent system of the form $A(c_1, \ldots, c_m)^T = 0$ where A has entries in $\mathrm{Const}(k)$. The solution of the initial problem is then $q = \sum_{i=1}^{m} d_i h_i$ where the additional equations $d_i = c_i$ for $1 \leq i \leq m$ are added to A, *i.e.* an $m \times 2m$ block of the form

$$
\begin{pmatrix}
1 & 0 & \cdots & \cdots & 0 & -1 & 0 & 0 & \cdots & 0 \\
0 & 1 & 0 & \cdots & \cdots & 0 & -1 & 0 & \cdots & 0 \\
\vdots & \ddots & \ddots & \ddots & & & \ddots & \ddots & & \vdots \\
\vdots & & \ddots & \ddots & \ddots & & & \ddots & \ddots & \vdots \\
0 & \cdots & \cdots & 0 & 1 & 0 & \cdots & \cdots & 0 & -1
\end{pmatrix}
\tag{7.18}
$$

is concatenated to the bottom of A, as well as a zero block to its right. The final system of linear constraints is then $A(c_1, \ldots, c_m, d_1, \ldots, d_m)^T = 0$.

ParamPolyRischDENoCancel1(b, c, D, n)
(* Parametric Poly Risch d.e. – no cancellation *)

(* Given a derivation D on $k[t]$, $n \in \mathbb{Z}$ and $b, q_1, \ldots, q_m \in k[t]$ with $b \neq 0$ and either $D = d/dt$ or $\deg(b) > \max(0, \delta(t) - 1)$, returns $h_1, \ldots, h_r \in k[t]$ and a matrix A with coefficients in $\mathrm{Const}(k)$ such that if $c_1, \ldots, c_m \in \mathrm{Const}(k)$ and $q \in k[t]$ satisfy $\deg(q) \leq n$ and $Dq + bq = \sum_{i=1}^{m} c_i$ then $q = \sum_{j=1}^{r} d_j h_j$ where $d_1, \ldots, d_r \in \mathrm{Const}(k)$ and $A(c_1, \ldots, c_m, d_1, \ldots, d_r)^T = 0$. *)

$d_b \leftarrow \deg(b)$, $b_d \leftarrow \mathrm{lc}(b)$
for $i \leftarrow 1$ **to** m **do** $h_i \leftarrow 0$
while $n \geq 0$ **do**
    for $i \leftarrow 1$ **to** m **do**
        $s_i \leftarrow$ **coefficient**$(q_i, t^{n+d_b})/b_d$
        $h_i \leftarrow h_i + s_i t^n$
        $q_i \leftarrow q_i - D(s_i t^n) - b s_i t^n$
    $n \leftarrow n - 1$
(* The remaining linear constraints are $\sum_{i=1}^{m} c_i q_i = 0$ *)
if $q_1 = \ldots = q_m = 0$ **then** $d_c \leftarrow -1$ **else** $d_c \leftarrow \max_{1 \leq i \leq m}(\deg(q_i))$
for $i \leftarrow 0$ **to** d_c **do for** $j \leftarrow 1$ **to** m **do** $M_{i+1,j} \leftarrow$ **coefficient**(q_j, t^i)
$(A, u) \leftarrow$ **ConstantSystem(M,0)**   (* $u = 0$ *)
(* Add the constraints $c_i - d_i = 0$ for $1 \leq i \leq m$ *)
$n_{eq} \leftarrow$ **number of rows**(A)
for $i \leftarrow 1$ **to** m **do** $A_{i+n_{eq},i} \leftarrow 1$, $A_{i+n_{eq},m+i} \leftarrow -1$
return$(h_1, \ldots, h_m, A)$

Example 7.1.5. Continuing example 7.1.3, let $k = \mathbb{Q}$, t be a monomial over k satisfying $Dt = 1$, *i.e.* $D = d/dt$, N be a positive integer, and consider the solutions $y \in k[t]$ of degree at most N of (7.13). We have $b = 1$, $m = 2$, $q_1 = t^N$ and $q_2 = 1$. Then,

1. $d_b = 0$, $b_d = 1$, $h_1 = h_2 = 0$
2. $n = N$, $s_1 = 1$, $h_1 = t^N$, $q_1 = -N t^{N-1}$, $s_2 = 0$, $h_2 = 0$, $q_2 = 1$

3. $n = N - 1$, $s_1 = -N$, $h_1 = t^N - Nt^{N-1}$, $q_1 = N(N-1)t^{N-2}$,
   $s_2 = 0$, $h_2 = 0$, $q_2 = 1$
4. ...

It is easy to prove by induction that after r steps through the loop $(r \leq N)$ we get

$$h_1 = \sum_{j=0}^{r-1} (-1)^j N^{\underline{j}} t^{N-j}, q_1 = (-1)^r N^{\underline{r}} t^{N-r}, h_2 = 0 \text{ and } q_2 = 1$$

where $N^{\underline{j}} = \prod_{i=0}^{j-1}(N-i)$. Thus, after N iterations we get $n = 0$,

$$h_1 = \sum_{j=0}^{N-1} (-1)^j N^{\underline{j}} t^{N-j},$$

$q_1 = (-1)^N N^{\underline{N}}$, $h_2 = 0$ and $q_2 = 1$. The last iteration then gives

1. $s_1 = (-1)^N N^{\underline{N}}$
2. $h_1 = h_1 + s_1 = \sum_{j=0}^{N}(-1)^j N^{\underline{j}} t^{N-j}$
3. $q_1 = (-1)^N N^{\underline{N}} - Ds_1 - s_1 = 0$
4. $s_2 = 1$, $h_2 = 1$, $q_2 = 0$

Proceeding with the algorithm, we get

1. $d_c = -1$, M and A are 0 by 0 matrices
2. $n_{eq} = 0$
3.
$$A = \begin{pmatrix} 1 & 0 & -1 & 0 \\ 0 & 1 & 0 & -1 \end{pmatrix}.$$

So the algorithm returns the above matrix, $h_2 = 1$ and

$$h_1 = \sum_{j=0}^{N} (-1)^j N^{\underline{j}} t^{N-j} = t^N - Nt^{N-1} + N(N-1)t^{N-2} + \ldots + (-1)^N N!.$$

The general solution of (7.13) is $y = d_1 h_1 + d_2$ where

$$\begin{pmatrix} 1 & 0 & -1 & 0 \\ 0 & 1 & 0 & -1 \end{pmatrix} \begin{pmatrix} c_1 \\ c_2 \\ d_1 \\ d_2 \end{pmatrix} = 0.$$

When deg(b) is Small Enough. Suppose that $\deg(b) < \delta(t) - 1$ and either $D = d/dt$, which implies that $b = 0$, or $\delta(t) \geq 2$. Let $q = y_n t^n + \ldots + y_0 \in k[t]$ be a solution of (7.16).

If $n > 0$, then Lemma 6.5.1 implies that $\deg(Dq + bq) \leq n + \delta(t) - 1$ and equating the coefficients of $t^{n+\delta(t)-1}$ on both sides yields

$$n\,\lambda(t)\,y_n = \sum_{i=1}^{m} c_i \operatorname{coefficient}(q_i, t^{n+\delta(t)-1}).$$

Replacing q by $h + \sum_{i=1}^{m} c_i s_{in} t^n$ in (7.16), where

$$s_{in} = \frac{\operatorname{coefficient}(q_i, t^{n+\delta(t)-1})}{n\,\lambda(t)} \quad \in k, \tag{7.19}$$

we get

$$Dh + bh = \sum_{i=1}^{m} c_i \left(q_i - D(s_{in} t^n) - b s_{in} t^n\right)$$

which is an equation of the same type as (7.16) with the same b as before. Hence the hypotheses of part (ii) of Lemma 6.5.1 are satisfied again, so we can repeat this process, but with a bound of $n - 1$ on $\deg(h)$. Note that although b remains the same, the right-hand side changes at every pass, so we must recompute the q_i's that appear in (7.19). The bound on $\deg(q)$ will decrease at every pass through this process, until we reach $n = 0$, *i.e.* we are looking for solutions $q = y_0 \in k$. At this point, the algorithm proceeds differently for $\deg(b) > 0$ and for $b \in k$.

If $\deg(b) > 0$, then equating the coefficients of $t^{\deg(b)}$ on both sides yields

$$\operatorname{lc}(b)\,y_0 = \sum_{i=1}^{m} c_i \operatorname{coefficient}(q_i, t^{\deg(b)}),$$

so any solution $y_0 \in k$ must be of the form $y_0 = \sum_{i=1}^{m} c_i s_{i0}$ where

$$s_{i0} = \frac{\operatorname{coefficient}(q_i, t^{\deg(b)})}{\operatorname{lc}(b)} \quad \in k.$$

This implies that any solution $q \in k[t]$ of the initial equation with $\deg(q) \leq n$ must be of the form $q = \sum_{i=1}^{m} c_i h_i$ where

$$h_i = \sum_{j=0}^{n} s_{ij} t^j \quad \in k[t].$$

Replacing q by that form in (7.16) with the original q_i's yields

$$\sum_{i=1}^{m} c_i(q_i - Dh_i - bh_i) = 0.$$

As we have seen earlier, this can be converted to a homogeneous system of the form $A(c_1, \ldots, c_m)^T = 0$ where A has entries in $\mathrm{Const}(k)$. The solution of the initial problem is then $q = \sum_{i=1}^{m} d_i h_i$ where the additional equations $d_i = c_i$ for $1 \le i \le m$ are added to A as earlier. The final system of linear constraints is then $A(c_1, \ldots, c_m, d_1, \ldots, d_m)^T = 0$.

If $b \in k$, then any solution $y_0 \in k$ of (7.16) satisfies

$$Dy_0 + by_0 = \sum_{i=1}^{m} c_i \, q_i(0) \, . \tag{7.20}$$

This is a parametric Risch differential equation of type (7.1) over k. Assuming that we can solve such problems over k, we obtain $f_1, \ldots, f_r \in k$ and a matrix B with coefficients in $\mathrm{Const}(k)$ such that any solution $y_0 \in k$ of (7.20) is of the form

$$y_0 = \sum_{j=1}^{r} d_j f_j$$

where $d_1, \ldots, d_j \in \mathrm{Const}(k)$ and $B(c_1, \ldots, c_m, d_1, \ldots, d_r)^T = 0$. This implies that any solution $q \in k[t]$ of the initial equation with $\deg(q) \le n$ must be of the form $q = \sum_{j=1}^{r} d_j f_j + \sum_{i=1}^{m} c_i h_i$ where

$$h_i = \sum_{j=1}^{n} s_{ij} t^j \quad \in k[t] \, .$$

Replacing q by that form in (7.16) with the original q_i's yields

$$\sum_{i=1}^{m} c_i(q_i - Dh_i - bh_i) - \sum_{j=1}^{r} d_j(Df_j + bf_j) = 0 \, .$$

In a similar way than in the previous cases, this can be converted to a homogeneous system of the form $A(c_1, \ldots, c_m, d_1, \ldots, d_r)^T = 0$ where A has entries in $\mathrm{Const}(k)$. The solution of the initial problem is then

$$q = \sum_{j=1}^{r} d_j f_j + \sum_{i=1}^{m} e_i h_i$$

where the additional equations $B(c_1, \ldots, c_m, d_1, \ldots, d_r)^T = 0$ are added to A, as well as the equations $e_i = c_i$ for $1 \le i \le m$. The final system of linear constraints is then $A(c_1, \ldots, c_m, d_1, \ldots, d_r, e_1, \ldots, e_m)^T = 0$.

ParamPolyRischDENoCancel2(b, c, D, n)
(* Parametric Poly Risch d.e. – no cancellation *)

  (* Given a derivation D on $k[t]$, $n \in \mathbb{Z}$ and $b, q_1, \ldots, q_m \in k[t]$ with
  $\deg(b) < \delta(t) - 1$ and either $D = d/dt$ or $\delta(t) \geq 2$, returns $h_1, \ldots, h_r \in$
  $k[t]$ and a matrix A with coefficients in $\text{Const}(k)$ such that if $c_1, \ldots, c_m \in$
  $\text{Const}(k)$ and $q \in k[t]$ satisfy $\deg(q) \leq n$ and $Dq + bq = \sum_{i=1}^{m} c_i$ then $q =$
  $\sum_{j=1}^{r} d_j h_j$ where $d_1, \ldots, d_r \in \text{Const}(k)$ and $A(c_1, \ldots, c_m, d_1, \ldots, d_r)^T =$
  0. *)

  $\delta \leftarrow \delta(t)$, $\lambda \leftarrow \lambda(t)$
  for $i \leftarrow 1$ **to** m **do** $h_i \leftarrow 0$
  while $n > 0$ **do**
    for $i \leftarrow 1$ **to** m **do**
      $s_i \leftarrow$ **coefficient**$(q_i, t^{n+\delta-1})/(n\lambda)$
      $h_i \leftarrow h_i + s_i t^n$, $q_i \leftarrow q_i - D(s_i t^n) - b s_i t^n$
    $n \leftarrow n - 1$
  if $\deg(b) > 0$ **then**
    for $i \leftarrow 1$ **to** m **do**
      $s_i \leftarrow$ **coefficient**$(q_i, t^{\deg(b)})/\text{lc}(b)$
      $h_i \leftarrow h_i + s_i$, $q_i \leftarrow q_i - D s_i - b s_i$
    if $q_1 = \ldots = q_m = 0$ **then** $d_c \leftarrow -1$ **else** $d_c \leftarrow \max_{1 \leq i \leq m}(\deg(q_i))$
    for $i \leftarrow 0$ **to** d_c **do for** $j \leftarrow 1$ **to** m **do** $M_{i+1,j} \leftarrow$ **coefficient**(q_j, t^i)
    $(A, u) \leftarrow$ **ConstantSystem**$(\mathbf{M}, \mathbf{0})$ (* $u = 0$ *)
    $n_{eq} \leftarrow$ **number of rows**(A)
    for $i \leftarrow 1$ **to** m **do** $A_{i+n_{eq},i} \leftarrow 1$, $A_{i+n_{eq},m+i} \leftarrow -1$
    return$(h_1, \ldots, h_m, A)$
  else   (* $b \in k$ *)
    $(f_1, \ldots, f_r, B) \leftarrow$ **ParamRischDE**$(b, q_1(0), \ldots, q_m(0))$
    if $q_1 = \ldots = q_m = 0$ **then**
        if $Df_1 + bf_1 = \ldots = Df_r + bf_r = 0$ **then** $d_c \leftarrow -1$ **else** $d_c \leftarrow 0$
    else $d_c \leftarrow \max_{1 \leq i \leq m}(\deg(q_i))$
    for $i \leftarrow 0$ **to** d_c **do for** $j \leftarrow 1$ **to** m **do** $M_{i+1,j} \leftarrow$ **coefficient**(q_j, t^i)
    for $j \leftarrow 1$ **to** r **do** $M_{1,j+m} \leftarrow -Df_j - bf_j$
    $(A, u) \leftarrow$ **ConstantSystem**$(\mathbf{M}, \mathbf{0})$ (* $u = 0$ *)
    $A \leftarrow A \cup B$ (* vertical concatenation *)
    (* Add the constraints $c_i - e_i = 0$ for $1 \leq i \leq m$ *)
    $n_{eq} \leftarrow$ **number of rows**(A)
    for $i \leftarrow 1$ **to** m **do** $A_{i+n_{eq},i} \leftarrow 1$, $A_{i+n_{eq},m+r+i} \leftarrow -1$
    return$(f_1, \ldots, f_r, h_1, \ldots, h_m, A)$

Example 7.1.6. Continuing example 7.1.1, let $k = \mathbb{Q}$, t be a monomial over k satisfying $Dt = 1$, *i.e.* $D = d/dt$, and consider the solutions $p \in k[t]$ of (7.11). We have $a = 1$, $b = 0$ and $q_1 = 2x$, so the degree bound algorithm for the base case yields an upper bound of $n = 2$ on $\deg(p)$. Then,

1. $\delta = 0$, $\lambda = 1$, $h_1 = 0$
2. $s_1 = 2/2 = 1$, $h_1 = t^2$, $q_1 = 0$, $n = 1$
3. $s_1 = 0$, $h_1 = t^2$, $q_1 = 0$, $n = 0$
4. $s_1 = 0$, $h_1 = t^2$, $q_1 = 0$, $n = -1$

At this point, since $b \in k$, we recursively find the solutions $y \in k$ of $Dy = 0$. This returns $f_1 = 1$ and the linear constraint $d_1 = \lambda$, i.e. $B = (1 \quad -1)$. We then have

1. $q_1 = Df_1 = 0$ so $d_c = -1$, hence M and A are 0 by 0 matrices
2. $A = A \cup B = (1 \quad -1)$
3. $n_{eq} = 1$
4.

$$A = \begin{pmatrix} 1 & -1 & 0 \\ 1 & 0 & -1 \end{pmatrix}.$$

So the algorithm returns the above matrix, $f_1 = 1$ and $h_1 = t^2$. The general solution of (7.11) is $p = d_1 + e_1 t^2$ where

$$\begin{pmatrix} 1 & -1 & 0 \\ 1 & 0 & -1 \end{pmatrix} \begin{pmatrix} \lambda \\ d_1 \\ e_1 \end{pmatrix} = 0.$$

Going back to example 7.1.1, Since we had $(c_1, c_2, c_3) = (\lambda, -3\lambda, -2\lambda)$, the general solution of (7.9) is $p = d_1 + e_1 t^2$ where

$$\begin{pmatrix} 3 & 1 & 0 & 0 & 0 \\ 2 & 0 & 1 & 0 & 0 \\ 1 & 0 & 0 & -1 & 0 \\ 1 & 0 & 0 & 0 & -1 \end{pmatrix} \begin{pmatrix} c_1 \\ c_2 \\ c_3 \\ d_1 \\ e_1 \end{pmatrix} = 0.$$

Since the above constraints imply that $d_1 = e_1 = c_1$, the general solution can also be simplified to $p = c_1(1 + t^2)$ subject to the constraints (7.10).

When $\delta(t) \geq 2$ and $\deg(b) = \delta(t) - 1$. In that case, we have cancellation only when $\deg(q) = -\mathrm{lc}(b)/\lambda(t)$, which implies in particular that $-\mathrm{lc}(b)/\lambda(t)$ is an integer between 1 and our degree bound n. Let $q = y_n t^n + \ldots y_0 \in k[t]$ be a solution of (7.16). If $n \neq -\mathrm{lc}(b)/\lambda(t)$, then Lemma 6.5.1 implies that $\deg(Dq + bq) \leq n + \delta(t) - 1$ and equating the coefficients of $t^{n+\deg(b)}$ on both sides yields

$$(n\,\lambda(t) + \mathrm{lc}(b))\,y_n = \sum_{i=1}^{m} c_i\,\mathrm{coefficient}(q_i, t^{n+\delta(t)-1}).$$

Replacing q by $h + \sum_{i=1}^{m} c_i s_{in} t^{n+\delta(t)-1}$ in (7.16), where

$$s_{in} = \frac{\mathrm{coefficient}(q_i, t^{n+\delta(t)-1})}{n\,\lambda(t) + \mathrm{lc}(b)} \in k, \tag{7.21}$$

we get

$$Dh + bh = \sum_{i=1}^{m} c_i\,(q_i - D(s_{in}t^n) - bs_{in}t^n)$$

which is an equation of the same type as (7.16) with the same b as before, but with a bound of $n-1$ on $\deg(h)$. As long as the degree bound is not equal to $-\text{lc}(b)/\lambda(t)$, the hypotheses of part (iii) of Lemma 6.5.1 are satisfied again, so we can repeat this process until either we have finished the case $n=0$, or we reach the case $n=-\text{lc}(b)/\lambda(t)$.

If we have finished the case $n=0$, then any solution $q \in k[t]$ of the initial equation with $\deg(q) \leq n$ must be of the form $q = \sum_{i=1}^{m} c_i h_i$ where

$$h_i = \sum_{j=0}^{n} s_{ij} t^j \quad \in k[t].$$

Replacing q by that form in (7.16) with the original q_i's yields

$$\sum_{i=1}^{m} c_i(q_i - Dh_i - bh_i) = 0.$$

As we have seen earlier, this is equivalent to homogeneous system of the form $A(c_1, \ldots, c_m)^T = 0$ where A has entries in $\text{Const}(k)$. The solution of the initial problem is then $q = \sum_{i=1}^{m} d_i h_i$ where the additional equations $d_i = c_i$ for $1 \leq i \leq m$ are added to A as earlier.

If we reach the case $n = -\text{lc}(b)/\lambda(t) > 0$, then the algorithm of the next section on the cancellation cases produce $f_1, \ldots, f_r \in k[t]$ and a matrix B with coefficients in $\text{Const}(k)$ such that any solution $q \in k[t]$ of degree at most n must be of the form

$$q = \sum_{j=1}^{r} d_j f_j$$

where $d_1, \ldots, d_j \in \text{Const}(k)$ and $B(c_1, \ldots, c_m, d_1, \ldots, d_r)^T = 0$. This implies that the solutions of the initial equation must be of the form

$$q = \sum_{j=1}^{r} d_j f_j + \sum_{i=1}^{m} c_i h_i$$

where

$$h_i = \sum_{j=1-\text{lc}(b)/\lambda(t)}^{n} s_{ij} t^j \quad \in k[t].$$

Replacing q by that form in (7.16) with the original q_i's yields

$$\sum_{i=1}^{m} c_i(q_i - Dh_i - bh_i) - \sum_{j=1}^{r} d_j(Df_j + bf_j) = 0.$$

As we have seen earlier, this is equivalent to homogeneous system of the form $A(c_1, \ldots, c_m, d_1, \ldots, d_r)^T = 0$ where A has entries in $\text{Const}(k)$. The solution of the initial problem is then

$$q = \sum_{j=1}^{r} d_j f_j + \sum_{i=1}^{m} e_i h_i$$

where the additional equations $B(c_1, \ldots, c_m, d_1, \ldots, d_r)^T = 0$ are added to A, as well as the equations $e_i = c_i$ for $1 \leq i \leq m$.

If $\deg(q) > 0$ and $\deg(q) \neq -\mathrm{lc}(b)/\lambda(t)$, then $\deg(q) + \delta(t) - 1 = \deg(c)$, so $\deg(q) = \deg(c) + 1 - \delta(t)$ and $(\deg(q)\lambda(t) + \mathrm{lc}(b))\mathrm{lc}(q) = \mathrm{lc}(c)$. This yields the leading monomial $u t^n$ of q, and replacing q by $u t^n + h$ in the equation yields a similar equation with a lower degree bound on its solution. We can repeat this as long as the new degree bound is not equal to $-\mathrm{lc}(b)/\lambda(t)$.

If $q \in k$, then the leading term of $Dq + bq$ is $q\,\mathrm{lc}(b)t^{\delta(t)-1}$, so either $\deg(c) = \delta(t) - 1$, in which case $q = \mathrm{lc}(c)/\mathrm{lc}(b)$ is the only potential solution, or $\deg(c) \neq \delta(t) - 1$ and (6.19) has no solution in k, hence in $k[t]$.

The Cancellation Cases

We finally study equation (7.16) whenever the non-cancellation cases do not hold, *i.e.* in one of the following cases:

1. $\delta(t) \leq 1$, $b \in k$ and $D \neq d/dt$.
2. $\delta(t) \geq 2$, $\deg(b) = \delta(t) - 1$, and $\deg(q) = -\mathrm{lc}(b)/\lambda(t)$.

The Liouvillian Case. If $D \neq d/dt$ and $Dt \in k$ or $Dt/t \in k$, then $\delta(t) \leq 1$, so the only case not handled by Lemma 6.5.1 is $b \in k$. Then, for any solution $q = y_n t^n + \ldots + y_0 \in k[t]$ of (7.16), Lemma 5.1.2 implies that $\deg(Dq + bq) \leq n$ and equating the coefficients of t^n on both sides yields

$$Dy_n + by_n = \sum_{i=1}^{m} c_i \, \mathrm{coefficient}(q_i, t^n) \tag{7.22}$$

if $Dt \in k$, and

$$Dy_n + \left(b + n\frac{Dt}{t}\right) y_n = \sum_{i=1}^{m} c_i \, \mathrm{coefficient}(q_i, t^n) \tag{7.23}$$

if $Dt/t \in k$. Both (7.22) and (7.23) are parametric Risch differential equations of type (7.1) over k. Assuming that we can solve such problems over k, we obtain $f_{1n}, \ldots, f_{r_n,n} \in k$ and a matrix A_n with coefficients in $\mathrm{Const}(k)$ such that y_n is of the form

$$y_n = \sum_{j=1}^{r_n} d_{jn} f_{jn}$$

where $d_{1n}, \ldots, d_{r_n,n} \in \mathrm{Const}(k)$ and $A_n(c_1, \ldots, c_m, d_{1n}, \ldots, d_{r_n,n})^T = 0$. Replacing q by $h + \sum_{j=1}^{r_n} d_{jn} f_{jn} t^n$ in (7.16), we get

$$Dh + bh = \sum_{i=1}^{m} c_i q_i - \sum_{j=1}^{r} d_{jn}(D(f_{jn}t^n) - bf_{jn}t^n)$$

which is an equation of the same type as (7.16) with the same b as before. Hence, we can repeat this process, but with a bound of $n-1$ on $\deg(h)$. Note that although b remains the same, the right side of (7.16) changes at every pass, so we must recompute the q_i's that appear in (7.22) or (7.23). Note also that the number of undetermined constants in the right side increases at each step. The bound on $\deg(q)$ will decrease at every pass through this process, guaranteeing termination. After finishing the case $n = 0$, we get that any solution $q \in k[t]$ of the initial equation with $\deg(q) \leq n$ must be of the form

$$q = \sum_{i=0}^{n} \sum_{j=1}^{r_i} d_{ji} h_{ji} \quad \text{where} \quad h_{ji} = f_{ji} t^i \,. \tag{7.24}$$

Replacing q by that form in (7.16) with the original q_i's yields

$$\sum_{i=1}^{m} c_i q_i - \sum_{i=0}^{n} \sum_{j=1}^{r_i} d_{ji}(Dh_{ji} - bh_{ji}) = 0 \,.$$

As we have seen in the non-cancellation cases, this can be converted to a homogeneous system of the form $A(c_1, \ldots, c_m, d_{11}, \ldots, d_{r_n, n})^T = 0$ where A has entries in $\mathrm{Const}(k)$. The solution of the initial problem is then given by (7.24) where the additional equations $A_i(c_1, \ldots, c_m, d_{1i}, \ldots, d_{r_i, i})^T = 0$ are added to A for $0 \leq i \leq n$.

The Nonlinear Case. If $\delta(t) \geq 2$, then we must have $\deg(b) = \delta(t) - 1$ and $\mathrm{lc}(b) = -n\lambda(t)$ where $n > 0$ is the bound on $\deg(q)$. As in the nonparametric case, there is no general algorithm for solving equation (7.16) in this case. If however $\mathcal{S}^{\mathrm{irr}} \neq \emptyset$, then projecting (7.16) to $k[t]/(p)$ for $p \in \mathcal{S}^{\mathrm{irr}}$ can be done as in the nonparametric case. Since $k[t]/(p)$ is a finite algebraic extension of k, $\mathrm{Const}(k[t]/(p))$ is a finite algebraic extension of $\mathrm{Const}(k)$ by Corollary 3.3.1, so let $b_1, \ldots, b_s$ be a vector space basis for $\mathrm{Const}(k[t]/(p))$ over $\mathrm{Const}(k)$. Now, with D^* being the induced derivation on $k[t]/(p)$, we get

$$D^* q^* + \pi_p(b) q^* = \sum_{i=1}^{m} c_i \pi_p(q_i) \tag{7.25}$$

where $q^* = \pi_p(q)$. Assuming that we have an algorithm for solving (7.25) in $k[t]/(p)$, we obtain $h_1, \ldots, h_r$ in $k[t]/(p)$ and a matrix B with coefficients in $\mathrm{Const}(k[t]/(p))$ such that any solution in $k[t]/(p)$ of (7.25) must be of the form $q^* = \sum_{j=1}^{r} d_j h_j$ where $d_1, \ldots, d_r \in \mathrm{Const}(k[t]/(p))$ and $B(c_1, \ldots, c_m, d_1, \ldots, d_r)^T = 0$. Expanding formally the constants $d_1, \ldots, d_r$ and the entries of B with respect to the basis $b_1, \ldots, b_s$ we obtain a matrix A with coefficients in $\mathrm{Const}(k)$ such that the system $B(c_1, \ldots, c_m, d_1, \ldots, d_r)^T = 0$ is equivalent to $A(c_1, \ldots, c_m, d_{11}, \ldots, d_{rs})^T = 0$ where

$$d_j = \sum_{l=1}^{s} d_{jl} b_l \quad \text{for} \quad 1 \le j \le r$$

and any solution in $k[t]/(p)$ of (7.25) must now be of the form

$$q^* = \sum_{j=1}^{r} \left(\sum_{l=1}^{s} d_{jl} b_l \right) h_j = \sum_{j=1}^{r} \sum_{l=1}^{s} d_{jl} h_{jl} \tag{7.26}$$

where $h_{jl} = b_l h_j \in k[t]/(p)$. For each j and l, let $r_{jl} \in k[t]$ be such that $\deg(r_{jl}) < \deg(p)$ and $\pi_p(r_{jl}) = h_{jl}$, and let

$$u = \sum_{j=1}^{r} \sum_{l=1}^{s} d_{jl} r_{jl} \in k[t].$$

We have $\deg(u) < \deg(p)$ and (7.26) implies that $\pi_p(q) = \pi_p(u)$, hence that $h = (q - u)/p \in k[t]$. Replacing q by $ph + u$ in (7.16) we get

$$\sum_{i=1}^{m} c_i q_i = Dq + bq = p \left(Dh + \left(b + \frac{Dp}{p} \right) h \right) + Du + bu$$

so h is a solution in $k[t]$ of degree at most $\deg(q) - \deg(p)$ of

$$Dh + \left(b + \frac{Dp}{p} \right) h = \frac{\sum_{i=1}^{m} c_i q_i - \sum_{j=1}^{r} \sum_{l=1}^{s} d_{jl}(Dr_{jl} + br_{jl})}{p}. \tag{7.27}$$

Write now $q_i = p\overline{q_i} + \widehat{q_i}$ and $Dr_{jl} + br_{jl} = p\overline{r_{jl}} + \widehat{r_{jl}}$ where $\widehat{q_i}, \widehat{r_{jl}} \in k[t]$, $\deg(\widehat{q_i}) < \deg(p)$ and $\deg(\widehat{r_{jl}}) < \deg(p)$. The right hand side of (7.27) becomes

$$\frac{\sum_{i=1}^{m} c_i q_i - \sum_{j=1}^{r} \sum_{l=1}^{s} d_{jl}(Dr_{jl} + br_{jl})}{p} =$$

$$\sum_{i=1}^{m} c_i \overline{q_i} - \sum_{j=1}^{r} \sum_{l=1}^{s} d_{jl} \overline{r_{jl}} + \frac{\sum_{i=1}^{m} c_i \widehat{q_i} - \sum_{j=1}^{r} \sum_{l=1}^{s} d_{jl} \widehat{r_{jl}}}{p}.$$

Since $\pi_p(u) = q^*$ is a solution of (7.25), we have

$$0 = \pi_p \left(\sum_{i=1}^{m} c_i q_i - Du - bu \right) = \pi_p \left(\sum_{i=1}^{m} c_i \widehat{q_i} - \sum_{j=1}^{r} \sum_{l=1}^{s} d_{jl} \widehat{r_{jl}} \right).$$

Since $\deg(\widehat{q_i}) < \deg(p)$ and $\deg(\widehat{r_{jl}}) < \deg(p)$, it follows that

$$\sum_{i=1}^{m} c_i \widehat{q_i} - \sum_{j=1}^{r} \sum_{l=1}^{s} d_{jl} \widehat{r_{jl}} = 0$$

so (7.27) becomes

$$Dh + \left(b + \frac{Dp}{p}\right)h = \sum_{i=1}^{m} c_i \overline{q}_i - \sum_{j=1}^{r} \sum_{l=1}^{s} d_{jl} \overline{r}_{jl}$$

which is an equation of type (7.16), but with a lower bound on the degree of its solution. Repeating this process until the lower bound becomes negative, and grouping all the linear constraints obtained at each step yields a complete solution of the initial parametric problem.

The remarks made in the nonparametric case about when (7.25) can be solved, for example when we can find an element of degree one, or an element with constant coefficients, in S^{irr}, remain valid in the parametric case.

The Hypertangent Case. If $Dt/(t^2 + 1) = \eta \in k$, then $\delta(t) = 2$, so the only case not handled by Lemma 6.5.1 is $b = b_0 - n\eta t$ where $b_0 \in k$ and $n > 0$ is the bound on $\deg(q)$. In such extensions, the method outlined above provides a complete algorithm: taking $p = t^2 + 1 \in S^{\mathrm{irr}}$, (7.25) becomes

$$Dq^* + (b_0 - n\eta\sqrt{-1})q^* = \sum_{i=1}^{m} c_i q_i(\sqrt{-1}) \qquad (7.28)$$

where D is extended to $k[t]/(p) \simeq k(\sqrt{-1})$ by $D\sqrt{-1} = 0$. One possibility is to view (7.28) as a parametric Risch differential equation in $k(\sqrt{-1})$ and to solve it recursively. After expanding the result with respect to the basis $\{1, \sqrt{-1}\}$ (only if $\sqrt{-1} \notin k$), we obtain $h_1, \ldots, h_r \in k(\sqrt{-1})$ and a matrix A with entries in $\mathrm{Const}(k)$ such that all the solutions in $k(\sqrt{-1})$ of (7.28) must be of the form $q^* = \sum_{j=1}^{r} d_j h_j$ where $d_j \in \mathrm{Const}(k)$ and $A(c_1, \ldots, c_m, d_1, \ldots, d_r)^T = 0$. Write $h_j = h_{j0} + h_{j1}\sqrt{-1}$ for each j with $h_{j0}, h_{j1} \in k$. Then, letting $r_j = h_{j0} + h_{j1}t \in k[t]$, $\overline{q}_i$ be the quotient of q_i by p and $\overline{r}_j$ be the quotient of $Dr_j + br_j$ by p, $h = (q - \sum_{j=1}^{m} d_j r_j)/p$ is a solution in $k[t]$ of degree at most $n - 2$ of

$$Dh + (b_0 - (n - 2)\eta t)h = \sum_{i=1}^{m} c_i \overline{q}_i - \sum_{j=1}^{r} d_j \overline{r}_j.$$

If $\sqrt{-1} \notin k$, it is possible to avoid introducing $\sqrt{-1}$ by considering the real and imaginary parts of (7.28): writing $q^* = u + v\sqrt{-1}$, we get

$$\begin{pmatrix} Du \\ Dv \end{pmatrix} + \begin{pmatrix} b_0 & n\eta \\ -n\eta & b_0 \end{pmatrix} = \sum_{i=1}^{m} c_i \begin{pmatrix} q_{i0} \\ q_{i1} \end{pmatrix} \qquad (7.29)$$

where $q_{i0} + q_{i1}t$ is the remainder of q_i by $t^2 + 1$. This is the parametric version of the coupled differential system introduced in Sect. 5.10, and the algorithm of Chap. 8 can be generalized to the parametric case, in a manner similar to what is done for the Risch differential equation in this chapter.

7.2 The Limited Integration Problem

We describe in this section a solution to the limited integration problem, *i.e.* given a differential field K of characteristic 0 and $f, w_1, \ldots, w_m \in K$, to decide whether there are constants $c_1, \ldots, c_m \in \mathrm{Const}(K)$ such that

$$f = Dv + c_1 w_1 + \ldots + c_m w_m \tag{7.30}$$

has a solution $v \in K$, and to find one such solution if there are solutions. As we have seen in Chap. 5, this problem arises from integrating polynomials in primitive extensions. There are several possible approaches to this problem:

– If all the w_i's are logarithmic derivatives of elements of K, then the existence of a solution of (7.30) implies that f has an integral in an elementary extension of K, so equation (7.30) can be seen as an elementary integration problem, and the algorithm of Chap. 5 can be used, followed by a step that attempts to rewrite the resulting integral in terms of the w_i's.

– Equation (7.30) can be considered a parametric Risch differential equation for v and can be solved by the algorithm of Sect. 7.1.

The first approach is applicable only when integrating elementary functions, since the only primitive monomials appearing in the integrand are then logarithms, and it is in fact the approach originally taken by Risch [60] and in most computer algebra systems and texts [27, 29, 31]. How to rewrite an elementary integral in terms of the w_i's is however never made explicit[1] and adds new difficulties and complexities to the algorithm. The second approach is applicable for arbitrary w_i's, so it allows arbitrary primitives in the integrand. Furthermore, algorithms for integrating in terms of some nonelementary functions, like Erf, Ei, Li and dilogarithms [20, 21, 38, 39], first produce candidate special functions and then solve the limited integration problem for those special functions. Because of those advantages, we essentially use that method here. However, the parametric Risch differential equation algorithm is made simpler by the fact that cancellation at the poles of v (including infinity) cannot occur since only Dv appears in the equation and no multiple of v, so bounding orders and degrees is significantly easier. We present in this section an simplified version of the algorithm of Sect. 7.1 that takes advantage of this fact.

We only study equation (7.30) in the transcendental case, *i.e.* when K is a simple monomial extension of a differential subfield k, so for the rest of this section, let k be a differential field of characteristic 0 and t be a monomial over k. We assume in addition that $\mathrm{Const}(k(t)) = \mathrm{Const}(k)$. We suppose that the coefficients f and $w_1, \ldots, w_m$ of our equation are in $k(t)$ and look for solutions $c_1, \ldots, c_m \in \mathrm{Const}(k)$ and $v \in k(t)$.

---

[1] In [60], Risch only has the following remark about the hypothesis that we can integrate elements of k: ``...we assume that the simpler variants, which occur when some of the c_i and v_i are given, have been established.''.

Because of the special form of equation (7.30), Theorem 7.1.1 can be strengthened to yield not only the normal part of the denominator, but also its special part whenever $S_1^{\mathrm{irr}} = S^{\mathrm{irr}}$, and the degree bound whenever t is either a Liouvillian or nonlinear monomial.

Theorem 7.2.1. *Let* $v, f, w_1, \ldots, w_m \in k(t)$ *and* $c_1, \ldots, c_m \in \mathrm{Const}(k)$ *be such that* $f = Dv + c_1 w_1 + \ldots + c_m w_m$. *Let* $d = d_s d_n$ *be a splitting factorization of the denominator of* f, *and* $e_i = e_{s,i} e_{n,i}$ *be splitting factorizations of the denominators of the* w_i's. *Let* $c = \mathrm{lcm}(d_n, e_{n,1}, \ldots, e_{n,m})$, $h_s = \mathrm{lcm}(d_s, e_{s,1}, \ldots, e_{s,m})$, *and*

$$h_n = \gcd\left(c, \frac{dc}{dt}\right).$$

Then,

(i) $vh_n \in k\langle t \rangle$,
(ii) *If* $S_1^{\mathrm{irr}} = S^{\mathrm{irr}}$, *then* $vh_n h_s \in k[t]$.
(iii) *If* t *is nonlinear or Liouvillian over* k, *then either* $\nu_\infty(v) = 0$ *or*

$$\nu_\infty(v) \geq \min\left(\nu_\infty(f), \nu_\infty(w_1), \ldots, \nu_\infty(w_m)\right) + \delta(t) - 1.$$

Proof. (i) Let $q = vh_n \in k(t)$. In order to show that $q \in k\langle t\rangle$, we need to show that $\nu_p(q) \geq 0$ for any normal irreducible $p \in k[t]$. We have $\nu_p(q) = \nu_p(v) + \nu_p(h_n)$ by Theorem 4.1.1. If $\nu_p(v) \geq 0$, then $\nu_p(q) \geq \nu_p(h_n) \geq 0$ since $h_n \in k[t]$. So suppose that $n = \nu_p(v) < 0$ and let $w = c_1 w_1 + \ldots + c_m w_m$. Then $\nu_p(Dv) = n - 1$ by Theorem 4.4.2, so $\nu_p(f - w) = n - 1$, which implies that $\nu_p(f) \leq n - 1$ or $\nu_p(w_i) \leq n - 1$ for some i. Hence $p^{1-n} \mid c$, so $\nu_p(c) \geq 1 - n$, which implies that $\nu_p(h_n) = \nu_p(c) - 1 \geq -n$, hence that $\nu_p(q) = n + \nu_p(h_n) \geq 0$.

(ii) Suppose that $S_1^{\mathrm{irr}} = S^{\mathrm{irr}}$ and let $q = vh_n h_s \in k(t)$. Since $h_s \in k[t] \subseteq k\langle t\rangle$, and we have from (i) that $vh_n \in k\langle t\rangle$, we get that $q \in k\langle t\rangle$, so in order to show that $q \in k[t]$, we need to show that $\nu_p(q) \geq 0$ for any $q \in S^{\mathrm{irr}}$. We have $\nu_p(q) = \nu_p(v) + \nu_p(h_n) + \nu_p(h_s)$ by Theorem 4.1.1. If $\nu_p(v) \geq 0$, then $\nu_p(q) \geq \nu_p(h_n) + \nu_p(h_s) \geq 0$ since $h_n, h_s \in k[t]$. So suppose that $n = \nu_p(v) < 0$ and let $w = c_1 w_1 + \ldots + c_m w_m$. Since $S_1^{\mathrm{irr}} = S^{\mathrm{irr}}$, $p \in S_1^{\mathrm{irr}}$, so $\nu_p(Dv) = n$ by Theorem 4.4.2, so $\nu_p(f - w) = n$, which implies that $\nu_p(f) \leq n$ or $\nu_p(w_i) \leq n$ for some i. Hence $p^{-n} \mid h_s$, so $\nu_p(h_s) \geq -n$, which implies that

$$\nu_p(q) = n + \nu_p(h_n) + \nu_p(h_s) \geq \nu_p(h_n) \geq 0.$$

(iii) Let $\mu = \min\left(\nu_\infty(f), \nu_\infty(w_1), \ldots, \nu_\infty(w_m)\right)$. Then $\nu_\infty(Dv) = \nu_\infty(f - \sum_{i=1}^m c_i v_i) \geq \mu$ by Theorem 4.3.1. Suppose now that $\nu_\infty(v) \neq 0$ and that t is either nonlinear or Liouvillian over k. If t is nonlinear, then $\nu_\infty(Dv) = \nu_\infty(v) - \delta(t) + 1$ by Theorem 4.4.4. If t is hyperexponential over k, then $\delta(t) = 1$, so $\nu_\infty(Dv) = \nu_\infty(v) = \nu_\infty(v) - \delta(t) + 1$ by Lemma 5.1.2. If t is primitive over k, then $\delta(t) = 0$ and $\nu_\infty(Dv) \in \{\nu_\infty(v), \nu_\infty(v) + 1\}$ by Lemma 5.1.2. Hence $\nu_\infty(Dv) \leq \nu_\infty(v) - \delta(t) + 1$ in all cases, so

$$\nu_\infty(v) \geq \nu_\infty(Dv) + \delta(t) - 1 \geq \mu + \delta - 1.$$

$\square$

Corollary 7.2.1. *Let $f, w_1, \ldots, w_m, c, h_n$ and h_s be as in Theorem 7.2.1. Then,*

(i) *For any solution $c_1, \ldots, c_m \in \mathrm{Const}(k)$ and $v \in k(t)$ of (7.30), $q = vh_n \in k\langle t \rangle$ and q is a solution of*

$$h_n Dq - q Dh_n = h_n^2 f - \sum_{i=1}^m c_i h_n^2 w_i. \tag{7.31}$$

Conversely, for any solution with $q \in k\langle t \rangle$ of (7.31), $v = q/h_n$ yield a solution of (7.30).

(ii) *If $S_1^{\mathrm{irr}} = S^{\mathrm{irr}}$, then for any solution $c_1, \ldots, c_m \in \mathrm{Const}(k)$ and $v \in k(t)$ of (7.30), $p = vh_n h_s \in k[t]$ and p is a solution of*

$$h_n h_s Dp - \left(Dh_n + h_n \frac{Dh_s}{h_s} \right) p = h_n^2 h_s f - \sum_{i=1}^m c_i h_n^2 h_s w_i. \tag{7.32}$$

In addition, if t is nonlinear or Liouvillian over k, then either

$$\deg(p) = \deg(h_n) + \deg(h_s),$$

or

$$\deg(p) \leq \deg(h_n) + \deg(h_s) + 1 - \delta(t) - \min\left(\nu_\infty(f), \nu_\infty(w_1), \ldots, \nu_\infty(w_m) \right).$$

Conversely, for any solution with $p \in k[t]$ of (7.32), $v = p/(h_n h_s)$ yield a solution of (7.30).

Proof. (i) Let $v \in k(t)$, $c_1, \ldots, c_m$ be a solution of (7.30), and let $q = vh_n$. $q \in k\langle t \rangle$ by Theorem 7.2.1, and

$$f = Dv + \sum_{i=1}^m c_i w_i = \frac{Dq}{h_n} - q \frac{Dh_n}{h_n^2} + \sum_{i=1}^m c_i w_i.$$

Multiplying through by h_n^2 yields (7.31). Conversely, the same calculation shows that for any solution with $q \in k\langle t \rangle$ of (7.31), $v = q/h_n$ yield a solution of (7.30).

(ii) Let $v \in k(t)$, $c_1, \ldots, c_m$ be a solution of (7.30), and let $p = vh_n h_s$. Since $S_1^{\mathrm{irr}} = S^{\mathrm{irr}}$, $p \in k[t]$ by Theorem 7.2.1, and

$$f = Dv + \sum_{i=1}^m c_i w_i = \frac{Dp}{h_n h_s} - p \frac{Dh_n}{h_n^2 h_s} - p \frac{Dh_s}{h_n h_s^2} + \sum_{i=1}^m c_i w_i.$$

Multiplying through by $h_n^2 h_s$ yields (7.32). Conversely, the same calculation shows that for any solution with $p \in k[t]$ of (7.32), $v = q/(h_n h_s)$ yield a solution of (7.30).

Suppose additionally that t is nonlinear or Liouvillian over k, and let $\mu = \min(\nu_\infty(f), \nu_\infty(w_1), \ldots, \nu_\infty(w_m))$. We have $\nu_\infty(v) = \nu_\infty(p/(h_n h_s)) = \deg(h_n) + \deg(h_s) - \deg(p)$. So if $\nu_\infty(v) = 0$, then $\deg(p) = \deg(h_n) + \deg(h_s)$. And if $\nu_\infty(v) \neq 0$, then $\nu_\infty(v) \geq \mu + \delta(t) - 1$ by Theorem 7.2.1, so $\deg(p) \leq \deg(h_n) + \deg(h_s) + 1 - \delta(t) - \mu$.   □

This gives us an algorithm that reduces a limited integration problem to one over $k\langle t \rangle$, or $k[t]$ if $\mathcal{S}_1^{\mathrm{irr}} = \mathcal{S}^{\mathrm{irr}}$.

LimitedIntegrateReduce$(f, w_1, \ldots, w_m, D)$
(* Reduction to a polynomial problem *)

(* Given a derivation D on $k(t)$ and $f, w_1, \ldots, w_m \in k(t)$, return $(a, b, h, N, g, v_1, \ldots, v_m)$ such that $a, b, h \in k[t]$, $N \in \mathbb{N}$, $g, v_1, \ldots, v_m \in k(t)$, and for any solution $v \in k(t)$, $c_1, \ldots, c_m \in C$ of $f = Dv + c_1 w_1 + \ldots c_m w_m$, $p = vh \in k\langle t \rangle$, and p and the c_i satisfy $aDp + bp = g + c_1 v_1 + \ldots + c_m v_m$). Furthermore, if $\mathcal{S}_1^{\mathrm{irr}} = \mathcal{S}^{\mathrm{irr}}$, then $p \in k[t]$, and if t is nonlinear or Liouvillian over k, then $\deg(p) \leq N$. *)

$(d_n, d_s) \leftarrow$ **SplitFactor**(denominator$(f), D)$
for $i \leftarrow 1$ **to** m **do** $(e_{n,i}, e_{s,i}) \leftarrow$ **SplitFactor**(denominator$(w_i), D)$
$c \leftarrow \mathrm{lcm}(d_n, e_{n,1}, \ldots, e_{n,m})$
$h_n \leftarrow \gcd(c, dc/dt)$
$a \leftarrow h_n, b \leftarrow -Dh_n, N \leftarrow 0$
if $\mathcal{S}_1^{\mathrm{irr}} = \mathcal{S}^{\mathrm{irr}}$ **then**
    $h_s \leftarrow \mathrm{lcm}(d_s, e_{s,1}, \ldots, e_{s,m})$
    $a \leftarrow h_n h_s, b \leftarrow -Dh_n - h_n Dh_s/h_s$ (* exact division *)
    $\mu \leftarrow \min(\nu_\infty(f), \nu_\infty(w_1), \ldots, \nu_\infty(w_m))$
    $N \leftarrow \deg(h_n) + \deg(h_s) + \max(0, 1 - \delta(t) - \mu)$
return$(a, b, a, N, ah_n f, -ah_n w_1, \ldots, -ah_n w_m)$

Example 7.2.1. Let $k = \mathbb{Q}(x)$ with $D = d/dx$ and let t be a monomial over k satisfying $Dt = 1/x$, *i.e.* $t = \log(x)$, and consider the limited integration problem
$$\frac{x}{t^2} = Dv + c_1 \frac{x}{t} \tag{7.33}$$
which arises when asking whether $\int x/\log(x)^2 dx$ is expressible in terms of $x, \log(x)$ and $\mathrm{Li}(x^2)$. We have $f = x/t^2$ and $w_1 = x/t$, so:

1. $(d_n, d_s) = $ **SplitFactor**$(t^2, D) = (t^2, 1)$
2. $(e_{n,1}, e_{s,1}) = $ **SplitFactor**$(t, D) = (t, 1)$
3. $c = \mathrm{lcm}(t^2, t) = t^2$
4. $h_n = \gcd(t^2, 2t/x) = t$
5. $a = t, b = -Dt = -1/x$

6. $h_s = \text{lcm}(1,1) = 1$
7. $\mu = \min(2,1) = 1$
8. $N = \deg(t) + \deg(1) + \max(0, 1 - \mu) = 1$
9. $ah_n f = x, \; -ah_n w_1 = -xt$

so any solution $c_1 \in \text{Const}(k)$ and $v \in k(t)$ of (7.33) must be of the form $v = q/t$ where $q \in k[t]$ has degree at most one and is a solution of

$$tDq - \frac{1}{x} q = x - c_1 xt . \tag{7.34}$$

In the case of Liouvillian or hypertangent monomials, we are reduced to finding solutions $c_1, \ldots, c_m \in \text{Const}(k)$ and $p \in k[t]$ of (7.32), which we rewrite as:

$$aDp + bp = g_0 + \sum_{i=1}^{m} c_i g_i \tag{7.35}$$

where $a, b \in k[t]$, $a \neq 0$ and $g_0, \ldots, g_m \in k(t)$. In addition, we have an upper bound n on $\deg(p)$ by Corollary 7.2.1. This is equivalent to looking for solutions $c_0, c_1, \ldots, c_m \in \text{Const}(k)$ and $p \in k[t]$ of

$$aDp + bp = \sum_{i=0}^{m} c_i g_i \tag{7.36}$$

with the additional constraint $c_0 = 1$. Since (7.36) is an equation of type (7.5), we can use the algorithms of the previous section to find all its solutions. This produces $h_1, \ldots, h_r \in k[t]$ and a matrix A with coefficients in $\text{Const}(k)$ such that any solution of (7.36) must be of the form

$$p = \sum_{j=1}^{r} d_j h_j$$

where $d_1, \ldots, d_r \in \text{Const}(k)$ and $A(c_0, \ldots, c_m, d_1, \ldots, d_r)^T = 0$. If this linear system has no solution with $c_0 = 1$, then (7.35) and the original limited integration problem have no solution, otherwise any solution with $c_0 = 1$ yields a solution of (7.35), hence of the original limited integration problem.

Several modifications can be made to the parametric Risch differential equation algorithm for the equations that arise from limited integration problems: if the linear constraints algorithm produces a 0-dimensional nullspace, then there are no solutions with $c_0 = 1$ and we can stop. If it produces a 1-dimensional nullspace, then there is a unique solution with $c_0 = 1$, so replacing $c_1, \ldots, c_m$ by that unique solution in (7.35) yields a nonparametric problem to which the SPDE algorithm of Sect. 6.4 is applicable. It is also possible to replace $c_1, \ldots, c_m$ by that unique solution in (7.30) and apply the in-field integration algorithm of Sect. 5.12, but that would imply recomputing the denominator of v. Finally, we can use the upper bound on $\deg(p)$ provided by Corollary 7.2.1 rather than recomputing it.

Example 7.2.2. Continuing example 7.2.1, let $k = \mathbb{Q}(x)$ with $D = d/dx$ and let t be a monomial over k satisfying $Dt = 1/x$, *i.e.* $t = \log(x)$, and consider the solutions $q \in k[t]$ with degree at most 1 of (7.34). Solving that equation is equivalent to finding a solution with $c_0 = 1$ of the parametric Risch differential equation

$$tDq - \frac{1}{x}q = c_0 x - c_1 x t.$$

That equation was solved in example 7.1.4, the general solution being

$$(q = \mu t - \lambda x^2, c_0 = \lambda, c_1 = 2\lambda).$$

Setting $c_0 = 1$, we find that (7.34) has a 1-parameter solution space, namely $c_1 = 2$ and $q = \mu t - x^2$ for any $\mu \in \text{Const}(k)$. This means that the solutions of (7.33) are

$$c_1 = 2 \quad \text{and} \quad v = \frac{q}{t} = \mu - \frac{x^2}{t}$$

for any $\mu \in \text{Const}(k)$ (it is of course normal for v to be defined up to an additive constant). As a consequence, we get

$$\int \frac{x\,dx}{\log(x)^2} = -\frac{x^2}{\log(x)} + 2\,\text{Li}(x^2).$$

7.3 The Parametric Logarithmic Derivative Problem

We describe in this section a solution to the parametric logarithmic derivative problem, *i.e.* given a differential field K of characteristic 0, an hyperexponential monomial θ over K and $f \in K$, to decide whether there are integers $n, m \in \mathbb{Z}$ with $n \neq 0$ such that

$$nf = \frac{Dv}{v} + m\frac{D\theta}{\theta} \tag{7.37}$$

has a solution $v \in K$, and to find one such solution if there are solutions. As we have seen in Chap. 5, this problem arises from determining whether elements of $K(\theta)$ are logarithmic derivatives of elements of $K(\theta)$ or logarithmic derivatives of $K(\theta)$-radicals. We can thus assume that we are able to determine recursively whether elements of K are logarithmic derivatives of K-radicals. Even though equation (7.37) is very similar to (7.30), the limited integration algorithm of Sect. 7.2 is not directly applicable to this problem. However, because the unknown constants are restricted to be integers, the structure theorems of Chap. 9 provide a complete solution to this problem whenever they are applicable. In fact they provide the only known complete solution, but we present first a variant of the linear constraints algorithm, which is often able to yield a unique potential solution for m/n, thereby solving the problem. This method is not a complete algorithm since it may fail to determine m/n, in which case we must revert to the structure theorems and their associated algorithms.

As previously, we only study equation (7.37) in the transcendental case, *i.e.* when K is a simple monomial extension of a differential subfield k, so for the rest of this section, let k be a differential field of characteristic 0 and t be a monomial over k. We assume in addition that $\mathrm{Const}(k(t)) = \mathrm{Const}(k)$. We suppose that the coefficients f and $D\theta/\theta$ of our equation are in $k(t)$ and look for solutions $n, m \in \mathbb{Z}$ and $v \in k(t)$.

Lemma 7.3.1. *Let $u, v, w \in k(t)$ and $c, \bar{c} \in \mathrm{Const}(k)$ be such that $v \neq 0$, $\bar{c} \neq 0$, and*

$$u = \bar{c}\frac{Dv}{v} + cw. \tag{7.38}$$

Write $u = p + a/d$ and $w = q + b/e$ where $p, q, a, b, d, e \in k[t]$, $d \neq 0$, $e \neq 0$, $\gcd(a, d) = \gcd(b, e) = 1$, $\deg(a) < \deg(d)$ and $\deg(b) < \deg(e)$. Then,

$$\deg(p - cq) \leq \max(0, \delta(t) - 1). \tag{7.39}$$

Furthermore, let $l = l_n l_s$ be a splitting factorization of $l = \mathrm{lcm}(d, e)$, and l_n^- be the deflation of l_n (Definition 1.6.2). Then,

$$lu - clw \equiv 0 \pmod{l_s l_n^-}.$$

Proof. Since $u = \bar{c}Dv/v + cw$, it follows that $u - cw = \bar{c}Dv/v$, hence that

$$\nu_\infty(u - cw) = \nu_\infty\left(\bar{c}\frac{Dv}{v}\right) = \nu_\infty\left(\frac{Dv}{v}\right) \geq -\max(0, \delta(t) - 1)$$

by Theorem 4.4.4. Since $u - cw = p - cq + a/d - cb/e$, either $\deg(p - cq) = 0$, or $\nu_\infty(u - cw) = -\deg(p - cq)$, in which case $\deg(p - cq) \leq \max(0, \delta(t) - 1)$. Since $l = \mathrm{lcm}(d, e)$, we have $lu - clw \in k[t]$. Let $p \in \mathcal{S}^{\mathrm{irr}}$ be any special irreducible factor of l_s. From $u - cw = \bar{c}Dv/v$ we get

$$\nu_p(u - cw) = \nu_p\left(\bar{c}\frac{Dv}{v}\right) = \nu_p\left(\frac{Dv}{v}\right) \geq 0$$

by Theorem 4.4.2. Therefore, $\nu_p(lu - clw) = \nu_p(l) + \nu_p(u - cw) \geq \nu_p(l) = \nu_p(l_s)$ since p is special. Since this holds for every irreducible factor of l_s, $l_s \mid lu - clw$. Let $p \in k[t]$ be any irreducible factor of l_n^-. From $u - cw = \bar{c}Dv/v$ we get

$$\nu_p(u - cw) = \nu_p\left(\bar{c}\frac{Dv}{v}\right) = \nu_p\left(\frac{Dv}{v}\right) \geq -1$$

by Corollary 4.4.2. Therefore,

$$\nu_p(lu - clw) = \nu_p(l) + \nu_p(u - cw) \geq \nu_p(l) - 1 = \nu_p(l_n) - 1 = \nu_p(l_n^-)$$

since $p \mid l_n$. Since this holds for every irreducible factor of l_n^-, $l_n^- \mid lu - clw$. Finally, $l_s l_n^- \mid lu - clw$ since $\gcd(l_s, l_n^-) = 1$. $\square$

Given $u, w \in k(t)$, Lemma 7.3.1 either proves that (7.38) has no solution $v \in k(t)$ and $c, \bar{c} \in \mathrm{Const}(k)$, or it yields a unique candidate $c \in \mathrm{Const}(k)$ for the solution in the following cases:

- If $\deg(q) > \max(0, \delta(t) - 1)$: then equating all the terms of $p - cq$ of degree higher than $\max(0, \delta(t) - 1)$ to 0 yields an overdetermined linear algebraic system for c. If this system has no solution in $\mathrm{Const}(k)$, then (7.38) has no solution, otherwise we get a unique candidate for c.
- If $\deg(p) > \max(0, \delta(t) - 1) \geq \deg(q)$: then (7.39) is never satisfied, so (7.38) has no solution.
- If $\deg(l_s l_n^-) > 0$, let then $r \in k[t]$ be the remainder of $lu - clw$ modulo $l_s l_n^-$. If r is identically 0, then $lu = lw \pmod{l_s l_n^-}$, which implies that $l_n^* u \in k[t]$ and $l_n^* w \in k[t]$ where l_n^* is the squarefree part of l_n, hence that d and e are normal, in contradiction with $\deg(l_s l_n^-) > 0$. Therefore r is not identically 0, so equating all its coefficients to 0 yields an overdetermined linear algebraic system for c. If this system has no solution in $\mathrm{Const}(k)$, then (7.38) has no solution, otherwise we get a unique candidate for c.

This in turns yields a method for solving (7.37): given f and θ, applying Lemma 7.3.1 to $u = f$ and $w = D\theta/\theta$, we can either prove that (7.38) has no solution with $c \in \mathbb{Q}$, in which case (7.37) has no solution, or get a unique candidate for $c = m/n$, or fail to get information about c if none of the above conditions is satisfied. If we get a unique candidate $c \in \mathbb{Q}$, write $c = M/N$ where $M, N \in \mathbb{Z}, N > 0$ and $\gcd(M, N) = 1$. Then for any solution of (7.37), we must have $n = QN$ and $m = QM$ for some nonzero integer Q, which implies that $QNf = Dv/v + QMD\theta/\theta$, hence that $Nf - MD\theta/\theta$ is the logarithmic derivative of a $k(t)$-radical, something that we can test recursively.

Note that if θ is an exponential over $k(t)$, then $D\theta/\theta = D\eta$ for some $\eta \in k(t)$. If $\nu_\infty(\eta) < 0$, then $\nu_\infty(D\eta) < -\max(0, \delta(t) - 1)$ by Theorem 4.4.4, so $\deg(q) > \max(0, \delta(t) - 1)$ and the above method succeeds. If $\nu_p(\eta) < 0$ for any normal irreducible $p \in k[t]$, then $\nu_p(D\eta) < -1$ by Theorem 4.4.2, so $p \mid l_n^-$ and the above method succeeds. If $\nu_p(\eta) < 0$ for any special $p \in S_1^{\mathrm{irr}}$, then $\nu_p(D\eta) < 0$ by Theorem 4.4.2, so $p \mid l_s$ and the above method succeeds. Thus, the only cases where the above method can fail when θ is an exponential over $k(t)$ and $S^{\mathrm{irr}} = S_1^{\mathrm{irr}}$ is if $\eta \in k$, i.e. θ is an exponential over k.

In a similar fashion, we see that if $f = Dg$ for some $g \in k(t)$, then the above method succeeds when $S^{\mathrm{irr}} = S_1^{\mathrm{irr}}$, unless $g \in k$. Thus if $S^{\mathrm{irr}} = S_1^{\mathrm{irr}}$, $f = Dg$ and θ is an exponential over $k(t)$, then the above method fails only if $f \in k$ and θ is an exponential over k, in which case an analysis similar to the one made in Sect. 5.12 shows that for any solution of (7.37), v must be in k^* if $Dt \in k$, in which case we are reduced to solving a similar problem over k, or v must be of the form $v = wt^q$ for $w \in k^*$ and an integer q if $Dt/t \in k$. In that latter case, we are reduced to solving an equation of the form

$$nDg = \frac{Dw}{w} + m\frac{D\theta}{\theta} + q\frac{Dt}{t} \tag{7.40}$$

where both θ and t are exponentials over k. Lemma 7.3.1 can be generalized to an arbitrary number of w's (Exercise 7.1) and applied to (7.40). This process stops when we reach the constant field, since $Dw = 0$ at that point, and (7.40) becomes a linear algebraic equation with integer unknowns. In practice however, it is preferable to use the structure theorems if Lemma 7.3.1 fails to produce c the first time around.

ParametricLogarithmicDerivative(f, θ, D)
(* Parametric Logarithmic Derivative Heuristic *)

(* Given a derivation D on $k[t]$, $f \in k(t)$ and a hyperexponential monomial θ over $k(t)$, returns either "failed", or "no solution", in which case $nf = Dv/v + mD\theta/\theta$ has no solution $v \in k(t)^*$ and $n, m \in \mathbb{Z}$ with $n \neq 0$, or a solution (n, m, v) of that equation. *)

$w \leftarrow D\theta/\theta$ $\qquad\qquad\qquad\qquad\qquad$ (* $w \in k(t)$ *)
$d \leftarrow \mathrm{denominator}(f)$, $e \leftarrow \mathrm{denominator}(w)$
$(p, a) \leftarrow$ **PolyDivide**$(\mathrm{numerator}(f), d)$ $\qquad$ (* $f = p + a/d$ *)
$(q, b) \leftarrow$ **PolyDivide**$(\mathrm{numerator}(w), e)$ $\qquad$ (* $w = q + b/e$ *)
$B \leftarrow \max(0, \deg(Dt) - 1)$
$C \leftarrow \max(\deg(p), \deg(q))$
if $\deg(q) > B$ **then**
$\quad s \leftarrow$ **solve**(coefficient$(p, t^i) = c$ coefficient$(q, t^i), B + 1 \leq i \leq C)$
$\quad$ **if** $s = \emptyset$ or $s \notin \mathbb{Q}$ **then return** "no solution"
$\quad N \leftarrow \mathrm{numerator}(s)$, $M \leftarrow \mathrm{denominator}(s)$ $\qquad$ (* $s \in \mathbb{Q}$ *)
$\quad$ **if** $Q(Nf - Mw) = Dv/v$ for some $Q \in \mathbb{Z}$ and $v \in k(t)$ with $Q \neq 0$
$\qquad$ and $v \neq 0$ **then return**(QN, QM, v) **else return** "no solution"
if $\deg(p) > B$ **then return** "no solution" $\qquad$ (* $\deg(q) \leq B$ *)
$l \leftarrow \mathrm{lcm}(d, e)$
$(l_n, l_s) \leftarrow$ **SplitFactor**(l, D)
$z \leftarrow l_s \gcd(l_n, dl_n/dt)$ $\qquad\qquad\qquad\qquad$ (* $z = l_s l_n^-$ *)
if $z \in k$ **then return** "failed"
$(u_1, r_1) \leftarrow$ **PolyDivide**(lf, z) $\qquad$ (* $r_1 \equiv lf \pmod{l_s l_n^-}$ *)
$(u_2, r_2) \leftarrow$ **PolyDivide**(lw, z) $\qquad$ (* $r_2 \equiv lw \pmod{l_s l_n^-}$ *)
$s \leftarrow$ **solve**(coefficient$(r_1, t^i) = c$ coefficient$(r_2, t^i), 0 \leq i < \deg(z))$
if $s = \emptyset$ or $s \notin \mathbb{Q}$ **then return** "no solution"
$M \leftarrow \mathrm{numerator}(s)$, $N \leftarrow \mathrm{denominator}(s)$ $\qquad$ (* $s \in \mathbb{Q}$ *)
if $Q(Nf - Mw) = Dv/v$ for some $Q \in \mathbb{Z}$ and $v \in k(t)$ with $Q \neq 0$ and
$v \neq 0$ **then return**(QN, QM, v) **else return** "no solution"

Example 7.3.1. Let $k = \mathbb{Q}(x)$ with $D = d/dx$, t be a monomial over k satisfying $Dt = 1/x$ and θ be an exponential monomial over $k(t)$ satisfying $D\theta = -\theta/(xt^2)$, *i.e.* $t = \log(x)$ and $\theta = e^{1/\log(x)}$, and consider the parametric logarithmic derivative problem

$$n \frac{5t^2 + t - 6}{2xt^2} = \frac{Dv}{v} + m \frac{D\theta}{\theta} \qquad (7.41)$$

for $n, m \in \mathbb{Z}$ and $v \in k(t)$. We get

1. $w = D\theta/\theta = -1/(xt^2)$
2. $(p, a) = \mathbf{PolyDivide}(5t^2 + t - 6, 2xt^2) = (5/(2x), t - 6)$
3. $(q, b) = \mathbf{PolyDivide}(-1, xt^2) = (0, -1)$
4. $B = \max(0, \deg_t(1/x) - 1) = 0$, $C = \max(\deg_t(5/(2x)), \deg_t(0)) = 0$
5. Since $\deg(p) \leq B$ and $\deg(q) \leq B$, $l = \mathrm{lcm}(2xt^2, xt^2) = xt^2$
6. $(l_n, l_s) = \mathbf{SplitFactor}(xt^2, D) = (xt^2, 1)$
7. $z = \gcd_t(xt^2, 2xt) = t$
8. $(u_1, r_1) = \mathbf{PolyDivide}(5/2t^2 + t/2 - 3, t) = (5/2t + 1/2, -3)$
9. $(u_2, r_2) = \mathbf{PolyDivide}(-1, t) = (0, -1)$
10. $s = \mathbf{solve}(-3 = -c) = 3$, so $M = 3$ and $N = 1$
11. $f - 3w = (5t + 1)/(2tx)$

Using the algorithm of Sect. 5.12, we find that $f - 3w$ is the logarithmic derivative of a $k(t)$-radical, namely

$$2(f - 3w) = \frac{5t + 1}{tx} = \frac{D(x^5 t)}{x^5 t}$$

so (7.41) has the solution $n = 2N = 2$, $m = 2M = 6$ and $v = x^5 t$. Note that it has in fact no solution with n and m coprime.

Example 7.3.2. Let $k = \mathbb{Q}$ and t be a monomial over k satisfying $Dt = 1$, θ be an exponential monomial over $k(t)$ satisfying $D\theta = \theta$, *i.e.* $D = d/dt$ and $\theta = e^t$, and consider the parametric logarithmic derivative problem

$$11 = \frac{Dv}{v} + m\frac{D\theta}{\theta} \tag{7.42}$$

for $m \in \mathbb{Z}$ and $v \in k(t)$. Even though this problem is trivial, it arises from bounding the degree in θ of the solutions $q \in k(t)[\theta]$ of

$$(t^2 + 2t + 1)Dq - (11t^2 + 22t + 10)q = \frac{234662231}{3628800}t + \frac{1255151}{28512} \tag{7.43}$$

which itself arises from computing the nontrivial integral[2]

$$\int \frac{2581284541e^t + 1757211400}{39916800e^{3t} + 119750400e^{2t} + 119750400e^t + 39916800} e^{1/(e^t + 1) - 10t} dt \ .$$

Despite its triviality, (7.42) is not solved by Lemma 7.3.1 because both 11 and $D\theta/\theta$ do not involve t. Since t is a primitive over k, (7.42) has a solution with $v \in k(t)$ if and only if it has a solution with $v \in k$. At this point, $Dv = 0$, so (7.42) becomes $11 = m$, whose solution yields the degree bound 11 on the solutions of (7.43). There happens to be a solution of degree 11, and the above integral is an elementary function.

---

[2] This example is attributed to M. Rothstein in [28].

The structure theorems of Chap. 9 provide an efficient alternative to solving (7.37): suppose first that f has an elementary integral over K, which turns out always to be the case in the parametric logarithmic problems that arise from the integration of elementary functions. Let then F be an elementary extension of $K(\theta)$ and $g \in F$ be such that $f = Dg$. Then, if (7.37) has a solution with $n \neq 0$, we get

$$nf = \frac{Dv}{v} + m\frac{D\theta}{\theta} = \frac{D(v\theta^m)}{v\theta^m}$$

which implies that $f = Dg$ is the logarithmic derivative of an F-radical. If $F = C(x)(t_1, \ldots, t_n)$ where $C = \mathrm{Const}(K)$, $Dx = 1$, and each t_i is either algebraic, or an elementary or real elementary, or a nonelementary primitive monomial over $C(x)(t_1, \ldots, t_{i-1})$, then it can be proven (Chap. 9) that f is the logarithmic derivative of an F-radical if and only if there are $r_i \in \mathbb{Q}$ such that

$$\sum_{i \in L} r_i Dt_i + \sum_{i \in E} r_i \frac{Dt_i}{t_i} = f \tag{7.44}$$

where

$$E = \{i \in \{1, \ldots, n\} \text{ s.t. } t_i \text{ exponential monomial over } C(x)(t_1, \ldots, t_{i-1})\}$$

and

$$L = \{i \in \{1, \ldots, n\} \text{ s.t. } t_i \text{ logarithmic monomial over } C(x)(t_1, \ldots, t_{i-1})\}.$$

Finding the rational solutions of (7.44) can be done by considering it a system of one linear equation for the r_i's with coefficients in F, then applying Lemma 7.1.2 to obtain a system with coefficients in C and the same constant solutions. Assuming that we have a vector space basis containing 1 for C over $\mathbb{Q}$, projecting that system on 1 yields a system with coefficients in $\mathbb{Q}$ and the same rational solutions as (7.44). This method is also applicable to equations of the form (7.40) with an arbitrary number of terms in the right hand side, since the existence of a solution implies that f is the logarithmic derivative of an F-radical.

Exercises

Exercise 7.1. Prove the following generalization of Lemma 7.3.1: let k be a differential field of characteristic 0, t be a monomial over k with $\mathrm{Const}(k(t)) = \mathrm{Const}(k)$, $v, w_0, \ldots, w_n \in k(t)$ and $c_1, \ldots, c_n, \bar{c} \in \mathrm{Const}(k)$ be such that $v \neq 0$, $\bar{c} \neq 0$, and

$$w_0 = \bar{c}\frac{Dv}{v} + \sum_{i=1}^{n} c_i w_i.$$

Write $w_i = p_i + a_i/d_i$ for $0 \leq i \leq n$ where $p_i, a_i, d_i \in k[t]$, $d_i \neq 0$, $\gcd(a_i, d_i) = 1$ and $\deg(a_i) < \deg(d_i)$. Then,

$$\deg\left(p_0 - \sum_{i=1}^{n} c_i p_i\right) \leq \max(0, \delta(t) - 1).$$

Furthermore, let $l = l_n l_s$ be a splitting factorization of $l = \mathrm{lcm}(d_0, \ldots, d_n)$, and l_n^- be the deflation of l_n. Then,

$$l w_0 - \sum_{i=1}^{n} c_i l w_i \equiv 0 \pmod{l_s l_n^-}.$$

Exercise 7.2. Solve the parametric logarithmic derivative problems (7.41) and (7.42) using the structure theorem approach.

8. The Coupled Differential System

We describe in this chapter the solution to the coupled differential system problem, *i.e.* given a differential field K of characteristic 0 and f_1, f_2, g_1, g_2 in K, to decide whether the system of equations

$$\begin{pmatrix} Dy_1 \\ Dy_2 \end{pmatrix} + \begin{pmatrix} f_1 & -f_2 \\ f_2 & f_1 \end{pmatrix} \begin{pmatrix} y_1 \\ y_2 \end{pmatrix} = \begin{pmatrix} g_1 \\ g_2 \end{pmatrix} \tag{8.1}$$

has a solution in $K \times K$, and to find one if there are some. It turns out that (8.1) is not really a second order equation, but the coupled system for the real and imaginary parts of a Risch differential equation. Indeed, suppose that $(y_1, y_2) \in K \times K$ is a solution of the slightly more general system

$$\begin{pmatrix} Dy_1 \\ Dy_2 \end{pmatrix} + \begin{pmatrix} f_1 & af_2 \\ f_2 & f_1 \end{pmatrix} \begin{pmatrix} y_1 \\ y_2 \end{pmatrix} = \begin{pmatrix} g_1 \\ g_2 \end{pmatrix} \tag{8.2}$$

for an arbitrary $a \in \mathrm{Const}_D(K)$. Then, since $D\sqrt{a} = 0$ by Lemma 3.3.2, writing $y = y_1 + y_2\sqrt{a}$ we have

$$
\begin{aligned}
Dy + (f_1 + f_2\sqrt{a})y &= Dy_1 + D(y_2)\sqrt{a} + (f_1 + f_2\sqrt{a})(y_1 + y_2\sqrt{a}) \\
&= Dy_1 + f_1y_1 + af_2y_2 + (Dy_2 + f_2y_1 + f_1y_2)\sqrt{a} \\
&= g_1 + g_2\sqrt{a}
\end{aligned}
$$

which implies that y is a solution in $K(\sqrt{a})$ of the Risch differential equation

$$Dy + (f_1 + f_2\sqrt{a})y = g_1 + g_2\sqrt{a}. \tag{8.3}$$

Conversely, if $\sqrt{a} \notin K$ and $y = y_1 + y_2\sqrt{a}$ satisfies (8.3) for $y_1, y_2 \in K$, then the above calculation shows that

$$Dy_1 + f_1y_1 + af_2y_2 + (Dy_2 + f_2y_1 + f_1y_2)\sqrt{a} = g_1 + g_2\sqrt{a}$$

hence that $Dy_1 + f_1y_1 + af_2y_2 = g_1$ and $Dy_2 + f_2y_1 + f_1y_2 = g_2$, since $\{1, \sqrt{a}\}$ is a vector space basis for $K(\sqrt{a})$ over K. Therefore, (y_1, y_2) is a solution of (8.2). Since coupled differential systems are generated by the integration algorithm only when $\sqrt{-1} \notin K$, the above remarks yield a trivial algorithm for finding the solutions in $K \times K$ of (8.1): find the solutions $y \in K(\sqrt{-1})$ of

$$Dy + (f_1 + f_2\sqrt{-1})y = g_1 + g_2\sqrt{-1}$$

and let y_1 and y_2 be the real and imaginary parts of y. While this approach is feasible, it has the following inconvenient: when $K = k(t)$ and t is a monomial over k, the Risch differential equation algorithms of Chap. 6 can generate Risch differential equations to be solved recursively over k. If we adjoin $\sqrt{-1}$ to K, then the equations to be solved recursively are over $k(\sqrt{-1})$, which means that any other hypertangent monomial in k has to be rewritten in terms of complex hyperexponentials. This cause the eventual solutions y_1, y_2 to be in a differential field isomorphic to K rather than in K itself, something that we would like to prevent. In order to avoid this problem, we present in this chapter direct algorithms corresponding to the cancellation cases of the Risch differential equation. When $K = k(t)$ and t is a monomial over k, these algorithms generate recursively coupled systems of the form (8.1) over k rather than Risch differential equations over $k(\sqrt{-1})$.

As in Chaps. 6 and 7, we only study (8.2) in the transcendental case, *i.e.* when K is a simple monomial extension of a differential subfield k, so for the rest of this chapter, let k be a differential field of characteristic 0 and t be a monomial over k. We assume in addition that $\mathrm{Const}(k(t)) = \mathrm{Const}(k)$. We suppose that the coefficients f_1, f_2, g_1 and g_2 of our system are in $k(t)$, that $a \in \mathrm{Const}(k)$ and $\sqrt{a} \notin k(t)$, and look for solutions $(y_1, y_2) \in k(t) \times k(t)$ of (8.2). In the first stage, we let $f = f_1 + f_2\sqrt{a}$, $g = g_1 + g_2\sqrt{a}$ and apply the algorithms of Chap. 6 to the Risch differential equation $Dy + fy = g$ up to and including the SPDE algorithm. At this point, either we have proven that (8.3) has no solution in $k(\sqrt{a})(t)$, in which case (8.2) has no solution in $k(t) \times k(t)$, or we have computed $b, c, d, \alpha, \beta \in k(\sqrt{a})[t]$ such that any solution in $k(\sqrt{a})(t)$ of (8.3) must be of the form $y = (\alpha q + \beta)/d$ where $q \in k(\sqrt{a})[t]$ is a solution of (6.19), *i.e.* $Dq + bq = c$. Furthermore, we have an upper bound n on $\deg(q)$. Although it may have been necessary to solve various problems over $k(\sqrt{a})$ recursively during the reduction of (8.3) to (6.19), those problems only occur when we compute various bounds on the poles of y, so after those integer bounds are computed we are again computing in $k(\sqrt{a})(t)$, even though we may have used isomorphic fields during the computation. If we are in one of the non-cancellation cases of Sect. 6.5, then we can apply the corresponding algorithms to the reduced equation $Dq + bq = c$ since they do not generate any recursive problem over $k(\sqrt{a})$. Thus, in the non-cancellation cases, we can either prove that (8.3) has no solution in $k(\sqrt{a})(t)$, or compute such a solution, thereby solving (8.2). We can therefore assume for the rest of this chapter that we are in one of the cancellation cases of Sect. 6.6, *i.e.* in one of the following cases:

1. $\delta(t) \leq 1$, $b \in k(\sqrt{a})$ and $D \neq d/dt$,
2. $\delta(t) \geq 2$, $\deg(b) = \delta(t) - 1$, and $\deg(q) = -\mathrm{lc}(b)/\lambda(t)$.

By our previous remarks, (6.19) is equivalent to

$$\begin{pmatrix} Dq_1 \\ Dq_2 \end{pmatrix} + \begin{pmatrix} b_1 & ab_2 \\ b_2 & b_1 \end{pmatrix} \begin{pmatrix} q_1 \\ q_2 \end{pmatrix} = \begin{pmatrix} c_1 \\ c_2 \end{pmatrix} \qquad (8.4)$$

where $q = q_1 + q_2\sqrt{a}$, $b = b_1 + b_2\sqrt{a}$, $c = c_1 + c_2\sqrt{a}$ and $q_1, q_2, b_1, b_2, c_1, c_2$ are in $k[t]$. Since $\sqrt{a} \notin k(t)$, $\deg(q) = \max(\deg(q_1), \deg(q_2))$, so $\deg(q_1) \le n$ and $\deg(q_2) \le n$. In addition, $\deg(b) = \max(\deg(b_1), \deg(b_2))$, so $b \in k(\sqrt{a})$ if and only if $b_1 \in k$ and $b_2 \in k$.

8.1 The Primitive Case

If $Dt \in k$, then $\delta(t) = 0$, so the only cancellation case for (8.4) is $b_1, b_2 \in k$.

If $b_1 = b_2 = 0$, then (8.4) becomes $Dq_1 = c_1$ and $Dq_2 = c_2$ for c_1, c_2 in $k[t]$, which are integration problems in $k[t]$, and deciding whether they have solutions in $k[t]$ can be done by the in-field integration algorithm (Sect. 5.12), so suppose now that $b_1 \in k^*$ or $b_2 \in k^*$.

If $b_1 + b_2\sqrt{a} = Du/u$ for some $u \in k(\sqrt{a})^*$, which can also be checked by a variant of the integration algorithm (Sect. 5.12), let then

$$\begin{pmatrix} p_1 \\ p_2 \end{pmatrix} = \begin{pmatrix} u_1 & au_2 \\ u_2 & u_1 \end{pmatrix} \begin{pmatrix} q_1 \\ q_2 \end{pmatrix} \tag{8.5}$$

where $u = u_1 + u_2\sqrt{a}$ with $u_1, u_2 \in k$. We have

$$\begin{aligned} p = p_1 + p_2\sqrt{a} &= (u_1 q_1 + au_2 q_2) + (u_2 q_1 + u_1 q_2)\sqrt{a} \\ &= (u_1 + u_2\sqrt{a})(q_1 + q_2\sqrt{a}) = uq \end{aligned}$$

so

$$Dp = D(uq) = uDq + qDu = uDq + ubq = u(Dq + bq) = uc$$

which implies that (8.4) becomes

$$\begin{pmatrix} Dp_1 \\ Dp_2 \end{pmatrix} = \begin{pmatrix} u_1 & au_2 \\ u_2 & u_1 \end{pmatrix} \begin{pmatrix} c_1 \\ c_2 \end{pmatrix}$$

which is, as earlier, a pair of integration problems in $k[t]$. The change of variable (8.5) is invertible since

$$\begin{vmatrix} u_1 & au_2 \\ u_2 & u_1 \end{vmatrix} = u_1^2 - au_2^2 = (u_1 + u_2\sqrt{a})(u_1 - u_2\sqrt{a}) \ne 0 \,.$$

Hence, solutions of the integration problems yield solutions q_1, q_2 of (8.4). Note that a necessary condition for $b_1 + b_2\sqrt{a} = Du/u$ is $2b_1 = Dv/v$ for some $v \in k^*$, and that condition can be tested in k rather than $k(\sqrt{a})$.

If $b_1 + b_2\sqrt{a}$ is not of the form Du/u for some $u \in k(\sqrt{a})^*$, then $D(\mathrm{lc}(q)) + b\,\mathrm{lc}(q) \ne 0$, so the leading monomial of $Dq + bq$ is

$$(D(\mathrm{lc}(q)) + b\,\mathrm{lc}(q))t^{\deg(q)} \,.$$

This implies that $\deg(q) = \deg(c)$, and that $\mathrm{lc}(q)$ is a solution in $k(\sqrt{a})^*$ of (6.23), i.e. $Dy + by = \mathrm{lc}(c)$. Since $\deg(q) = \max(\deg(q_1), \deg(q_2))$ and

$\deg(c) = \max(\deg(c_1), \deg(c_2))$, we get that $\deg(q_1) \leq n$ and $\deg(q_2) \leq n$ where $n = \max(\deg(c_1), \deg(c_2))$. Furthermore, $\mathrm{lc}(q) = y_1 + y_2\sqrt{a}$ and $\mathrm{lc}(c) = z_1 + z_2\sqrt{a}$ where y_1, y_2, z_1 and z_2 are the coefficients of t^n in q_1, q_2, c_1 and c_2. Therefore (6.23) is equivalent to

$$\begin{pmatrix} Dy_1 \\ Dy_2 \end{pmatrix} + \begin{pmatrix} b_1 & ab_2 \\ b_2 & b_1 \end{pmatrix} \begin{pmatrix} y_1 \\ y_2 \end{pmatrix} = \begin{pmatrix} z_1 \\ z_2 \end{pmatrix}$$

which is a coupled differential system in k. If it has no solution in k, then (8.4) has no solution in $k[t]$. Otherwise, since it is equivalent to (6.23), Lemma 5.9.1 implies that it has a unique solution y_1, y_2, which must be the coefficients of t^n in q_1 and q_2. Replacing each q_i by $y_i t^n + h_i$ in (8.4), we get

$$\begin{pmatrix} Dh_1 \\ Dh_2 \end{pmatrix} + \begin{pmatrix} b_1 & ab_2 \\ b_2 & b_1 \end{pmatrix} \begin{pmatrix} h_1 \\ h_2 \end{pmatrix} = \begin{pmatrix} c_1 - D(y_1 t^n) \\ c_2 - D(y_2 t^n) \end{pmatrix} - \begin{pmatrix} b_1 & ab_2 \\ b_2 & b_1 \end{pmatrix} \begin{pmatrix} y_1 t^n \\ y_2 t^n \end{pmatrix}$$

which is a system of the same type as (8.4) with the same b_1 and b_2 as before, but a bound of $n - 1$ on $\deg(h_1)$ and $\deg(h_2)$. We can therefore repeat this process, decreasing the bound each time until we have solved (8.4).

CoupledDECancelPrim$(a, b_1, b_2, c_1, c_2, D, n)$
(* Cancellation – primitive case *)

(* Given a derivation D on $k[t]$, n either an integer or $+\infty$, $a \in \mathrm{Const}(k)$, $b_1, b_2 \in k$ and $c_1, c_2 \in k[t]$ with $Dt \in k$, $\sqrt{a} \notin k(t)$ and $b_1 \neq 0$ or $b_2 \neq 0$, return either "no solution", in which case the system (8.4) has no solution with both degrees at most n in $k[t]$, or a solution $q_1, q_2 \in k[t] \times k[t]$ of this system with $\deg(q_1) \leq n$ and $\deg(q_2) \leq n$. *)

if $b = Dz/z$ for $z \in k(\sqrt{a})^*$ **then**
$\quad z = z_1 + z_2\sqrt{a}$ $\qquad\qquad\qquad\qquad\qquad$ (* $z_1, z_2 \in k$ *)
$\quad$ **if** $z_1 c_1 + a z_2 c_2 = Dp_1$ and $z_2 c_1 + z_1 c_2 = Dp_2$ for $p_1, p_2 \in k[t]$ with
$\quad\quad \deg(p_1) \leq n$ and $\deg(p_2) \leq n$
$\quad\quad$ **then return** $((z_1 p_1 - a z_2 p_2)/(z_1^2 - a z_2^2), (z_1 p_2 - z_2 p_1)/(z_1^2 - a z_2^2))$
$\quad$ **else return** "no solution"
if $c_1 = 0$ and $c_2 = 0$ **then return** $(0, 0)$
if $n < \max(\deg(c_1), \deg(c_2))$ **then return** "no solution"
$q_1 \leftarrow 0, q_2 \leftarrow 0$
while $c_1 \neq 0$ or $c_2 \neq 0$ **do**
$\quad m \leftarrow \max(\deg(c_1), \deg(c_2))$ $\qquad$ (* m becomes smaller at each pass *)
$\quad$ **if** $n < m$ **then return** "no solution"
$\quad (s_1, s_2) \leftarrow$ **CoupledDESystem**$(b_1, b_2,$
$\quad\quad\quad\quad\quad\quad\quad\quad\quad\quad \mathrm{coefficient}(c_1, t^m), \mathrm{coefficient}(c_2, t^m))$
$\quad$ **if** $(s_1, s_2) = $ "no solution" **then return** "no solution"
$\quad q_1 \leftarrow q_1 + s_1 t^m, q_2 \leftarrow q_2 + s_2 t^m$
$\quad n \leftarrow m - 1$
$\quad c_1 \leftarrow c_1 - D(s_1 t^m) - (b_1 s_1 + ab_2 s_2)t^m$
$\quad c_2 \leftarrow c_2 - D(s_2 t^m) - (b_2 s_1 + b_1 s_2)t^m$
return (q_1, q_2)

8.2 The Hyperexponential Case

If $Dt/t = \eta \in k$, then $\delta(t) = 1$, so the only cancellation case for (8.4) is $b_1, b_2 \in k$.

If $b_1 = b_2 = 0$, then (8.4) becomes $Dq_1 = c_1$ and $Dq_2 = c_2$ for c_1, c_2 in $k[t]$, which are integration problems in $k[t]$, and deciding whether they have solutions in $k[t]$ can be done by the in-field integration algorithm (Sect. 5.12), so suppose now that $b_1 \in k^*$ or $b_2 \in k^*$.

If $b_1 + b_2\sqrt{a} = Du/u + m\eta$ for some $u \in k(\sqrt{a})^*$ and $m \in \mathbb{Z}$, let then

$$
\begin{pmatrix} p_1 \\ p_2 \end{pmatrix} = \begin{pmatrix} u_1 & au_2 \\ u_2 & u_1 \end{pmatrix} \begin{pmatrix} q_1 t^m \\ q_2 t^m \end{pmatrix}
\tag{8.6}
$$

where $u = u_1 + u_2\sqrt{a}$ with $u_1, u_2 \in k$. We have

$$
\begin{aligned}
p = p_1 + p_2\sqrt{a} &= (u_1 q_1 t^m + au_2 q_2 t^m) + (u_2 q_1 t^m + u_1 q_2 t^m)\sqrt{a} \\
&= (u_1 + u_2\sqrt{a})(q_1 + q_2\sqrt{a})t^m = uq t^m
\end{aligned}
$$

so

$$
Dp = D(uq t^m) = (uDq + qDu + m\eta uq)t^m = u(Dq + bq)t^m = uct^m
$$

which implies that (8.4) becomes

$$
\begin{pmatrix} Dp_1 \\ Dp_2 \end{pmatrix} = \begin{pmatrix} u_1 & au_2 \\ u_2 & u_1 \end{pmatrix} \begin{pmatrix} c_1 t^m \\ c_2 t^m \end{pmatrix}
$$

which is a pair of integration problems in $k\langle t \rangle$, and deciding whether they have solutions in $k\langle t \rangle$ can be done by a variant of the integration algorithm (Sect. 5.12). As in the primitive case, the change of variable (8.6) is invertible, so solutions of the integration problems yield solutions of (8.4). Note that a necessary condition for $b_1 + b_2\sqrt{a} = Du/u + m\eta$ is $2b_1 = Dv/v + 2m\eta$ for some $v \in k^*$, and that condition can be tested in k rather than $k(\sqrt{a})$, yielding a unique potential candidate for the integer m.

Suppose finally that $b_1 + b_2\sqrt{a}$ is not of the form $Du/u + m\eta$ for some $u \in k(\sqrt{a})^*$ and $m \in \mathbb{Z}$. Then $D(\mathrm{lc}(q)) + \deg(q)\,\eta\,\mathrm{lc}(q) + b\,\mathrm{lc}(q) \neq 0$, so the leading monomial of $Dq + bq$ is

$$
(D(\mathrm{lc}(q)) + \deg(q)\,\eta\,\mathrm{lc}(q) + b\,\mathrm{lc}(q))\, t^{\deg(q)}.
$$

This implies that $\deg(q) = \deg(c)$, and that $\mathrm{lc}(q)$ is a solution in $k(\sqrt{a})^*$ of (6.24), i.e. $Dy + (b + \deg(q)\eta)y = \mathrm{lc}(c)$. Since $\deg(c) = \max(\deg(c_1), \deg(c_2))$ and $\deg(q) = \max(\deg(q_1), \deg(q_2))$, we get that $\deg(q_1) \leq n$ and $\deg(q_2) \leq n$ where $n = \max(\deg(c_1), \deg(c_2))$. Furthermore, $\mathrm{lc}(q) = y_1 + y_2\sqrt{a}$ and $\mathrm{lc}(c) = z_1 + z_2\sqrt{a}$ where y_1, y_2, z_1 and z_2 are the coefficients of t^n in q_1, q_2, c_1 and c_2. Therefore (6.24) is equivalent to

$$
\begin{pmatrix} Dy_1 \\ Dy_2 \end{pmatrix} + \begin{pmatrix} b_1 + n\eta & ab_2 \\ b_2 & b_1 + n\eta \end{pmatrix} \begin{pmatrix} y_1 \\ y_2 \end{pmatrix} = \begin{pmatrix} z_1 \\ z_2 \end{pmatrix}
$$

which is a coupled differential system in k. If it has no solution in k, then (8.4) has no solution in $k[t]$. Otherwise, since it is equivalent to (6.24), Lemma 5.9.1 implies that it has a unique solution y_1, y_2, which must be the coefficients of t^n in q_1 and q_2. As in the primitive case, replacing q_1 by $y_1 t^n + h_1$ and q_2 by $y_2 t^n + h_2$ in (8.4) yields a system of the same type with a lower degree bound on its solutions, and a lower degree right hand side.

CoupledDECancelExp$(a, b_1, b_2, c_1, c_2, D, n)$
(* Cancellation – hyperexponential case *)

(* Given a derivation D on $k[t]$, n either an integer or $+\infty$, $a \in \mathrm{Const}(k)$, $b_1, b_2 \in k$ and $c_1, c_2 \in k[t]$ with $Dt/t \in k$, $\sqrt{a} \notin k(t)$ and $b_1 \neq 0$ or $b_2 \neq 0$, return either "no solution", in which case the system (8.4) has no solution with both degrees at most n in $k[t]$, or a solution $q_1, q_2 \in k[t] \times k[t]$ of this system with $\deg(q_1) \leq n$ and $\deg(q_2) \leq n$. *)

if $b = Dz/z + mDt/t$ for $z \in k(\sqrt{a})^*$ and $m \in \mathbb{Z}$ **then**
$\quad z = z_1 + z_2\sqrt{a}$ $\hspace{4cm}$ (* $z_1, z_2 \in k$ *)
$\quad$**if** $(z_1 c_1 + a z_2 c_2)t^m = Dp_1$ and $(z_2 c_1 + z_1 c_2)t^m = Dp_2$ for $p_1, p_2 \in k\langle t\rangle$
$\quad$**then**
$\qquad q_1 \leftarrow (z_1 p_1 - a z_2 p_2)t^{-m}/(z_1^2 - a z_2^2)$
$\qquad q_2 \leftarrow (z_1 p_2 - z_2 p_1)t^{-m}/(z_1^2 - a z_2^2)$
$\qquad$**if** $q_1 \in k[t]$ and $q_2 \in k[t]$ and $\deg(q_1) \leq n$ and $\deg(p_2) \leq n$
$\qquad\qquad$**then return**(q_1, q_2) **else return** "no solution"
$\quad$**else return** "no solution"
if $c_1 = 0$ and $c_2 = 0$ **then return** $(0, 0)$
if $n < \max(\deg(c_1), \deg(c_2))$ **then return** "no solution"
$q_1 \leftarrow 0$, $q_2 \leftarrow 0$
while $c_1 \neq 0$ or $c_2 \neq 0$ **do**
$\quad m \leftarrow \max(\deg(c_1), \deg(c_2))$ $\hspace{1.5cm}$ (* m becomes smaller at each pass *)
$\quad$**if** $n < m$ **then return** "no solution"
$\quad (s_1, s_2) \leftarrow$ **CoupledDESystem**$(b_1 + mDt/t, b_2,$
$\hspace{4.5cm}$ **coefficient**(c_1, t^m), **coefficient**$(c_2, t^m))$
$\quad$**if** $(s_1, s_2) =$ "no solution" **then return** "no solution"
$\quad q_1 \leftarrow q_1 + s_1 t^m$, $q_2 \leftarrow q_2 + s_2 t^m$
$\quad n \leftarrow m - 1$
$\quad c_1 \leftarrow c_1 - D(s_1 t^m) - (b_1 s_1 + a b_2 s_2)t^m$
$\quad c_2 \leftarrow c_2 - D(s_2 t^m) - (b_2 s_1 + b_1 s_2)t^m$
return (q_1, q_2)

8.3 The Nonlinear Case

If $\delta(t) \geq 2$, then the only cancellation case for (8.4) is $\max(\deg(b_1), \deg(b_2)) = \delta(t) - 1$ and $\mathrm{lc}(b) = -n\lambda(t)$ where $n > 0$ is the bound on $\deg(q)$. Since $\mathrm{lc}(b) = \beta_1 + \beta_2\sqrt{a}$ where β_1 and β_2 are the coefficients of $t^{\delta(t)-1}$ in b_1 and b_2, and $\lambda(t) \in k$, we must have $\beta_2 = 0$ and $\beta_1 = -n\lambda(t)$, i.e. $\deg(b_1) = \delta(t) - 1$,

$\deg(b_2) < \delta(t) - 1$ and $\mathrm{lc}(b_1) = -n\lambda(t)$. As in the Risch differential equation case, there is no general algorithm for solving the system (8.4) in this case. If however $\mathcal{S}^{\mathrm{irr}} \neq \emptyset$, then projecting (8.4) to $k[t]/(p)$ for $p \in \mathcal{S}^{\mathrm{irr}}$ can be done: with D^* being the induced derivation on $k[t]/(p)$, applying π_p to (8.4) we get

$$\begin{pmatrix} D^*q_1^* \\ D^*q_2^* \end{pmatrix} + \begin{pmatrix} \pi_p(b_1) & \pi_p(a)\pi_p(b_2) \\ \pi_p(b_2) & \pi_p(b_1) \end{pmatrix} \begin{pmatrix} q_1^* \\ q_2^* \end{pmatrix} = \begin{pmatrix} \pi_p(c_1) \\ \pi_p(c_2) \end{pmatrix} \qquad (8.7)$$

where $q_1^* = \pi_p(q_1)$ and $q_2^* = \pi_p(q_2)$. Assuming that we have an algorithm for solving (8.7) in $k[t]/(p)$, we can then solve (8.4) as follows: if (8.7) has no solution in $k[t]/(p)$, then (8.4) has no solution in $k[t]$. Otherwise, let $(q_1^*, q_2^*) \in k[t]/(p) \times k[t]/(p)$ be a solution of (8.7), and let $r_1, r_2 \in k[t]$ be such that $\deg(r_1) < \deg(p)$, $\deg(r_2) < \deg(p)$, $\pi_p(r_1) = q_1^*$ and $\pi_p(r_2) = q_2^*$. Note that $\pi_p(Dr_1 + b_1r_1 + ab_2r_2) = \pi_p(c_1)$, and $\pi_p(Dr_2 + b_2r_1 + b_1r_2) = \pi_p(c_2)$, so $p \mid c_1 - Dr_1 - b_1r_1 - ab_2r_2$ and $p \mid c_2 - Dr_2 - b_2r_1 - b_1r_2$. In addition, $\pi_p(q_1) = \pi_p(r_1)$ and $\pi_p(q_2) = \pi_p(r_2)$, so $h_1 = (q_1 - r_1)/p \in k[t]$, $h_2 = (q_2 - r_2)/p \in k[t]$ and we have

$$\begin{aligned} \begin{pmatrix} c_1 \\ c_2 \end{pmatrix} &= \begin{pmatrix} Dq_1 \\ Dq_2 \end{pmatrix} + \begin{pmatrix} b_1 & ab_2 \\ b_2 & b_1 \end{pmatrix} \begin{pmatrix} q_1 \\ q_2 \end{pmatrix} \\ &= p\left(\begin{pmatrix} Dh_1 \\ Dh_2 \end{pmatrix} + \begin{pmatrix} b_1 + \frac{Dp}{p} & ab_2 \\ b_2 & b_1 + \frac{Dp}{p} \end{pmatrix} \begin{pmatrix} h_1 \\ h_2 \end{pmatrix} \right) \\ &\quad + \begin{pmatrix} Dr_1 \\ Dr_2 \end{pmatrix} + \begin{pmatrix} b_1 & ab_2 \\ b_2 & b_1 \end{pmatrix} \begin{pmatrix} r_1 \\ r_2 \end{pmatrix} \end{aligned}$$

so (h_1, h_2) is a solution in $k[t] \times k[t]$ of

$$\begin{pmatrix} Dh_1 \\ Dh_2 \end{pmatrix} + \begin{pmatrix} b_1 + \frac{Dp}{p} & ab_2 \\ b_2 & b_1 + \frac{Dp}{p} \end{pmatrix} \begin{pmatrix} h_1 \\ h_2 \end{pmatrix} =$$
$$\frac{1}{p}\left(\begin{pmatrix} c_1 - Dr_1 \\ c_2 - Dr_2 \end{pmatrix} - \begin{pmatrix} b_1 & ab_2 \\ b_2 & b_1 \end{pmatrix} \begin{pmatrix} r_1 \\ r_2 \end{pmatrix} \right) \qquad (8.8)$$

which is a coupled system of type (8.4), but with a lower bound on the degree of its solutions since $\deg(h_1) \leq \deg(q_1) - \deg(p)$ and $\deg(h_2) \leq \deg(q_2) - \deg(p)$.

As was the case for Risch differential equations, there are cases when (8.7) can be solved, for example when we can find an element of degree one, or an element with constant coefficients, in $\mathcal{S}^{\mathrm{irr}}$. Although $\sqrt{a} \notin k(t)$, it may happen that $\sqrt{\pi_p(a)} \in k[t]/(p)$, in which case two new difficulties arise:

- p is then reducible over $k(\sqrt{a})$, so we must use an irreducible factor $\overline{p}$ of p in $k(\sqrt{a})[t]$ rather than p.
- (8.7) is not equivalent to a Risch differential equation over $k[t]/(\overline{p})$ anymore, so we must revert to solving (6.25), taking care to generate coupled systems over k recursively, rather than Risch differential equations over $k(\sqrt{a})$.

An example of those difficulties is provided by the hypertangent case with $a = -1$ and $\mathcal{S}^{\mathrm{irr}} = \{t^2 + 1\}$.

8.4 The Hypertangent Case

If $Dt/(t^2 + 1) = \eta \in k$, then $\delta(t) = 2$, so the only cancellation case for (8.4) is $b_2 \in k$ and $b_1 = b_0 - n\eta t$, where $b_0 \in k$ and $n > 0$ is the bound on $\deg(q)$. In such extensions, the method outlined above is not applicable for $a = -1$, since $S^{\text{irr}} = \{t^2 + 1\}$, and $\sqrt{-1} \in k[t]/(t^2 + 1)$. Since $\sqrt{-1} \notin k(t)$ by assumption, (8.4) is equivalent to (6.19), in this case

$$Dq + (b_0 - n\eta t + b_2\sqrt{-1})q = c_1 + c_2\sqrt{-1}$$

and we can use the method of Sect. 6.6: taking

$$p = t - \sqrt{-1} \quad \in S^{\text{irr}}_{k(\sqrt{-1})[t]:k(\sqrt{-1})}$$

(6.25) becomes

$$Dq^* + (b_0 + (b_2 - n\eta)\sqrt{-1})q^* = c_1(\sqrt{-1}) + c_2(\sqrt{-1})\sqrt{-1} \qquad (8.9)$$

where D is extended to $k(\sqrt{-1})$ by $D\sqrt{-1} = 0$ and $q^* = q(\sqrt{-1}) \in k(\sqrt{-1})$. Writing

$$q^* = y_1 + y_2\sqrt{-1} \quad \text{and} \quad c_1(\sqrt{-1}) + c_2(\sqrt{-1})\sqrt{-1} = z_1 + z_2\sqrt{-1}$$

where $y_1, y_2, z_1, z_2 \in k$, (8.9) is equivalent to

$$\begin{pmatrix} Dy_1 \\ Dy_2 \end{pmatrix} + \begin{pmatrix} b_0 & n\eta - b_2 \\ b_2 - n\eta & b_0 \end{pmatrix} \begin{pmatrix} y_1 \\ y_2 \end{pmatrix} = \begin{pmatrix} z_1 \\ z_2 \end{pmatrix} \qquad (8.10)$$

which is a coupled differential system in k. If it has no solution in k, then (8.9) has no solution in $k(\sqrt{-1})$, which implies that (8.4) has no solution in $k[t]$. Otherwise, if $y_1, y_2 \in k \times k$ is a solution of (8.10), letting $r = y_1 + y_2\sqrt{-1}$, $h = (q - r)/(t - \sqrt{-1})$ is a solution in $k(\sqrt{-1})[t]$ of degree at most $n - 1$ of (6.26), which in this case becomes

$$Dh + \left(b_0 - (n-1)\eta t + (b_2 + \eta)\sqrt{-1}\right)h = \frac{c - Dr - br}{t - \sqrt{-1}}$$

$$= \frac{c_1 - z_1 + n\eta(y_1 t + y_2) + (c_2 - z_2 + n\eta(y_2 t - y_1))\sqrt{-1}}{t - \sqrt{-1}}$$

where the right hand side is an exact quotient in $k(\sqrt{-1})[t]$. Repeating this process we either prove that (8.4) has no solution in $k[t]$, or obtain a solution $q \in k(\sqrt{-1})[t]$ of (6.19). Writing $q = q_1 + q_2\sqrt{-1}$ with $q_1, q_2 \in k[t]$, we get that q_1, q_2 is a solution of (8.4).

CoupledDECancelTan$(b_0, b_2, c_1, c_2, D, n)$
(* Cancellation – tangent case *)

(* Given a derivation D on $k[t]$, n either an integer or $+\infty$, $b_0, b_2 \in k$ and $c_1, c_2 \in k[t]$ with $Dt/(t^2 + 1) = \eta \in k$, $\sqrt{-1} \notin k(t)$ and $b_0 \neq 0$ or $b_2 \neq 0$, return either "no solution", in which case the system

$$\begin{pmatrix} Dq_1 \\ Dq_2 \end{pmatrix} + \begin{pmatrix} b_0 - n\eta t & -b_2 \\ b_2 & b_0 + n\eta t \end{pmatrix} \begin{pmatrix} q_1 \\ q_2 \end{pmatrix} = \begin{pmatrix} c_1 \\ c_2 \end{pmatrix}$$

has no solution with both degrees at most n in $k[t]$, or a solution $q_1, q_2 \in k[t] \times k[t]$ of this system with $\deg(q_1) \leq n$ and $\deg(q_2) \leq n$. *)

if $n = 0$ **then**
    if $c_1 \in k$ and $c_2 \in k$ **then return CoupledDESystem**(b_0, b_2, c_1, c_2)
    else return "no solution"
$p \leftarrow t - \sqrt{-1}$
$\eta \leftarrow Dt/(t^2 + 1)$ (* $t = \tan(\int \eta)$ *)
$c_1(\sqrt{-1}) + c_2(\sqrt{-1})\sqrt{-1} = z_1 + z_2\sqrt{-1}$ (* $z_1, z_2 \in k$ *)
$(s_1, s_2) \leftarrow$ **CoupledDESystem**$(b_0, b_2 - n\eta, z_1, z_2)$
if $(s_1, s_2) = $ "no solution" **then return** "no solution"
$c \leftarrow (c_1 - z_1 + n\eta(s_1 t + s_2) + (c_2 - z_2 + n\eta(s_2 t - s_1))\sqrt{-1})/p$
$c = d_1 + d_2\sqrt{-1}$ (* $d_1, d_2 \in k[t]$ *)
$(h_1, h_2) \leftarrow$ **CoupledDECancelTan**$(b_0, b_2 + \eta, d_1, d_2, D, n - 1)$
if $(h_1, h_2) = $ "no solution" **then return** "no solution"
return$(h_1 t + h_2 + s_1, h_2 t - h_1 + s_2)$

Example 8.4.1. Let $k = \mathbb{Q}(x)$ with $D = d/dx$, and let t be a monomial over k satisfying $Dt = 1 + t^2$, *i.e.* $t = \tan(x)$, and consider the coupled system

$$\begin{pmatrix} Dy_1 \\ Dy_2 \end{pmatrix} + \begin{pmatrix} 0 & -4x \\ 4x & 0 \end{pmatrix} \begin{pmatrix} y_1 \\ y_2 \end{pmatrix} = \begin{pmatrix} -(t^2 - 2t + 8x^2 - 1)/(t^2 + 1) \\ 2(1 - 2x)/(t^2 + 1) \end{pmatrix} \quad (8.11)$$

which arises from computing

$$\int -\frac{(\tan(x)^2 - 2\tan(x) + 8x^2 - 1)\tan(x^2) + 4x - 2}{(\tan(x)^2 + 1)(\tan(x^2)^2 + 1)} \, dx. \quad (8.12)$$

The system (8.11) is equivalent to the Risch differential equation

$$Dy + 4xy\sqrt{-1} = -\frac{(t^2 - 2t + 8x^2 - 1) + 2(2x - 1)\sqrt{-1}}{t^2 + 1} \quad (8.13)$$

over $k(\sqrt{-1})(t)$. Since $4x\sqrt{-1}$ is weakly normalized w.r.t. t and the denominator of the right hand side of (8.13) is special, Theorem 6.1.2 implies that any solution in $k(\sqrt{-1})(t)$ of (8.13) must be in $k(\sqrt{-1})\langle t \rangle$. With $a = 1$, $b = 4x\sqrt{-1}$ and $\epsilon = \pm 1$, we have

$$-\frac{b(\epsilon\sqrt{-1})}{a(\epsilon\sqrt{-1})} = -4x\sqrt{-1}$$

which is not of the form $2m\epsilon\sqrt{-1} + Du/u$ for $m \in \mathbb{Z}$ and $u \in k(\sqrt{-1})^*$, so $\nu_{t-\epsilon\sqrt{-1}}(Dy + 4xy\sqrt{-1}) = \nu_{t-\epsilon\sqrt{-1}}(y)$ for $\epsilon = \pm 1$ and any $y \in k(\sqrt{-1})(t)$ by Lemma 6.2.1 (see also Exercise 6.1). Hence, $\nu_{t-\epsilon\sqrt{-1}}(y) \geq -1$ for any solution $y \in k(\sqrt{-1})(t)$ of (8.13), which implies that any such solution must be of the form $y = q/(t^2 + 1)$ where $q \in k(\sqrt{-1})[t]$. Making that substitution in (8.13) we obtain

$$Dq + (4x\sqrt{-1} - 2t)q = -(t^2 - 2t + 8x^2 - 1) + 2(1 - 2x)\sqrt{-1} \qquad (8.14)$$

which is equivalent to the system

$$\begin{pmatrix} Dq_1 \\ Dq_2 \end{pmatrix} + \begin{pmatrix} -2t & -4x \\ 4x & -2t \end{pmatrix} \begin{pmatrix} q_1 \\ q_2 \end{pmatrix} = \begin{pmatrix} -t^2 + 2t - 8x^2 + 1 \\ 2(1 - 2x) \end{pmatrix} . \qquad (8.15)$$

With $a = 1$, $b = 4x\sqrt{-1} - 2t$ and $c = (-t^2 + 2t - 8x^2 + 1) + 2(1 - 2x)$ we have

$$-\frac{\mathrm{lc}(b)}{\mathrm{lc}(a)} = 2 > \deg(c) - \deg(b)$$

so any solution $q \in k(\sqrt{-1})[t]$ of (8.14) has degree at most 2 by Lemma 6.3.5. Since $\deg(b) = 1$ and $\mathrm{lc}(b) = -2$, we are in the cancellation case of this section. Applying **CoupledDECancelTan** to $b_0 = 0$, $b_2 = 4x$, $c_1 = -t^2 + 2t - 8x^2 + 1$, $c_2 = 2(1 - 2x)$ and $n = 2$, we get

1. $p = t - \sqrt{-1}$
2. $\eta = Dt/(t^2 + 1) = 1$
3.
$$c_1(\sqrt{-1}) + c_2(\sqrt{-1})\sqrt{-1} = 2(1 - 4x^2) + 4(1 - x)\sqrt{-1}$$

   so $z_1 = 2(1 - 4x^2)$ and $z_2 = 4(1 - x)$
4. Since

$$\begin{pmatrix} Ds_1 \\ Ds_2 \end{pmatrix} + \begin{pmatrix} 0 & 2(1 - 2x) \\ 2(2x - 1) & 0 \end{pmatrix} \begin{pmatrix} s_1 \\ s_2 \end{pmatrix} = \begin{pmatrix} 2(1 - 4x^2) \\ 4(1 - x) \end{pmatrix}$$

   has the solution $s_1 = -1$ and $s_2 = 2x + 1$,
   $(s_1, s_2) = $ **CoupledDESystem**$(0, 4x - 2, 2 - 8x^2, 4 - 4x) = (-1, 2x + 1)$
5.
$$c = -\frac{t^2 - 2(2x + 1)t\sqrt{-1} - 4x - 1}{t - \sqrt{-1}} = -t + (4x + 1)\sqrt{-1}$$

   so $d_1 = -t$ and $d_2 = 4x + 1$
6. recursive call, **CoupledDECancelTan**$(0, 4x + 1, -t, 4x + 1, D, 1)$:
   a) $c_1(\sqrt{-1}) + c_2(\sqrt{-1})\sqrt{-1} = 4x\sqrt{-1}$, so $z_1 = 0$ and $z_2 = 4x$
   b) Since

$$\begin{pmatrix} Ds_1 \\ Ds_2 \end{pmatrix} + \begin{pmatrix} 0 & -4x \\ 4x & 0 \end{pmatrix} \begin{pmatrix} s_1 \\ s_2 \end{pmatrix} = \begin{pmatrix} 0 \\ 4x \end{pmatrix}$$

   has the solution $s_1 = 1$ and $s_2 = 0$,
   $(s_1, s_2) = $ **CoupledDESystem**$(0, 4x, 0, 4x) = (1, 0)$

c) $c = (-t + t + (4x + 1 - 4x - 1)\sqrt{-1})/(t - \sqrt{-1}) = 0$, so $d_1 = d_2 = 0$

d) **CoupledDECancelTan**$(0, 4x + 2, 0, 0, D, 0)$ returns $(0, 0)$, so we return $h_1 t + h_2 + s_1 = 1$ and $h_2 t - h_1 + s_2 = 0$

7. We obtain $(h_1, h_2) = (1, 0)$ from the recursion, so we return the following solution of (8.15):

$$q_1 = h_1 t + h_2 + s_1 = t - 1 \quad \text{and} \quad q_2 = h_2 t - h_1 + s_2 = -1 + 2x + 1 = 2x$$

We conclude that a solution of (8.11) is

$$y_1 = \frac{t - 1}{t^2 + 1} \quad \text{and} \quad y_2 = \frac{2x}{t^2 + 1}$$

hence that

$$\int -\frac{(\tan(x)^2 - 2\tan(x) + 8x^2 - 1)\tan(x^2) + 4x - 2}{(\tan(x)^2 + 1)(\tan(x^2)^2 + 1)} \, dx \ =$$

$$\frac{(\tan(x) - 1)\tan(x^2) + 2x}{(\tan(x)^2 + 1)(\tan(x^2)^2 + 1)} + \int 2x \frac{\tan(x) - 1}{\tan(x)^2 + 1} \, dx$$

and the latter integral does not involve $\tan(x^2)$ anymore. Applying the algorithm of Sect. 5.10 to it, which involves solving a coupled differential system over $\mathbb{Q}(x)$, we find

$$\int 2x \frac{\tan(x) - 1}{\tan(x)^2 + 1} \, dx = \frac{x(1 - x)\tan(x)^2 + (1 - 2x)\tan(x) - x^2 - x - 1}{2(\tan(x)^2 + 1)}$$

which yields a complete formula for (8.12).

9. Structure Theorems

We present in this chapter proofs of the various structure theorems that were used in Chap. 7 for solving the parametric logarithmic derivative problem. Although they are used in the integration algorithm, the main application of structure theorems is to determine algebraic dependencies between functions.

9.1 The Module of Differentials

We first need to slightly generalize the concept of derivation we used previously.

Definition 9.1.1. *Let $S \subseteq R$ be commutative rings and M be an R-module. An S-derivation of R into M is a map $D : R \to M$ such that for any $x, y \in R$:*

(i) $D(x + y) = Dx + Dy$.
(ii) $D(xy) = xDy + yDx$.
(iii) $Dc = 0$ for any $c \in S$.

Note that a derivation of R in the sense of Chap. 3 is an S-derivation of R into R for any subring S of $\text{Const}_D(R)$, in particular for $S = \mathbb{Z}$.

The usual properties of derivation (Theorem 3.1.1) are easily generalized to S-derivations.

Theorem 9.1.1. *Let $S \subseteq R$ be commutative rings, M be an R-module, and $D : R \to M$ be an S-derivation. Then,*

(i) $D(cx) = cDx$ for any $c \in S$ and $x \in R$.
(ii) If R is a field, then

$$D\frac{x}{y} = \frac{yDx - xDy}{y^2}$$

for any $x, y \in R$, $y \neq 0$.
(iii) $Dx^n = nx^{n-1}Dx$ for any $x \in R \setminus \{0\}$ and any integer $n > 0$ (any integer n if R is a field).
(iv) Logarithmic derivative identity:

$$\frac{D(u_1^{e_1} \dots u_n^{e_n})}{u_1^{e_1} \dots u_n^{e_n}} = e_1\frac{Du_1}{u_1} + \dots + e_n\frac{Du_n}{u_n}$$

for any $u_1, \dots, u_n \in R^*$ and any integers $e_1, \dots, e_n$.

(v)

$$DP(x_1, \ldots, x_n) = \sum_{i=1}^{n} \frac{\partial P}{\partial X_i}(x_1, \ldots, x_n) Dx_i$$

for any $x_1, \ldots, x_n \in R$ and any polynomial P with coefficients in S.

Proof. The proofs are similar to the proofs of the corresponding statements in Theorem 3.1.1 and are left as exercises. □

Let R be a commutative ring and Φ_R be the free R-module generated by the symbols δx for all $x \in R$. Its elements are all the finite sums $\sum_i a_i \delta x_i$ with $a_i, x_i \in R$. For any subring S of R, let $\Psi_{R/S}$ be the submodule of Φ_R generated by $\delta(x+y) - \delta x - \delta y$ and $\delta(xy) - x\delta y - y\delta x$ for all $x, y \in R$ and δc for all $c \in S$, and let $\Omega_{R/S}$ be the quotient module $\Phi_R/\Psi_{R/S}$. It is easily checked that the map $d_{R/S} : R \to \Omega_{R/S}$ that sends $x \in R$ to the equivalence class of δx is an S-derivation of R into $\Omega_{R/S}$.

The pair $(\Omega_{R/S}, d_{R/S})$ is called the *module of S-differentials of R*, and we omit the subscript on d when the context is clear. The S-differentials of R are all the finite sums $\sum_i a_i dx_i$ with $a_i, x_i \in R$, subject to the relations

$$d(x+y) = dx + dy, \quad d(xy) = x\,dy + y\,dx \quad \text{for all } x, y \in R$$

and $dc = 0$ for all $c \in S$. If $S \subseteq T \subseteq R$ are commutative rings, then $\Psi_{R/S}$ is a submodule of $\Psi_{R/T}$. This implies the existence of a canonical projection $\pi : \Omega_{R/S} \to \Omega_{R/T}$, which is the surjective R-linear map given by $\pi\left(\sum_i a_i d_{R/S}\, x_i\right) = \sum_i a_i d_{R/T}\, x_i$.

We first show that $\Omega_{R/S}$ is a universal object, *i.e.* that every S-derivation can be factored as in the following diagram:

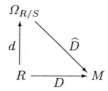

Lemma 9.1.1. *Let $S \subseteq R$ be commutative rings, M be an R-module, and $D : R \to M$ be an S-derivation. Then, there is a unique R-linear map $\widehat{D} : \Omega_{R/S} \to M$ such that $D = \widehat{D}d$.*

Proof. Since Φ_R is a free R-module, let $\overline{D} : \Phi_R \to M$ be the R-linear map given by $\overline{D}(\delta x) = Dx$ for all $x \in R$. Since D is an S-derivation, $\overline{D}(\delta c) = Dc = 0$ for all $c \in S$. Furthermore, $\overline{D}(\delta(x+y) - \delta x - \delta y) = D(x+y) - Dx - Dy = 0$ and $\overline{D}(\delta(xy) - x\delta y - y\delta x) = D(xy) - xDy - yDx = 0$ for any $x, y \in R$, which implies that $\Psi_{R/S} \subseteq \ker\overline{D}$, hence that $\overline{D}$ induces an R-linear map $\widehat{D} : \Omega_{R/S} \to M$. For any $x \in R$ we have $\widehat{D}dx = \overline{D}\delta x = Dx$, so $D = \widehat{D}d$.

Suppose that D_1 and D_2 are both R-linear maps from $\Omega_{R/S}$ into M such that $D = D_1 d = D_2 d$. Then, since any $\omega \in \Omega_{R/S}$ is a finite sum of the form $\omega = \sum_i a_i dx_i$ with $a_i, x_i \in R$, we get by linearity of D_1 and D_2 that

$$D_1\omega = \sum_i a_i D_1(dx_i) = \sum_i a_i Dx_i = \sum_i a_i D_2(dx_i) = D_2\omega$$

hence that $\widehat{D}$ is unique. $\square$

Any ring homomorphism induces a skew-linear map on the differentials such that the following diagram commutes:

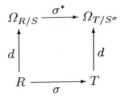

Lemma 9.1.2. *Let T and $S \subseteq R$ be commutative rings, and $\sigma : R \to T$ be a homomorphism. Then, there is a unique map $\sigma^* : \Omega_{R/S} \to \Omega_{T/S^\sigma}$ such that*

(i) $(\omega + \eta)^{\sigma^*} = \omega^{\sigma^*} + \eta^{\sigma^*}$ *for any* $\omega, \eta \in \Omega_{R/S}$
(ii) $(x\omega)^{\sigma^*} = x^\sigma \omega^{\sigma^*}$ *for any* $\omega \in \Omega_{R/S}$ *and* $x \in R$.
(iii) $\sigma^* d = d\sigma$

Proof. Let $\overline{\sigma} : \Phi_R \to \Phi_T$ be the map given by

$$\left(\sum_i a_i \delta x_i \right)^{\overline{\sigma}} = \sum_i a_i^\sigma \delta(x_i^\sigma)$$

for all finite sums with $a_i, x_i \in R$. $\overline{\sigma}$ is well-defined since Φ_R is free over R. Furthermore, $\overline{\sigma}$ is an abelian group homomorphism by definition. Since σ is a homomorphism, $1^\sigma = 1$, so $\overline{\sigma}\delta = \delta\sigma$. In addition we have $(\delta c)^{\overline{\sigma}} = \delta(c^\sigma) \in \Psi_{T/S^\sigma}$ for all $c \in S$. Furthermore,

$$
\begin{aligned}
(\delta(x + y) - \delta x - \delta y)^{\overline{\sigma}} &= \delta((x + y)^\sigma) - \delta(x^\sigma) - \delta(y^\sigma) \\
&= \delta(x^\sigma + y^\sigma) - \delta(x^\sigma) - \delta(y^\sigma) \quad \in \Psi_{T/S^\sigma}
\end{aligned}
$$

and

$$
\begin{aligned}
(\delta(xy) - x\delta y - y\delta x)^{\overline{\sigma}} &= \delta((xy)^\sigma) - x^\sigma \delta(y^\sigma) - y^\sigma \delta(x^\sigma) \\
&= \delta(x^\sigma y^\sigma) - x^\sigma \delta(y^\sigma) - y^\sigma \delta(x^\sigma) \quad \in \Psi_{T/S^\sigma}
\end{aligned}
$$

for any $x, y \in R$, which implies that $(\Psi_{R/S})^{\overline{\sigma}} \subseteq \Psi_{T/S^\sigma}$, hence that $\overline{\sigma}$ induces an abelian group homomorphism $\sigma^* : \Omega_{R/S} \to \Omega_{T/S^\sigma}$ that satisfies

$$\left(\sum_i a_i dx_i\right)^{\sigma^*} = \sum_i a_i^\sigma d(x_i^\sigma).$$

Furthermore, $\sigma^* d = d\sigma$ since $\overline{\sigma} d = d\sigma$. Let $x \in R$ and $\omega \in \Omega_{R/S}$, and write $\omega = \sum_i a_i dx_i$ with $a_i, x_i \in R$. Then,

$$(x\omega)^{\sigma^*} = \left(\sum_i a_i x dx_i\right)^{\sigma^*} = \sum_i (a_i x)^\sigma d(x_i^\sigma) = x^\sigma \sum_i a_i^\sigma d(x_i^\sigma) = x^\sigma \omega^{\sigma^*}.$$

Suppose that σ_1 and σ_2 are both maps from $\Omega_{R/S}$ into Ω_{T/S^σ} that satisfy the lemma. Then, since any $\omega \in \Omega_{R/S}$ is a finite sum of the form $\omega = \sum_i a_i dx_i$ with $a_i, x_i \in R$, we get

$$\omega^{\sigma_1} = \sum_i a_i^\sigma d(x_i^\sigma) = \omega^{\sigma_2}$$

hence that σ^* is unique. $\square$

In a similar manner, we can show that a derivation on R induces a skew-derivation on the differentials such that the following diagram commutes:

$$\begin{array}{ccc}
\Omega_{R/S} & \xrightarrow{\;D^*\;} & \Omega_{R/S} \\[2pt]
{\scriptstyle d}\Big\uparrow & & \Big\uparrow{\scriptstyle d} \\[2pt]
R & \xrightarrow[\;D\;]{} & R
\end{array}$$

Lemma 9.1.3. *Let (R, D) be a differential ring and $S \subseteq R$ be a differential subring. Then, there is a unique map $D^* : \Omega_{R/S} \to \Omega_{R/S}$ such that*

(i) $D^*(\omega + \eta) = D^*\omega + D^*\eta$ *for any $\omega, \eta \in \Omega_{R/S}$*
(ii) $D^*(x\omega) = (Dx)\omega + x D^*\omega$ *for any $\omega \in \Omega_{R/S}$ and $x \in R$.*
(iii) $D^* d = dD$

Proof. Let $\overline{D} : \Phi_R \to \Phi_R$ be the map given by

$$\overline{D}\left(\sum_i a_i \delta x_i\right) = \sum_i (Da_i)\delta x_i + a_i \delta(Dx_i)$$

for all finite sums with $a_i, x_i \in R$. $\overline{D}$ is well-defined since Φ_R is free over R. Furthermore, $\overline{D}$ is an abelian group homomorphism by definition. Since D is a derivation, $D1 = 0$, so $\overline{D}\delta = \delta D$. Since S is a differential subring, $DS \subseteq S$, so $\overline{D}(\delta c) = \delta(Dc) \in \Psi_{R/S}$ for all $c \in S$. Furthermore,

$$\overline{D}(\delta(x+y) - \delta x - \delta y) = \delta D(x+y) - \delta Dx - \delta Dy$$
$$= \delta(Dx + Dy) - \delta Dx - \delta Dy \quad \in \Psi_{R/S}$$

and

$$\overline{D}(\delta(xy) - x\delta y - y\delta x) = \delta D(xy) - (Dx)\delta y - x\delta Dy - (Dy)\delta x - y\delta Dx$$
$$= \delta(xDy + yDx) - \delta(xDy)$$
$$+ (\delta(xDy) - (Dx)\delta y - x\delta Dy)$$
$$- \delta(yDx) + (\delta(yDx) - (Dy)\delta x - y\delta Dx)$$
$$= (\delta(xDy + yDx) - \delta(xDy) - \delta(yDx))$$
$$+ (\delta(xDy) - (Dy)\delta x - x\delta Dy)$$
$$+ (\delta(yDx) - (Dx)\delta y - y\delta Dx) \quad \in \Psi_{R/S}$$

for any $x, y \in R$, which implies that $\overline{D}\Psi_{R/S} \subseteq \Psi_{R/S}$, hence that $\overline{D}$ induces an abelian group homorphism $D^* : \Omega_{R/S} \to \Omega_{R/S}$ that satisfies

$$D^*\left(\sum_i a_i dx_i\right) = \sum_i (Da_i)dx_i + a_i d(Dx_i).$$

Furthermore, $D^*d = dD$ since $\overline{D}d = dD$. Let $x \in R$ and $\omega \in \Omega_{R/S}$, and write $\omega = \sum_i a_i dx_i$ with $a_i, x_i \in R$. Then,

$$D^*(x\omega) = D^*\left(\sum_i a_i x dx_i\right) = \sum_i (Da_i x)dx_i + a_i x d(Dx_i)$$
$$= \sum_i a_i(Dx)dx_i + x(Da_i)dx_i + a_i x d(Dx_i) = (Dx)\omega + xD^*\omega.$$

Suppose that D_1 and D_2 are both maps from $\Omega_{R/S}$ into $\Omega_{R/S}$ that satisfy the lemma. Then, since any $\omega \in \Omega_{R/S}$ is a finite sum of the form $\omega = \sum_i a_i dx_i$ with $a_i, x_i \in R$, we get

$$D_1\omega = \sum_i (Da_i)dx_i + a_i D_1(dx_i) = \sum_i (Da_i)dx_i + a_i d(Dx_i)$$
$$= \sum_i (Da_i)dx_i + a_i D_2(dx_i) = D_2\omega$$

hence that D^* is unique. $\qquad\square$

Lemma 9.1.4. *Let $S \subseteq R$ be commutative rings, $(\Omega_{R/S}, d)$ be the module of S-differentials of R, and $\mathcal{B} \subseteq R$. Then,*

(i) $d(S[\mathcal{B}])$ and $\{db\}_{b \in \mathcal{B}}$ generate the same submodule of $\Omega_{R/S}$ over R.

(ii) If R and S are fields, then $d(S(\mathcal{B}))$ and $\{db\}_{b \in \mathcal{B}}$ generate the same submodule of $\Omega_{R/S}$ over R.

Proof. For any $S \subseteq \Omega_{R/S}$, we write $R\langle S \rangle$ for the submodule of $\Omega_{R/S}$ generated by S over R.

(i) Let $p \in S[\mathcal{B}]$. Then $p = P(x_1, \ldots, x_n)$ where $x_i \in \mathcal{B}$ and $P \in S[X_1, \ldots, X_n]$ is a polynomial with coefficients in S. Therefore,

$$dp = \sum_{i=1}^{n} \frac{\partial P}{\partial X_i}(x_1, \ldots, x_n)\, dx_i$$

by Theorem 9.1.1, so $R\langle d(S[\mathcal{B}]) \rangle \subseteq R\langle d(\mathcal{B}) \rangle$. Since $\mathcal{B} \subseteq S[\mathcal{B}]$, $R\langle d(\mathcal{B}) \rangle \subseteq R\langle d(S[\mathcal{B}]) \rangle$ so both submodules are equal.

(ii) Suppose that R and S are fields, let $x \in S(\mathcal{B})$ and write $x = p/q$ where $p, q \in S[\mathcal{B}]$ and $q \neq 0$. By Theorem 9.1.1,

$$dx = d\frac{p}{q} = \frac{q\, dp - p\, dq}{q^2} = \frac{1}{q}\, dp - \frac{p}{q^2}\, dq\,.$$

Since dp and dq are in the span of $\{db\}_{b \in \mathcal{B}}$ by (i), we conclude that $R\langle d(S(\mathcal{B})) \rangle \subseteq R\langle d(\mathcal{B}) \rangle$. Since $\mathcal{B} \subseteq S(\mathcal{B})$, $R\langle d(\mathcal{B}) \rangle \subseteq R\langle d(S(\mathcal{B})) \rangle$ so both submodules are equal.   □

We now determine the dimension of $\Omega_{R/S}$ over R when R and S are fields.

Lemma 9.1.5. *Let $k \subseteq K$ be fields of characteristic 0 and $\mathcal{B} \subseteq K$ be algebraically dependent over k. Then, $\{db\}_{b \in \mathcal{B}}$ is linearly dependent over K.*

Proof. Since $\mathcal{B}$ is algebraically dependent over k, there are $x_1, \ldots, x_n \in \mathcal{B}$ and a polynomial $P \in k[X_1, \ldots, X_n] \setminus \{0\}$ such that $P(x_1, \ldots, x_n) = 0$. Let Q be a nonzero polynomial with coefficients in k and of minimal total degree such that $Q(x_1, \ldots, x_n) = 0$. Applying d and Theorem 9.1.1 we get

$$0 = d0 = dQ(x_1, \ldots, x_n) = \sum_{i=0}^{n} \frac{\partial Q}{\partial X_i}(x_1, \ldots, x_n)\, dx_i\,.$$

Since $Q \notin k$ and k has characteristic 0, $\partial Q / \partial X_{i_0}$ is not indentically 0 for some i_0. By minimality of the total degree, $(\partial Q / \partial X_{i_0})(x_1, \ldots, x_n) \neq 0$, which implies that $dx_1, \ldots, dx_n$, and therefore $\{db\}_{b \in \mathcal{B}}$, are linearly dependent over K.   □

Theorem 9.1.2. *Let $k \subseteq K$ be fields and $\mathcal{B} \subseteq K$ be algebraically independent over k. If K is separable algebraic over $k(\mathcal{B})$, then $\{db\}_{b \in \mathcal{B}}$ is a basis for $\Omega_{K/k}$ over K.*

Proof. Let $x \in K$, then x is separable algebraic over $k(\mathcal{B})$, so let $P \in k(\mathcal{B})[X]$ be its minimal irreducible polynomial and write $P = \sum_{j=0}^{m} a_j X^j$ where $a_0, \ldots, a_m \in k(\mathcal{B})$. Since d is a k-derivation of K into $\Omega_{K/k}$ we get

$$0 = d0 = dP(x) = d\left(\sum_{j=0}^{m} a_j x^j\right) = \sum_{j=0}^{m} x^j da_j + j a_j x^{j-1} dx$$

$$= \sum_{j=0}^{m} x^j da_j + dx \sum_{j=1}^{m} j a_j x^{j-1}.$$

Let $a = \sum_{j=1}^{m} j a_j x^{j-1} \in K$. Since x is separable over $k(\mathcal{B})$, $a \neq 0$, so $dx = \sum_{j=0}^{m}(-x^j/a)da_j$ is in the subspace of $\Omega_{K/k}$ generated by $da_0, \ldots, da_m$, hence in the subspace generated by $d(k(\mathcal{B}))$. By Lemma 9.1.4, this implies that dx is in the subspace of $\Omega_{K/k}$ generated by $\{db\}_{b\in\mathcal{B}}$. Since this holds for any $x \in K$, $\{db\}_{b\in\mathcal{B}}$ generate $\Omega_{K/k}$.

Suppose that $\sum_{j=1}^{n} a_j dx_j = 0$ for some $a_1, \ldots, a_n \in K$ and $x_1, \ldots, x_n \in \mathcal{B}$. Since $\mathcal{B}$ is algebraically independent over k, $\partial/\partial x_1, \ldots, \partial/\partial x_n$ are derivations on $k(\mathcal{B})$ by Theorem 3.2.2. Those derivations can be extended to derivations of K by Theorem 3.2.3. Since $k \subseteq \mathrm{Const}_{D_i}(K)$, each $\partial/\partial x_i$ is a k-derivation of K into K, so let $\widehat{D_1}, \ldots, \widehat{D_n}$ be the induced K-linear maps from $\Omega_{K/k}$ into K given by Lemma 9.1.1. Applying $\widehat{D_i}$ and Lemma 9.1.1 we get

$$0 = \widehat{D_i}0 = \widehat{D_i}\left(\sum_{j=1}^{n} a_j dx_j\right) = \sum_{j=1}^{n} a_j \widehat{D_i}(dx_j) = \sum_{j=1}^{n} a_j \frac{\partial x_j}{\partial x_i} = a_i$$

which implies that $dx_1, \ldots, dx_n$ are linearly independent over K, hence that $\{db\}_{b\in\mathcal{B}}$ is linearly independent over K. $\qquad\square$

As a consequence, the dimension of $\Omega_{K/k}$ over K is exactly the transcendence degree of K over k. Another consequence is that in characteristic 0, algebraically independent elements yield linearly independent differentials.

Corollary 9.1.1. *Let $k \subseteq K$ be fields of characteristic 0. Then, $\mathcal{B} \subseteq K$ is algebraically independent over k if and only if $\{db\}_{b\in\mathcal{B}} \subseteq \Omega_{K/k}$ is linearly independent over K.*

Proof. Let $\mathcal{A}$ be a transcendence basis of K over k containing $\mathcal{B}$. Since the fields have characteristic 0, K is separable over $k(\mathcal{A})$, so $\{db\}_{b\in\mathcal{A}}$, and therefore $\{db\}_{b\in\mathcal{B}}$, is linearly independent over K by Theorem 9.1.2. Conversely, if $\{db\}_{b\in\mathcal{B}}$ is linearly independent over K, then $\mathcal{B}$ must be algebraically independent over k by Lemma 9.1.5. $\qquad\square$

Corollary 9.1.2. *Let $k \subseteq K$ be fields and $t \in K$ be transcendental over k. If K is separable algebraic over $k(t)$, then*

$$dx = \frac{\partial x}{\partial t} dt$$

in $\Omega_{K/k}$ for any $x \in K$, where $\partial/\partial t$ is the derivation on K that maps t to 1 and every element of k to 0.

Proof. Note that $\partial/\partial t$ is uniquely defined on K by Theorems 3.2.2 and 3.2.3. Since t is transcendental over k and K is separable algebraic over $k(t)$, dt is a basis for $\Omega_{K/k}$ over K by Theorem 9.1.2, so let $D : K \to K$ be the map given by $dx = (Dx)dt$ for every $x \in K$. Since d is a k-derivation we have

$$D(x + y)dt = d(x + y) = dx + dy = (Dx)dt + (Dy)dt$$

and

$$D(xy)dt = d(xy) = x\,dy + y\,dx = (xDy)dt + (yDx)dt$$

for any $x, y \in K$, which implies that D is a derivation on K. Since $Dc = 0$ for any $c \in k$ and $Dt = 1$, we have $D = \partial/\partial t$ by unicity of the differential extension. $\square$

Let $k \subseteq K \subseteq L$ be fields of characteristic 0. The restriction to K of the k-derivation $d_{L/k} : L \to \Omega_{L/k}$ is a k-derivation of K into $\Omega_{L/k}$, so by Lemma 9.1.1, it induces a K-linear map $\widehat{d} : \Omega_{K/k} \to \Omega_{L/k}$ such that $d_{L/k} = \widehat{d}\,d_{K/k}$. Let $\mathcal{B}$ be a transcendence basis of K over k. Then, $\{d_{K/k}b\}_{b\in\mathcal{B}}$ is a basis of $\Omega_{K/k}$ over K by Theorem 9.1.2. In addition, $\{d_{L/k}b\}_{b\in\mathcal{B}}$ is linearly independent over L by Corollary 9.1.1. Since $\widehat{d}\,d_{K/k}b = d_{L/k}b$ for any $b \in \mathcal{B}$, this implies that $\widehat{d}$ is injective, hence that it is an embedding of $\Omega_{K/k}$ into $\Omega_{L/k}$.

9.2 Rosenlicht's Theorem

We prove in this section a fundamental theorem of Rosenlicht, itself a generalization of a result of Ax [4] on Schanuel's conjecture for differential fields, that is used to prove the various structure theorems later. From now on, let all fields in this chapter have characteristic 0. We start with an analogue of Theorem 3.2.4 for differentials: the trace map in algebraic extensions induces a linear trace on the differentials.

Lemma 9.2.1. *Let $k \subseteq K$ be fields, E a finitely generated algebraic extension of K, and $Tr : E \to K$ and $N : E \to K$ be the trace and norm maps from E to K. Then, there is a K-linear map $Tr^* : \Omega_{E/k} \to \Omega_{K/k}$ such that $Tr^*\,d = d\,Tr$ and*

$$Tr^*\left(\frac{da}{a}\right) = \frac{dN(a)}{N(a)} \quad \text{for any } a \in E^*.$$

Proof. Let $\overline{K}$ be the algebraic closure of K and $\sigma_1, \ldots, \sigma_n$ be the distinct embeddings of E in $\overline{K}$ over K. Note that $k^{\sigma_i} = k$ for each i since σ_i is the identity on k, so let $\sigma_i^* : \Omega_{E/k} \to \Omega_{\overline{K}/k}$ be the induced map given by Lemma 9.1.2. Define $Tr^* : \Omega_{E/k} \to \Omega_{\overline{K}/k}$ by $Tr^* = \sum_{i=1}^{n} \sigma_i^*$. Since σ_i is the identity on K, σ_i^* is a K-linear map by Lemma 9.1.2, so Tr^* is K-linear. We have

$$Tr^*(da) = \sum_{i=1}^n (da)^{\sigma_i^*} = \sum_{i=1}^n d(a^{\sigma_i}) = d\left(\sum_{i=1}^n a^{\sigma_i}\right) = d(Tr(a))$$

for any $a \in E$, so $Tr^* d = d Tr$. Furthermore,

$$Tr^*\left(\frac{da}{a}\right) = \sum_{i=1}^n \left(\frac{da}{a}\right)^{\sigma_i^*} = \sum_{i=1}^n \frac{d(a^{\sigma_i})}{a^{\sigma_i}} = \frac{d\left(\prod_{i=1}^n a^{\sigma_i}\right)}{\prod_{i=1}^n a^{\sigma_i}} = \frac{d(N(a))}{N(a)}$$

for any $a \in E^*$. Let $\mathcal{B}$ be a transcendence basis for K over k. Then, $\{db\}_{b \in \mathcal{B}}$ is a basis for $\Omega_{K/k}$ over K by Theorem 9.1.2. But E is algebraic over K, so $\mathcal{B}$ is a transcendence basis for E over k and $\{db\}_{b \in \mathcal{B}}$ is a basis for $\Omega_{E/k}$ over E by Theorem 9.1.2. Write then $\omega \in \Omega_{E/k}$ as $\omega = \sum_{b \in \mathcal{B}} a_b db$ where the a_b are in E and only finitely many of them are nonzero. Then,

$$Tr^*(\omega) = \sum_{i=1}^n \omega^{\sigma_i^*} = \sum_{i=1}^n \sum_{b \in \mathcal{B}} a_b^{\sigma_i} db = \sum_{b \in \mathcal{B}} Tr(a_b) db$$

which is in the image of $\Omega_{K/k}$ under the natural embedding $\Omega_{K/k} \to \Omega_{\overline{K}/k}$, so $Tr^*(\Omega_{E/k}) \subseteq \Omega_{K/k}$. $\qquad\square$

Lemma 9.2.2. *Let $k \subseteq K$ be fields, $v \in K$, $u_1, \ldots, u_n \in K^*$, $c_1, \ldots, c_n \in k$ be linearly independent over $\mathbb{Q}$, and*

$$\omega = dv + \sum_{i=1}^n c_i \frac{du_i}{u_i} \quad \in \Omega_{K/k}. \tag{9.1}$$

Then,

$$\omega = 0 \iff u_1, \ldots, u_n, v \text{ are all algebraic over } k$$
$$\iff du_1 = \ldots = du_n = dv = 0.$$

Proof. Note that Corollary 9.1.1 implies that any $x \in K$ is algebraic over k if and only if $dx = 0$. Suppose first that $u_1, \ldots, u_n, v$ are all algebraic over k. Then, $du_1 = \ldots = du_n = dv = 0$ so $\omega = 0$. Conversely, suppose that $\omega = 0$ and that one of the u_i, say u_1, is transcendental over k. Let $\mathcal{B}$ be a transcendence basis of K over k containing u_1 and $E = k(\mathcal{B} \setminus \{u_1\})$. K is then algebraic over $E(u_1)$, so $F = E(u_1)(u_2, \ldots, u_n, v) \subseteq K$ is a finitely generated algebraic extension of $E(u_1)$. Identifying $\Omega_{F/k}$ with its image under the embedding $\Omega_{F/k} \to \Omega_{K/k}$ mentioned at the end of the previous section, we can consider ω and the differentials appearing in (9.1) as being elements of $\Omega_{F/k}$. Let $Tr : F \to E(u_1)$ and $N : F \to E(u_1)$ be the trace and norm maps from F to $E(u_1)$ and $Tr^* : \Omega_{F/k} \to \Omega_{E(u_1)/k}$ be the induced $E(u_1)$-linear map given by Lemma 9.2.1. Applying Tr^* and Lemma 9.2.1 to (9.1) we get

$$0 = Tr^*(0) = Tr^*(\omega) = Tr^*(dv) + \sum_{i=1}^{n} c_i \, Tr^* \frac{du_i}{u_i}$$

$$= d(Tr(v)) + \sum_{i=1}^{n} c_i \frac{dN(u_i)}{N(u_i)} = dw + mc_1 \frac{du_1}{u_1} + \sum_{i=2}^{n} c_i \frac{dv_i}{v_i}$$

in $\Omega_{E(u_1)/k}$, where $w = Tr(v) \in E(u_1)$, $m = [F : E(u_1)] > 0$ and $v_i = N(u_i) \in E(u_1)$ for $2 \leq i \leq n$. Applying the canonical projection $\pi : \Omega_{E(u_1)/k} \to \Omega_{E(u_1)/E}$ to the above, we get

$$0 = d_{E(u_1)/E} w + mc_1 \frac{d_{E(u_1)/E} u_1}{u_1} + \sum_{i=2}^{n} c_i \frac{d_{E(u_1)/E} v_i}{v_i}.$$

By Corollary 9.1.2, $d_{E(u_1)/E} x = (\partial x / \partial u_1) d_{E(u_1)/E} u_1$ for any $x \in E(u_1)$, so

$$0 = \frac{\partial w}{\partial u_1} d_{E(u_1)/E} u_1 + mc_1 \frac{d_{E(u_1)/E} u_1}{u_1} + \sum_{i=2}^{n} \frac{c_i}{v_i} \frac{\partial v_i}{\partial u_1} d_{E(u_1)/E} u_1.$$

Since u_1 is transcendental over E, $d_{E(u_1)/E} u_1 \neq 0$, so the above implies that

$$c_1 \frac{m}{u_1} + \sum_{i=2}^{n} \frac{c_i}{v_i} \frac{\partial v_i}{\partial u_1} = -\frac{\partial w}{\partial u_1}. \tag{9.2}$$

Note that u_1 is a monomial over E with respect to $\partial / \partial u_1$ and that u_1 is normal and irreducible as an element of $E[u_1]$. Furthermore, the left hand side of (9.2) is simple by Corollary 4.4.2, so $\partial w / \partial u_1$ must be simple too, which implies that $\nu_{u_1}(\partial w / \partial u_1) \geq -1$. But $\nu_{u_1}(\partial w / \partial u_1) \neq -1$ by Corollary 4.4.2, so $\text{residue}_{u_1}(\partial w / \partial u_1) = 0$ by Theorem 4.4.1. Applying Corollary 4.4.2 and residue_{u_1} to (9.2) we get

$$0 = \text{residue}_{u_1} \left(c_1 \frac{m}{u_1} + \sum_{i=2}^{n} \frac{c_i}{v_i} \frac{\partial v_i}{\partial u_1} \right) = mc_1 + \sum_{i=2}^{n} \nu_{u_1}(v_i) c_i$$

where ν_{u_1} is the order function at u_1. Since m is a positive integer, the above is a contradiction with $c_1, \ldots, c_n$ linearly independent over $\mathbb{Q}$, Therefore $u_1, \ldots, u_n$ are all algebraic over k, so $du_1 = \ldots = du_n = 0$, which implies that $0 = \omega = dv$, hence that v is also algebraic over k. $\qquad \square$

Lemma 9.2.3. *Let (K, D) be a differential field, $k \subseteq K$ be a differential subfield and $D^* : \Omega_{K/k} \to \Omega_{K/k}$ be the induced skew-derivation given by Lemma 9.1.3. For any $u, v \in K$, if u and v are algebraically dependent over $\text{Const}_D(k)$, then $D^*(u\,dv) = d(u\,Dv)$ in $\Omega_{K/k}$.*

Proof. Suppose that u and v are algebraically dependent over $C = \mathrm{Const}_D(k)$, and let $P \in C[X, Y]$ be a nonzero polynomial of minimal total degree such that $P(u, v) = 0$. Applying D we get

$$0 = D(P(u, v)) = \frac{\partial P}{\partial X}(u, v)Du + \frac{\partial P}{\partial Y}(u, v)Dv \qquad (9.3)$$

by Theorem 3.1.1. Applying d we get

$$0 = d(P(u, v)) = \frac{\partial P}{\partial X}(u, v)du + \frac{\partial P}{\partial Y}(u, v)dv \qquad (9.4)$$

by Theorem 9.1.1. Using (9.3) and (9.4), we obtain

$$
\begin{aligned}
\left(\frac{\partial P}{\partial X}(u, v) \frac{\partial P}{\partial Y}(u, v)Du \right) dv &= \left(\frac{\partial P}{\partial X}(u, v)Du \right) \left(\frac{\partial P}{\partial Y}(u, v)dv \right) \\
&= \left(-\frac{\partial P}{\partial Y}(u, v)Dv \right) \left(-\frac{\partial P}{\partial X}(u, v)du \right) \\
&= \left(\frac{\partial P}{\partial X}(u, v) \frac{\partial P}{\partial Y}(u, v)Dv \right) du \qquad (9.5)
\end{aligned}
$$

If $\frac{\partial P}{\partial X}(u, v) = 0$, then $\frac{\partial P}{\partial X}$ is identically 0 by the minimality of P, which implies that $P \in C[Y]$, hence that v is algebraic over C, *i.e.* that $dv = 0$ and $Dv = 0$ by Lemma 3.3.2. Therefore, $(Du)dv = (Dv)du = 0$. Similarly, $(Du)dv = (Dv)du = 0$ if $\frac{\partial P}{\partial Y}(u, v) = 0$. If $\frac{\partial P}{\partial X}(u, v) \neq 0$ and $\frac{\partial P}{\partial Y}(u, v) \neq 0$, then (9.5) implies that $(Du)dv = (Dv)du$. Using that equality together with Lemma 9.1.3 we get

$$D^*(udv) = (Du)dv + uD^*(dv) = (Dv)du + ud(Dv) = d(uDv).$$

$\square$

Lemma 9.2.4. *Let (K, D) be a differential field, $k \subseteq K$ be a differential subfield and $D^* : \Omega_{K/k} \to \Omega_{K/k}$ be the induced skew-derivation given by Lemma 9.1.3. Then, for any $v \in K$, $u_1, \ldots, u_n \in K^*$ and $c_1, \ldots, c_n \in \mathrm{Const}_D(k)$,*

$$d\left(Dv + \sum_{i=1}^n c_i \frac{Du_i}{u_i} \right) = D^*\left(dv + \sum_{i=1}^n c_i \frac{du_j}{u_j} \right) \qquad in \ \Omega_{K/k}.$$

Proof. Since d is k-linear, we have

$$d\left(Dv + \sum_{i=1}^n c_i \frac{Du_i}{u_i} \right) = d(Dv) + \sum_{i=1}^n c_i d\left(\frac{Du_i}{u_i} \right).$$

Since $u_i u_i^{-1} - 1 = 0$, u_i and u_i^{-1} are algebraically dependent over $\mathrm{Const}_D(k)$, so $d(Du_j/u_j) = D^*(du_j/u_j)$ by Lemma 9.2.3. In addition, $d(Dv) = D^*(dv)$ by Lemma 9.1.3, so

$$d\left(Dv + \sum_{i=1}^{n} c_i \frac{Du_i}{u_i}\right) = D^*(dv) + \sum_{i=1}^{n} c_i D^*\left(\frac{du_i}{u_i}\right) = D^*\left(dv + \sum_{i=1}^{n} c_i \frac{du_i}{u_i}\right).$$

□

Lemma 9.2.5. *Let (K, D) be a differential field, $k \subseteq K$ be a differential subfield, $D^* : \Omega_{K/k} \to \Omega_{K/k}$ be the induced skew-derivation given by Lemma 9.1.3 and $\omega_1, \ldots, \omega_n \in \Omega_{K/k}$ be such that $D^*\omega_i = 0$ for each i. If $\omega_1, \ldots, \omega_n$ are linearly dependent over K, then they are linearly dependent over $\mathrm{Const}_D(K)$.*

Proof. Suppose that $\omega_1, \ldots, \omega_n$ are linearly dependent over K, and let $a_1, \ldots, a_n \in K$ be not all 0, and such that $\sum_{i=1}^{n} a_i \omega_i = 0$ with the number of nonzero a_i's minimal over all such linear combinations. We can assume without loss of generality that $a_1 \neq 0$, and dividing by a_1 if needed, that $a_1 = 1$. Applying D^* we get

$$0 = D^*\left(\omega_1 + \sum_{i=2}^{n} a_i \omega_i\right) = D^*(\omega_1) + \sum_{i=2}^{n} D(a_i)\omega_i + a_i D^*(\omega_i) = \sum_{i=2}^{n} D(a_i)\omega_i.$$

Since $a_1 = 1$, the above is a linear combination of the ω_i with one less nonzero a_i, so by minimality we must have $Da_i = 0$ for $2 \leq i \leq n$, hence $a_1, \ldots, a_n \in \mathrm{Const}_D(K)$. Therefore $\omega_1, \ldots, \omega_n$ are linearly dependent over $\mathrm{Const}_D(K)$. □

Theorem 9.2.1 (Rosenlicht [66]). *Let (K, D) be a differential field, k be a differential subfield of K with $\mathrm{Const}_D(K) = \mathrm{Const}_D(k)$, and let $v_1, \ldots, v_n \in K$ and $u_1, \ldots, u_m \in K^*$. If there are constants c_{ij} in $\mathrm{Const}_D(k)$ such that*

$$Dv_i + \sum_{j=1}^{m} c_{ij} \frac{Du_j}{u_j} \quad \in k \quad for \; 1 \leq i \leq n$$

then either $k(u_1, \ldots, u_m, v_1, \ldots, v_n)$ has degree of transcendence at least n over k, or $\omega_1, \ldots, \omega_n$ are linearly dependent over $\mathrm{Const}_D(k)$, where

$$\omega_i = dv_i + \sum_{j=1}^{m} c_{ij} \frac{du_j}{u_j} \quad \in \Omega_{K/k}.$$

Proof. Let $C = \mathrm{Const}_D(K) = \mathrm{Const}_D(k)$, $E = k(u_1, \ldots, u_m, v_1, \ldots, v_n)$, $D^* : \Omega_{K/k} \to \Omega_{K/k}$ be the induced skew-derivation given by Lemma 9.1.3, $x_i = Dv_i + \sum_{j=1}^{m} c_{ij} Du_j/u_j$, and suppose that $x_i \in k$ for $1 \leq i \leq n$. Then $dx_i = 0$ for each i by the definition of $\Omega_{K/k}$, so $D^*\omega_i = dx_i = 0$ by Lemma 9.2.4.

Suppose that $\omega_1, \ldots, \omega_n$ are linearly independent over C. Then, they are linearly independent over K by Lemma 9.2.5, so considered as elements of $\Omega_{E/k}$, they are linearly independent over E. This implies that the dimension of $\Omega_{E/k}$ over E is at least n, hence by Theorem 9.1.2 that E has transcendence degree at least n over k. □

Corollary 9.2.1. *Let (K, D) be a differential field, $k \subseteq K$ be a differential subfield with $\mathrm{Const}_D(K) = \mathrm{Const}_D(k)$ and let $v_1, \ldots, v_n \in K$ and $u_1, \ldots, u_n \in K^*$ be such that*

$$Dv_i - \frac{Du_i}{u_i} \in k \quad \text{for } 1 \leq i \leq n.$$

Then, either $k(u_1, \ldots, u_n, v_1, \ldots, v_n)$ has degree of transcendence at least n over k, or there are $c_1, \ldots, c_n \in \mathrm{Const}_D(k)$ not all zero, and $e_1, \ldots, e_n \in \mathbb{Z}$ not all zero, such that $\sum_{i=1}^{n} c_i v_i$ and $\prod_{i=1}^{n} u_i^{e_i}$ are both algebraic over k.

Proof. Suppose that the degree of transcendence of $k(u_1, \ldots, u_n, v_1, \ldots, v_n)$ over k is stricly less than n. Taking $c_{ij} = -1$ if $i = j$ and 0 if $i \neq j$, we see that $u_1, \ldots, u_n, v_1, \ldots, v_n$ satisfy the hypothesis of Theorem 9.2.1. Therefore, $\omega_1, \ldots, \omega_n$ are linearly dependent over $\mathrm{Const}_D(k)$ where $\omega_i = dv_i - du_i/u_i \in \Omega_{K/k}$. Let then $c_1, \ldots, c_n \in \mathrm{Const}_D(k)$ be not all zero and such that $\sum_{i=1}^{n} c_i \omega_i = 0$. Since $\mathrm{Const}_D(k)$ contains $\mathbb{Q}$, it is a vector space over $\mathbb{Q}$, so there are $b_1, \ldots, b_r \in \mathrm{Const}_D(k)$ linearly independent over $\mathbb{Q}$, and $m_{ij} \in \mathbb{Z}$ not all zero such that $c_i = \sum_{j=1}^{r} m_{ij} b_j$. We then have

$$
\begin{aligned}
0 &= \sum_{i=1}^{n} c_i \omega_i = \sum_{i=1}^{n} c_i dv_i - \sum_{i=1}^{n} \sum_{j=1}^{r} m_{ij} b_j \frac{du_i}{u_i} \\
&= d\left(\sum_{i=1}^{n} c_i v_i \right) + \sum_{j=1}^{r} b_j \frac{d\left(\prod_{i=1}^{n} u_i^{-m_{ij}} \right)}{u_i^{-m_{ij}}}.
\end{aligned}
$$

By Lemma 9.2.2, this implies that $\sum_{i=1}^{n} c_i v_i$ is algebraic over k, and that $\prod_{i=1}^{n} u_i^{-m_{ij}}$ is algebraic over k for each j. Since at least one of the m_{ij}'s is nonzero, this proves the corollary. $\quad\square$

Corollary 9.2.2. *Let C be a field, x be transcendental over C, (K, D) be a differential extension of $(C(x), d/dx)$ with $\mathrm{Const}_D(K) = C$, and $v_1, \ldots, v_n \in K$ and $u_1, \ldots, u_n \in K^*$ be such that*

$$Dv_i - \frac{Du_i}{u_i} \in C \quad \text{for } 1 \leq i \leq n.$$

Then, either $C(x)(u_1, \ldots, u_n, v_1, \ldots, v_n)$ has degree of transcendence at least n over $C(x)$, or there are $e_1, \ldots, e_n \in \mathbb{Z}$ not all zero such that $\prod_{i=1}^{n} u_i^{e_i} \in C$.

Proof. Let $u_0 = 1$, $v_0 = x$ and suppose that the degree of transcendence of $C(x)(u_1, \ldots, u_n, v_1, \ldots, v_n)$ over $C(x)$ is stricly less than n. Then the transcendence degree of $C(u_0, u_1, \ldots, u_n, v_0, v_1, \ldots, v_n)$ over C is stricly less than $n + 1$, so by Theorem 9.2.1, $\omega_0, \ldots, \omega_n$ are linearly dependent over C where $\omega_i = dv_i - du_i/u_i \in \Omega_{K/C}$. Let then $c_0, \ldots, c_n \in C$ be not all zero and such that $\sum_{i=0}^{n} c_i \omega_i = 0$. Note that $\omega_0 = dx$ is linearly independent over K by Corollary 9.1.1, so there must be some $i_0 > 0$ such that $c_{i_0} \neq 0$. Expressing each c_i as $c_i = \sum_{j=1}^{r} m_{ij} b_j$ with $m_{ij} \in \mathbb{Z}$ as in the proof of Corollary 9.2.1, we get in a similar way that $\prod_{i=0}^{n} u_i^{-m_{ij}} = \prod_{i=1}^{n} u_i^{-m_{ij}}$ is algebraic over C for each j. Since $C = \mathrm{Const}_D(K)$, Lemma 3.3.2 implies that $\prod_{i=1}^{n} u_i^{-m_{ij}} \in C$ for each j. Since $c_{i0} \neq 0$, $m_{i_0 j} \neq 0$ for some j, which proves the corollary. $\square$

9.3 The Risch Structure Theorems

Recall (Definition 5.1.4) that a differential field (K, D) is an elementary extension of (k, D) if $K = k(t_1, \ldots, t_n)$ with each t_i elementary over $k(t_1, \ldots, t_{i-1})$. In that case we define the following index sets:

$$
E_{K/k} = \{i \in \{1, \ldots, n\} \text{ such that } t_i \text{ transcendental over } k(t_1, \ldots, t_{i-1})
$$
$$
\text{and } Dt_i/t_i = Da_i, a_i \in k(t_1, \ldots, t_{i-1})\} \tag{9.6}
$$

and

$$
L_{K/k} = \{i \in \{1, \ldots, n\} \text{ such that } t_i \text{ transcendental over } k(t_1, \ldots, t_{i-1})
$$
$$
\text{and } Dt_i = Da_i/a_i, a_i \in k(t_1, \ldots, t_{i-1})^*\} . \tag{9.7}
$$

Note that the cardinality of $E_{K/k} \cup L_{K/k}$ is exactly the transcendence degree of K over k and that it is at most n. In addition, if $\mathrm{Const}_D(K) = \mathrm{Const}_D(k)$, then $\deg_{t_i}(Dt_i)$ is 0 when $t_i \in L_{K/k}$ and 1 when $t_i \in E_{K/k}$ (since $Dt_i \neq 0$), so $E_{K/k} \cap L_{K/k} = \emptyset$.

Lemma 9.3.1. *Let (K, D) be an elementary extension of (k, D) satisfying $\mathrm{Const}_D(K) = \mathrm{Const}_D(k) = C$. Write $K = k(t_1, \ldots, t_n)$ with each t_i elementary over $k(t_1, \ldots, t_{i-1})$, and let $E_{K/k}$ and $L_{K/k}$ be given by (9.6) and (9.7) respectively. If there are integers $e_i \in \mathbb{Z}$ such that*

$$
\prod_{i \in E_{K/k}} t_i^{e_i} \prod_{i \in L_{K/k}} a_i^{e_i} \in C
$$

where $t_i = \log(a_i)$ for $i \in L_{K/k}$, then $e_i = 0$ for all i in $E_{K/k} \cup L_{K/k}$.

Proof. Suppose that $e_i \neq 0$ for some i and let then

$$
j = \max\{i \in E_{K/k} \cup L_{K/k} \text{ such that } e_i \neq 0\},
$$

$$
\alpha = \prod_{\substack{i \in E_{K/k} \\ i < j}} t_i^{e_i} \prod_{\substack{i \in L_{K/k} \\ i < j}} a_i^{e_i} \quad \text{and} \quad \beta = \sum_{\substack{i \in E_{K/k} \\ i < j}} e_i a_i + \sum_{\substack{i \in L_{K/k} \\ i < j}} e_i t_i
$$

where $t_i = \log(a_i)$ for $i \in L_{K/k}$ and $t_i = \exp(a_i)$ for $i \in E_{K/k}$. Note that $D\alpha/\alpha = D\beta$ by the logarithmic derivative identity.

If $j \in E_{K/k}$, then $t_j^{e_j} \alpha \in C$, in contradiction with t_j transcendental over $k(t_1, \ldots, t_{j-1})$ since $\alpha \in k(t_1, \ldots, t_{j-1})$.

If $j \in L_{K/k}$, then $a_j^{e_j} \alpha \in C$, so

$$0 = \frac{D\left(a_j^{e_j} \alpha\right)}{a_j^{e_j} \alpha} = e_j \frac{Da_j}{a_j} + \frac{D\alpha}{\alpha} = e_j Dt_j + D\beta = D(e_j t_j + \beta)$$

which implies that $e_j t_j + \beta \in C$, in contradiction with t_j transcendental over $k(t_1, \ldots, t_{j-1})$ since $\beta \in k(t_1, \ldots, t_{j-1})$.

Theorem 9.3.1 (Risch [62]). *Let C be a field, x be transcendental over C, and (K, D) be an elementary extension of $(C(x), d/dx)$ with $\mathrm{Const}_D(K) = C$. Write $K = C(x)(t_1, \ldots, t_n)$ with each t_i elementary over $C(x)(t_1, \ldots, t_{i-1})$, and let $E_{K/C(x)}$ and $L_{K/C(x)}$ be given by (9.6) and (9.7) respectively. If there are $v \in K$ and $u \in K^*$ such that $Dv = Du/u$, then there are $r_i \in \mathbb{Q}$ such that*

$$v + \sum_{i \in L_{K/C(x)}} r_i t_i + \sum_{i \in E_{K/C(x)}} r_i a_i \in C$$

where $t_i = \exp(a_i)$ for $i \in E_{K/C(x)}$.

Proof. Let $u_i = t_i$ and $v_i = a_i$ for $i \in E_{K/C(x)}$, $u_i = a_i$ and $v_i = t_i$ for $i \in L_{K/C(x)}$, $I = E_{K/C(x)} \cup L_{K/C(x)}$, m be the cardinality of I, and $F = C(x)(u, v, \{u_i\}_{i \in I}, \{v_i\}_{i \in I})$. Since the degree of transcendence of K over $C(x)$ is exactly m, the degree of transcendence of F over $C(x)$ is at most m, hence strictly less than $m + 1$. Since $Dv - Du/u = 0 \in C$ and $Dv_i - Du_i/u_i = 0 \in C$ for each $i \in I$, Corollary 9.2.2 implies that there are integers e and $\{e_i\}_{i \in I}$, not all zero, such that $u^e \prod_{i \in I} u_i^{e_i} \in C$. Note that $e \neq 0$ by Lemma 9.3.1. We then have

$$
\begin{aligned}
0 &= \frac{D\left(u^e \prod_{i \in I} u_i^{e_i}\right)}{u^e \prod_{i \in I} u_i^{e_i}} = e\frac{Du}{u} + \sum_{i \in I} e_i \frac{Du_i}{u_i} = eDv + \sum_{i \in I} e_i Dv_i \\
&= eDv + \sum_{i \in L_{K/C(x)}} e_i Dt_i + \sum_{i \in E_{K/C(x)}} e_i Da_i \\
&= D\left(ev + \sum_{i \in L_{K/C(x)}} e_i t_i + \sum_{i \in E_{K/C(x)}} e_i a_i\right)
\end{aligned}
$$

which implies that

$$ev + \sum_{i \in L_{K/C(x)}} e_i t_i + \sum_{i \in E_{K/C(x)}} e_i a_i \in C$$

and dividing by e proves the theorem. $\qquad \square$

As consequences we get algorithms for determining whether new logarithms or exponentials over a differential field are monomials over that field having the same constants.

Corollary 9.3.1. *Let $C, x, K, E_{K/C(x)}$ and $L_{K/C(x)}$ be as in Theorem 9.3.1, $a \in K^*$ and $b \in K$. Then,*

(i) Da/a is the derivative of an element of K if and only if there are $r_i \in \mathbb{Q}$ such that

$$\sum_{i \in L_{K/C(x)}} r_i Dt_i + \sum_{i \in E_{K/C(x)}} r_i \frac{Dt_i}{t_i} = \frac{Da}{a}. \tag{9.8}$$

(ii) Db is the logarithmic derivative of a K-radical if and only if there are $r_i \in \mathbb{Q}$ such that

$$\sum_{i \in L_{K/C(x)}} r_i Dt_i + \sum_{i \in E_{K/C(x)}} r_i \frac{Dt_i}{t_i} = Db. \tag{9.9}$$

Proof. (i) Suppose that $Da/a = Dv$ for some $v \in K$. By Theorem 9.3.1 there are $r_i \in \mathbb{Q}$ such that

$$v + \sum_{i \in L_{K/C(x)}} r_i t_i + \sum_{i \in E_{K/C(x)}} r_i a_i \in C$$

where $t_i = \exp(a_i)$ for $i \in E_{K/C(x)}$. Applying D yields (9.8). Conversely, if there are $r_i \in \mathbb{Q}$ satisfying (9.8), then

$$\frac{Da}{a} = D \left(\sum_{i \in L_{K/C(x)}} r_i t_i + \sum_{i \in E_{K/C(x)}} r_i a_i \right)$$

is the derivative of an element of K.

(ii) Suppose that $nDb = Du/u$ for some integer $n \neq 0$ and $u \in K^*$. By Theorem 9.3.1 applied to $v = nb$, there are $r_i \in \mathbb{Q}$ such that

$$nb + \sum_{i \in L_{K/C(x)}} r_i t_i + \sum_{i \in E_{K/C(x)}} r_i a_i \in C$$

where $t_i = \exp(a_i)$ for $i \in E_{K/C(x)}$. Applying D and dividing by n yields (9.9). Conversely, if there are $r_i \in \mathbb{Q}$ satisfying (9.9), then

$$Db = \sum_{i \in L_{K/C(x)}} r_i \frac{Da_i}{a_i} + \sum_{i \in E_{K/C(x)}} r_i \frac{Dt_i}{t_i}$$

where $t_i = \log(a_i)$ for $i \in L_{K/C(x)}$. Putting the r_i's over a common denominator $e \neq 0$, we get

$$eDb = \sum_{i \in L_{K/C(x)}} e_i \frac{Da_i}{a_i} + \sum_{i \in E_{K/C(x)}} e_i \frac{Dt_i}{t_i} = \frac{Du}{u}$$

where

$$u = \prod_{i \in L_{K/C(x)}} a_i^{e_i} \prod_{i \in E_{K/C(x)}} t_i^{e_i} \in K^*.$$

$\square$

The algorithms follow from Corollary 9.3.1 and Theorems 5.1.1 and 5.1.2: let (K, D) be given explicitly as an elementary extension of $(C(x), d/dx)$ where $C = \text{Const}_D(K)$ and suppose that the sets $E_{K/C(x)}$ and $L_{K/C(x)}$ are known (those can be computed by applying the algorithm to $t_1, t_2, \ldots, t_n$ in that order).

Let $a \in K^*$ and let t in a differential extension of K be such that $t = \log(a)$, i.e. $Dt = Da/a$. If (9.8) has a solution $r_i \in \mathbb{Q}$, then it provides $v \in K$ such that $Dv = Da/a$, hence $c = t - v \in \text{Const}_D(K(t))$ and $K(t) = C(c)(t_1, \ldots, t_n)$. Otherwise, Da/a is not the derivative of an element of K by Corollary 9.3.1, so t is a monomial over K and $\text{Const}_D(K(t)) = \text{Const}_D(K)$ by Theorem 5.1.1.

Let $b \in K$ and let t in a differential extension of K be such that $t = \exp(b)$, i.e. $Dt/t = Db$. If (9.9) has a solution $r_i \in \mathbb{Q}$, then it provides a nonzero integer e and $u \in K^*$ such that $eDb = Du/u$, hence $c = t^e/u \in \text{Const}_D(K(t))$ and $K(t)$ is algebraic over $C(c)(t_1, \ldots, t_n)$ since $t^e = cu$. Otherwise, Db is not the logarithmic derivative of a K-radical by Corollary 9.3.1, so t is a monomial over K and $\text{Const}_D(K(t)) = \text{Const}_D(K)$ by Theorem 5.1.2.

To determine whether (9.8) and (9.9) have solutions in $\mathbb{Q}$, we compute a linear system with coefficients in C and the same constant solutions by Lemma 7.1.2. Assuming[1] that we have a vector space basis $\mathcal{B}$ containing 1 for C over $\mathbb{Q}$, projecting that system on 1 yields a linear system with coefficients in $\mathbb{Q}$ for the r_i's.

Risch also gave a real version of his structure theorem, which is applicable to towers of logarithms, exponentials, arc-tangents and tangents over a real constant field. Recall (Definition 5.10.1) that t is a tangent over k if $Dt/(t^2 + 1) = Da$ for some $a \in k$. In a similar fashion, we say that t is an arc-tangent over k if $Dt = Da/(a^2 + 1)$ for some $a \in k$ such that $a^2 + 1 \neq 0$, and that t is real elementary over k if t is either algebraic, a logarithm, an exponential, an arc-tangent or a tangent over k. We say that (K, D) is a real elementary extension of (k, D) if $K = k(t_1, \ldots, t_n)$ with each t_i real elementary over $k(t_1, \ldots, t_{i-1})$. In that case, in addition to the index sets $E_{K/k}$ and $L_{K/k}$ defined by (9.6) and (9.7), we introduce the following index sets:

$$T_{K/k} = \{i \in \{1, \ldots, n\} \text{ such that } t_i \text{ transcendental over } k(t_1, \ldots, t_{i-1})$$
$$\text{and } Dt_i/(t_i^2 + 1) = Da_i, a_i \in k(t_1, \ldots, t_{i-1})\} \quad (9.10)$$

and

$$A_{K/k} = \{i \in \{1, \ldots, n\} \text{ such that } t_i \text{ transcendental over } k(t_1, \ldots, t_{i-1})$$
$$\text{and } Dt_i = Da_i/(a_i^2 + 1), a_i \in k(t_1, \ldots, t_{i-1})\}. \quad (9.11)$$

---

[1] This may cause undecidability problems in general, but we usually compute in cases where C is a finitely generated extension of $\mathbb{Q}$ and such a basis is available. The integration algorithm requires an explicitly computable constant field.

Note that the cardinality of $E_{K/k} \cup L_{K/k} \cup T_{K/k} \cup A_{K/k}$ is exactly the transcendence degree of K over k and that it is at most n. In addition, if $\mathrm{Const}_D(K) = \mathrm{Const}_D(k)$, then $\deg_{t_i}(Dt_i)$ is 0 when $t_i \in L_{K/k} \cup A_{K/t}$, 1 when $t_i \in E_{K/k}$ and 2 when $t_i \in T_{K/k}$, so the sets $A_{L/k} \cup L_{L/k}$, $E_{L/k}$ and $T_{K/k}$ are disjoints. It can be shown that $A_{L/k} \cap L_{L/k} = \emptyset$ when $\sqrt{-1} \notin L$ (Exercise 9.1), so the four index sets are disjoints.

Lemma 9.3.2. *Let K be a field, X be an indeterminate and $p, q \in K[X]$ be irreducible. Then,*

$$p \text{ irreducible over } K[X]/(q) \iff q \text{ irreducible over } K[X]/(p)$$

Proof. Let $\overline{K}$ be the algebraic closure of K and $\alpha, \beta \in \overline{K}$ be such that $p(\alpha) = q(\beta) = 0$. Then, $K[X]/(q) \simeq K(\beta)$ and $K[X]/(p) \simeq K(\alpha)$. We proceed by a degree argument in the following diagram:

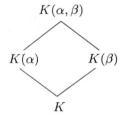

We have $[K(\alpha) : K] = \deg(p)$ and $[K(\beta) : K] = \deg(q)$ since p and q are irreducible over K. If p is irreducible over $K(\beta)$, then $[K(\alpha, \beta) : K(\beta)] = \deg(p)$, so

$$[K(\alpha, \beta) : K] = [K(\alpha, \beta) : K(\beta)][K(\beta) : K] = \deg(p)\deg(q) \,.$$

But

$$[K(\alpha, \beta) : K] = [K(\alpha, \beta) : K(\alpha)][K(\alpha) : K] = [K(\alpha, \beta) : K(\alpha)]\deg(p)$$

which implies that $[K(\alpha, \beta) : K(\alpha)] = \deg(q)$, hence that q is irreducible over $K(\alpha)$. The converse follows by symmetry.   □

Lemma 9.3.3. *Let C be a field, x be transcendental over C, and (K, D) be a real elementary extension of $(C(x), d/dx)$ with $\sqrt{-1} \notin K$ and $\mathrm{Const}_D(K) = C$. Then, $K(\sqrt{-1})$ is an elementary extension of $(C(\sqrt{-1})(x), d/dx)$. Furthermore, $\mathrm{Const}_\Delta(E) = C(\sqrt{-1})$, and the elementary tower from $C(\sqrt{-1})(x)$ to $K(\sqrt{-1})$ can be chosen so that $L_{K(\sqrt{-1})/C(\sqrt{-1})} = L_{K/C} \cup A_{K/C}$ and $E_{K(\sqrt{-1})/C(\sqrt{-1})} = E_{K/C} \cup T_{K/C}$.*

Proof. Note that $\mathrm{Const}_D(K(\sqrt{-1})) = C(\sqrt{-1})$ by Lemma 3.3.4. We write $K = C(x)(t_1, \ldots, t_n)$ where each t_i is real elementary over $C(x)(t_1, \ldots, t_{i-1})$ and proceed by induction on n. If $n = 0$, then $K = C(x)$ so $K(\sqrt{-1}) = C(\sqrt{-1})(x)$. Suppose now that $n > 0$ and that the lemma holds for $k =$

$C(x)(t_1, \ldots, t_{n-1})$. Then, $\mathrm{Const}_D(k(\sqrt{-1})) = C(\sqrt{-1})$, $k(\sqrt{-1})$ is an elementary extension of $(C(\sqrt{-1})(x), d/dx)$, and the elementary tower can be chosen so that $L_{k(\sqrt{-1})/C(\sqrt{-1})} = L_{k/C} \cup A_{k/C}$ and $E_{k(\sqrt{-1})/C(\sqrt{-1})} = E_{k/C} \cup T_{k/C}$. In addition, $K = k(t)$ where $t = t_n$ is real elementary over K.

Case 1: t is transcendental over k. Then, t is transcendental over $k(\sqrt{-1})$.

Case 1a: t is a logarithm over k. Then, t is a logarithm over $k(\sqrt{-1})$, so $K(\sqrt{-1}) = k(\sqrt{-1})(t)$ is elementary over $k(\sqrt{-1})$, hence over $C(\sqrt{-1})$. Furthermore,

$$L_{K(\sqrt{-1})/C(\sqrt{-1})} = \{n\} \cup L_{k(\sqrt{-1})/C(\sqrt{-1})} = \{n\} \cup L_{k/C} \cup A_{k/C} = L_{K/C} \cup A_{K/C}$$

and

$$E_{K(\sqrt{-1})/C(\sqrt{-1})} = E_{k(\sqrt{-1})/C(\sqrt{-1})} = E_{k/C} \cup T_{k/C} = E_{K/C} \cup T_{K/C}.$$

Case 1b: t is an exponential over k. Then, t is an exponential over $k(\sqrt{-1})$, so $K(\sqrt{-1}) = k(\sqrt{-1})(t)$ is elementary over $k(\sqrt{-1})$, hence over $C(\sqrt{-1})$. Furthermore,

$$E_{K(\sqrt{-1})/C(\sqrt{-1})} = \{n\} \cup E_{k(\sqrt{-1})/C(\sqrt{-1})} = \{n\} \cup E_{k/C} \cup T_{k/C} = E_{K/C} \cup T_{K/C}$$

and

$$L_{K(\sqrt{-1})/C(\sqrt{-1})} = L_{k(\sqrt{-1})/C(\sqrt{-1})} = L_{k/C} \cup A_{k/C} = L_{K/C} \cup A_{K/C}.$$

Case 1c: t is an arc-tangent over k. Let then $\theta = 2t\sqrt{-1} \in K(\sqrt{-1})$. We have $Dt = Da/(a^2 + 1)$ where $a \in k$ satisfies $a^2 + 1 \neq 0$, so

$$D\theta = 2Dt\sqrt{-1} = 2\sqrt{-1}\frac{Da}{a^2 + 1} = \frac{Db}{b}$$

where $b = (a - \sqrt{-1})/(a + \sqrt{-1}) \in k(\sqrt{-1})^*$, so θ is a logarithm over $k(\sqrt{-1})$, which implies that $K(\sqrt{-1}) = k(\sqrt{-1})(\theta)$ is elementary over $k(\sqrt{-1})$, hence over $C(\sqrt{-1})$. Furthermore,

$$L_{K(\sqrt{-1})/C(\sqrt{-1})} = \{n\} \cup L_{k(\sqrt{-1})/C(\sqrt{-1})} = \{n\} \cup L_{k/C} \cup A_{k/C} = L_{K/C} \cup A_{K/C}$$

and

$$E_{K(\sqrt{-1})/C(\sqrt{-1})} = E_{k(\sqrt{-1})/C(\sqrt{-1})} = E_{k/C} \cup T_{k/C} = E_{K/C} \cup T_{K/C}.$$

Case 1d: t is a tangent over k. Let then $\theta = (\sqrt{-1} - t)/(\sqrt{-1} + t) \in K(\sqrt{-1})$. We have $Dt = (t^2 + 1)Da$ for some $a \in k$, so

$$\frac{D\theta}{\theta} = -\frac{Dt}{\sqrt{-1} - t} - \frac{Dt}{\sqrt{-1} + t} = \frac{2\sqrt{-1}Dt}{t^2 + 1} = D(2a\sqrt{-1})$$

so θ is an exponential over $k(\sqrt{-1})$, which implies that $K(\sqrt{-1}) = k(\sqrt{-1})(\theta)$ is elementary over $k(\sqrt{-1})$, hence over $C(\sqrt{-1})$. Furthermore,

$$E_{K(\sqrt{-1})/C(\sqrt{-1})} = \{n\} \cup E_{k(\sqrt{-1})/C(\sqrt{-1})} = \{n\} \cup E_{k/C} \cup T_{k/C} = E_{K/C} \cup T_{K/C}$$

and

$$L_{K(\sqrt{-1})/C(\sqrt{-1})} = L_{k(\sqrt{-1})/C(\sqrt{-1})} = L_{k/C} \cup A_{k/C} = L_{K/C} \cup A_{K/C}.$$

Case 2: t is algebraic over k. Let then X be an indeterminate and $p \in k[X]$ be the minimal irreducible polynomial for t over k. Since $\sqrt{-1} \notin k(t)$, p remains irreducible over $k(\sqrt{-1})$ by Lemma 9.3.2, so $K(\sqrt{-1}) = k(\sqrt{-1})(t) \simeq k(\sqrt{-1})[X]/(p)$ is algebraic, hence elementary, over $k(\sqrt{-1})$. Furthermore,

$$E_{K(\sqrt{-1})/C(\sqrt{-1})} = E_{k(\sqrt{-1})/C(\sqrt{-1})} = E_{k/C} \cup T_{k/C} = E_{K/C} \cup T_{K/C}$$

and

$$L_{K(\sqrt{-1})/C(\sqrt{-1})} = L_{k(\sqrt{-1})/C(\sqrt{-1})} = L_{k/C} \cup A_{k/C} = L_{K/C} \cup A_{K/C}.$$

$\square$

We can now prove a slight generalization of the real Risch structure Theorem to arbitrary constant fields.

Theorem 9.3.2 (Risch [62]). *Let C be a field, x be transcendental over C, and (K, D) be a real elementary extension of $(C(x), d/dx)$ with $\mathrm{Const}_D(K) = C$ and $\sqrt{-1} \notin K$. Write $K = C(x)(t_1, \ldots, t_n)$ with each t_i real elementary over $C(x)(t_1, \ldots, t_{i-1})$, and let $E_{K/C(x)}$, $L_{K/C(x)}$, $T_{K/C(x)}$ and $A_{K/C(x)}$ be given by (9.6), (9.7), (9.10) and (9.11) respectively.*

(i) *If there are $v \in K$ and $u \in K^*$ such that $Dv = Du/u$, then there are $r_i \in \mathbb{Q}$ such that*

$$v + \sum_{i \in L_{K/C(x)}} r_i t_i + \sum_{i \in E_{K/C(x)}} r_i a_i \in C$$

   where $t_i = \exp(a_i)$ for $i \in E_{K/C(x)}$.

(ii) *If there are $u, v \in K$ such that $Dv = Du/(u^2 + 1)$, then there are $r_i \in \mathbb{Q}$ such that*

$$v + \sum_{i \in A_{K/C(x)}} r_i t_i + \sum_{i \in T_{K/C(x)}} r_i a_i \in C$$

   where $t_i = \tan(a_i)$ for $i \in T_{K/C(x)}$.

Proof. Let $F = C(\sqrt{-1})$. By Lemma 9.3.3, $E = K(\sqrt{-1})$ is elementary over $(F(x), d/dx)$, $\mathrm{Const}_D(E) = F$, $L_{E/F(x)} = L_{K/C(x)} \cup A_{K/C(x)}$ and $E_{E/F(x)} = E_{K/C(x)} \cup T_{K/C(x)}$.

(i) Suppose that $Dv = Du/u$ for $v \in K$ and $u \in K^*$. By Theorem 9.3.1 applied to E, there are $r_i \in \mathbb{Q}$ such that

$$v + \sum_{i \in L_{E/F(x)}} r_i \theta_i + \sum_{i \in E_{E/F(x)}} r_i \alpha_i \in C(\sqrt{-1})$$

where $\theta_i = \exp(\alpha_i)$ for $i \in E_{E/F(x)}$. Differentiating, we get

$$
\begin{aligned}
0 \;=\; & Dv + \sum_{i \in L_{K/C(x)}} r_i D\theta_i + \sum_{i \in A_{K/C(x)}} r_i D\theta_i \\
& + \sum_{i \in E_{K/C(x)}} r_i \frac{D\theta_i}{\theta_i} + \sum_{i \in T_{K/C(x)}} r_i \frac{D\theta_i}{\theta_i} \\
\;=\; & Dv + \sum_{i \in L_{K/C(x)}} r_i Dt_i + 2\sqrt{-1} \sum_{i \in A_{K/C(x)}} r_i Dt_i \\
& + \sum_{i \in E_{K/C(x)}} r_i \frac{Dt_i}{t_i} + 2\sqrt{-1} \sum_{i \in T_{K/C(x)}} r_i \frac{Dt_i}{t_i^2 + 1}.
\end{aligned}
$$

Since $\sqrt{-1} \notin K$, we get that

$$
Dv + \sum_{i \in L_{K/C(x)}} r_i Dt_i + \sum_{i \in E_{K/C(x)}} r_i \frac{Dt_i}{t_i} = 0
$$

hence that

$$
v + \sum_{i \in L_{K/C(x)}} r_i t_i + \sum_{i \in E_{K/C(x)}} r_i a_i \in C
$$

where $t_i = \exp(a_i)$ for $i \in E_{K/C(x)}$.

(ii) Suppose that $Dv = Du/(u^2 + 1)$ for $u, v \in K$. Then,

$$
D(2v\sqrt{-1}) = 2Dv\sqrt{-1} = 2\sqrt{-1} \frac{Du}{u^2 + 1} = \frac{Dz}{z}
$$

where $z = (u - \sqrt{-1})/(u + \sqrt{-1})$, so by Theorem 9.3.1 applied to E, there are $r_i \in \mathbb{Q}$ such that

$$
2v\sqrt{-1} + \sum_{i \in L_{E/F(x)}} r_i \theta_i + \sum_{i \in E_{E/F(x)}} r_i \alpha_i \in C(\sqrt{-1})
$$

where $\theta_i = \exp(\alpha_i)$ for $i \in E_{E/F(x)}$. Differentiating as in part (i), we get

$$
\begin{aligned}
0 \;=\; & 2Dv\sqrt{-1} + \sum_{i \in L_{K/C(x)}} r_i Dt_i + 2\sqrt{-1} \sum_{i \in A_{K/C(x)}} r_i Dt_i \\
& + \sum_{i \in E_{K/C(x)}} r_i \frac{Dt_i}{t_i} + 2\sqrt{-1} \sum_{i \in T_{K/C(x)}} r_i \frac{Dt_i}{t_i^2 + 1}.
\end{aligned}
$$

Since $\sqrt{-1} \notin K$, we get that

$$
Dv + \sum_{i \in A_{K/C(x)}} r_i Dt_i + \sum_{i \in T_{K/C(x)}} r_i \frac{Dt_i}{t_i^2 + 1} = 0
$$

hence that

$$
v + \sum_{i \in A_{K/C(x)}} r_i t_i + \sum_{i \in T_{K/C(x)}} r_i a_i \in C
$$

where $t_i = \tan(a_i)$ for $i \in T_{K/C(x)}$. $\qquad\square$

As consequences we get algorithms for determining whether new logarithms, exponentials, tangents and arc-tangents over a real differential field are monomials over that field having the same constants.

Corollary 9.3.2. *Let* $C, x, K, E_{K/C(x)}, L_{K/C(x)}, T_{K/C(x)}$ *and* $A_{K/C(x)}$ *be as in Theorem 9.3.2,* $a \in K^*$ *and* $b \in K$. *Then,*

(i) Da/a *is the derivative of an element of* K *if and only if there are* $r_i \in \mathbb{Q}$ *such that*

$$\sum_{i \in L_{K/C(x)}} r_i Dt_i + \sum_{i \in E_{K/C(x)}} r_i \frac{Dt_i}{t_i} = \frac{Da}{a}. \tag{9.12}$$

(ii) Db *is the logarithmic derivative of a* K-*radical if and only if there are* $r_i \in \mathbb{Q}$ *such that*

$$\sum_{i \in L_{K/C(x)}} r_i Dt_i + \sum_{i \in E_{K/C(x)}} r_i \frac{Dt_i}{t_i} = Db. \tag{9.13}$$

(iii) $Db/(b^2 + 1)$ *is the derivative of an element of* K *if and only if there are* $r_i \in \mathbb{Q}$ *such that*

$$\sum_{i \in A_{K/C(x)}} r_i Dt_i + \sum_{i \in T_{K/C(x)}} r_i \frac{Dt_i}{t_i^2 + 1} = \frac{Db}{b^2 + 1}. \tag{9.14}$$

(iv) $\sqrt{-1}Db$ *is the logarithmic derivative of a* $K(\sqrt{-1})$-*radical if and only if there are* $r_i \in \mathbb{Q}$ *such that*

$$\sum_{i \in A_{K/C(x)}} r_i Dt_i + \sum_{i \in T_{K/C(x)}} r_i \frac{Dt_i}{t_i^2 + 1} = Db. \tag{9.15}$$

Proof. The proofs of parts (i) and (iii) are similar to the proof of part (i) of Corollary 9.3.1, using Theorem 9.3.2 instead of 9.3.1, while the proofs of part (ii) is similar to the proof of part (ii) of Corollary 9.3.1.
(iv) Suppose that $n\sqrt{-1}Db = Dw/w$ for some integer $n \neq 0$ and w in $K(\sqrt{-1})^*$, and write $w = y + z\sqrt{-1}$ where $y, z \in K$. If $y = 0$, then $z \neq 0$ and $Dw/w = Dz/z = n\sqrt{-1}Db$, which implies that $Dz = Db = 0$, hence that (9.15) is satisfied with $r_i = 0$ for each i, so suppose from now on that $y \neq 0$. We then have,

$$n\sqrt{-1}Db = \frac{Dw}{w} = \frac{Dy + \sqrt{-1}Dz}{y + z\sqrt{-1}} = \frac{yDy + zDz}{y^2 + z^2} + \frac{yDz - zDy}{y^2 + z^2}\sqrt{-1}$$

so $yDy + zDz = 0$, which implies that $c = y^2 + z^2 \in C^*$. Let then $W = w^2/c$ and

$$\begin{aligned}
u &= \frac{1 - W}{1 + W}\sqrt{-1} = \frac{c - w^2}{c + w^2}\sqrt{-1} = \frac{c - y^2 + z^2 - 2yz\sqrt{-1}}{c + y^2 - z^2 + 2yz\sqrt{-1}}\sqrt{-1} \\
&= \frac{z^2\sqrt{-1} + yz}{y^2 + yz\sqrt{-1}} = \frac{z}{y}\frac{z\sqrt{-1} + y}{y + z\sqrt{-1}} = \frac{z}{y} \in K.
\end{aligned}$$

We have $W = (\sqrt{-1} - u)/(\sqrt{-1} + u)$, so by Lemma 5.10.1,

$$2\sqrt{-1}\frac{Du}{u^2 + 1} = \frac{DW}{W} = 2\frac{Dw}{w} - \frac{Dc}{c} = 2n\sqrt{-1}Db$$

which implies that $nDb = Du/(u^2 + 1)$. By Theorem 9.3.2 applied to $v = nb$, there are $r_i \in \mathbb{Q}$ such that

$$nb + \sum_{i \in A_{K/C(x)}} r_i t_i + \sum_{i \in T_{K/C(x)}} r_i a_i \in C$$

where $t_i = \tan(a_i)$ for $i \in T_{K/C(x)}$. Applying D and dividing by n yields (9.15). Conversely, if there are $r_i \in \mathbb{Q}$ satisfying (9.15), then

$$Db = \sum_{i \in A_{K/C(x)}} r_i \frac{Da_i}{a_i^2 + 1} + \sum_{i \in T_{K/C(x)}} r_i \frac{Dt_i}{t_i^2 + 1}.$$

so putting the r_i's over a common denominator $e \neq 0$ and multiplying by $2\sqrt{-1}$, we get

$$2e\sqrt{-1}Db = \sum_{i \in A_{K/C(x)}} 2e_i\sqrt{-1}\frac{Da_i}{a_i^2 + 1} + \sum_{i \in T_{K/C(x)}} 2e_i\sqrt{-1}\frac{Dt_i}{t_i^2 + 1}$$

where $t_i = \arctan(a_i)$ for $i \in A_{K/C(x)}$. Let $b_i = (\sqrt{-1} - a_i)/(\sqrt{-1} + a_i) \in K(\sqrt{-1})^*$ and $\theta_i = (\sqrt{-1} - t_i)/(\sqrt{-1} + t_i) \in K(\sqrt{-1})^*$. By Lemma 5.10.1, $Db_i/b_i = 2\sqrt{-1}Da_i/(a_i^2 + 1)$ and $D\theta_i/\theta_i = 2\sqrt{-1}Dt_i/(t_i^2 + 1)$, so

$$2e\sqrt{-1}Db = \sum_{i \in A_{K/C(x)}} e_i \frac{Db_i}{b_i} + \sum_{i \in T_{K/C(x)}} e_i \frac{D\theta_i}{\theta_i} = \frac{Dw}{w}$$

where

$$w = \prod_{i \in A_{K/C(x)}} b_i^{e_i} \prod_{i \in T_{K/C(x)}} \theta_i^{e_i} \in K(\sqrt{-1})^*. \tag{9.16}$$

$\square$

The algorithms follow from Corollary 9.3.2 and Theorems 5.1.1, 5.1.2 and 5.10.1: let (K, D) be given explicitly as a real elementary extension of $(C(x), d/dx)$ where $C = \mathrm{Const}_D(K)$, $\sqrt{-1} \notin K$, and suppose that the sets $E_{K/C(x)}, L_{K/C(x)}, T_{K/C(x)}$ and $A_{K/C(x)}$ are known (those can be computed by applying the algorithm to $t_1, t_2, \ldots, t_n$ in that order).
Let $a \in K^*$ and let t in a differential extension of K be such that $t = \log(a)$, i.e. $Dt = Da/a$. If (9.12) has a solution $r_i \in \mathbb{Q}$, then it provides $v \in K$ such that $Dv = Da/a$, hence $c = t - v \in \mathrm{Const}_D(K(t))$ and $K(t) = C(c)(t_1, \ldots, t_n)$. Otherwise, Da/a is not the derivative of an element of K by Corollary 9.3.1, so t is a monomial over K and $\mathrm{Const}_D(K(t)) = \mathrm{Const}_D(K)$ by Theorem 5.1.1.

Let $b \in K$ and let t in a differential extension of K be such that $t = \exp(b)$, i.e. $Dt/t = Db$. If (9.13) has a solution $r_i \in \mathbb{Q}$, then it provides a nonzero integer e and $u \in K^*$ such that $eDb = Du/u$, hence $c = t^e/u \in \mathrm{Const}_D(K(t))$ and $K(t)$ is algebraic over $C(c)(t_1, \ldots, t_n)$ since $t^e = cu$. Otherwise, Db is not the logarithmic derivative of a K-radical by Corollary 9.3.1, so t is a monomial over K and $\mathrm{Const}_D(K(t)) = \mathrm{Const}_D(K)$ by Theorem 5.1.2.

Let $b \in K$ and let t in a differential extension of K be such that $t = \arctan(b)$, i.e. $Dt = Db/(b^2 + 1)$. If (9.14) has a solution $r_i \in \mathbb{Q}$, then it provides $v \in K$ such that $Dv = Db/(b^2 + 1)$, hence $c = t - v \in \mathrm{Const}_D(K(t))$ and $K(t) = C(c)(t_1, \ldots, t_n)$. Otherwise, $Db/(b^2 + 1)$ is not the derivative of an element of K by Corollary 9.3.2, so t is a monomial over K and $\mathrm{Const}_D(K(t)) = \mathrm{Const}_D(K)$ by Theorem 5.1.1.

Let $b \in K$ and let t in a differential extension of K be such that $t = \tan(b)$, i.e. $Dt/(t^2 + 1) = Db$. If (9.15) has a solution $r_i \in \mathbb{Q}$, then it provides a nonzero integer e and $w \in K(\sqrt{-1})^*$ given by (9.16) such that $2e\sqrt{-1}Db = Dw/w$. Let

$$\theta = (\sqrt{-1} - t)/(\sqrt{-1} + t) \in K(\sqrt{-1})(t)^* \text{ and } c = \theta^e/w \in K(\sqrt{-1})(t)^*.$$

Using Lemma 5.10.1 we get

$$\frac{Dc}{c} = e\frac{D\theta}{\theta} - \frac{Dw}{w} = 2e\sqrt{-1}\frac{Dt}{t^2 + 1} - 2e\sqrt{-1}Db = 0$$

so $c \in \mathrm{Const}_D(K(\sqrt{-1})(t))$ and θ is algebraic over $C(c, \sqrt{-1})(t_1, \ldots, t_n)$ since $\theta^e = cw$. Since $t = (\theta - 1)\sqrt{-1}/(\theta + 1)$, this implies that $K(t)$ is algebraic over $C(c, \sqrt{-1})(t_1, \ldots, t_n)$, hence over $C(c)(t_1, \ldots, t_n)$. It is actually possible to compute the minimal polynomial for t over $C(c)(t_1, \ldots, t_n)$ directly from the solution of (9.15) without introducing $\sqrt{-1}$, see [10] for details. Otherwise, if (9.15) has no solution in $\mathbb{Q}$, then $\sqrt{-1}Db$ is not the logarithmic derivative of a $K(\sqrt{-1})$-radical by Corollary 9.3.2, so t is a monomial over K and $\mathrm{Const}_D(K(t)) = \mathrm{Const}_D(K)$ by Theorem 5.10.1.

9.4 The Rothstein–Caviness Structure Theorem

Rothstein and Caviness [69] have generalized the Risch structure theorem by allowing arbitrary primitives instead of logarithms in the tower of extensions. Since a hyperexponential extension can be embedded in an exponential extension of a primitive extension, this yields a structure theorem applicable to arbitrary Liouvillian extensions. In order to avoid having logarithms cancel with primitives, it is necessary to introduce the restriction that for a primitive t over a field F, either t is explicitly given as a logarithm over F, or Dt does not have an elementary integral over F.

Definition 9.4.1. *Let (K, D) be a differential extension of (k, D). $t \in K$ is nonsimple primitive over k if $Dt \in k$ and Dt does not have an integral in any elementary extension of k.*

Note that Theorem 5.1.1 implies that if t is nonsimple primitive over k, then t is transcendental over k and $\mathrm{Const}_D(k(t)) = \mathrm{Const}_D(k)$.

Definition 9.4.2. *(K, D) is a log-explicit Liouvillian extension of (k, D) if there are $t_1, \dots, t_n$ in K such that $K = k(t_1, \dots, t_n)$ and for $i \in \{1, \dots, n\}$, either t_i is elementary over $k(t_1, \dots, t_{i-1})$, or t_i is nonsimple primitive over $k(t_1, \dots, t_{i-1})$.*

Theorem 9.4.1 (Rothstein & Caviness [69]). *Let C be a field, x be transcendental over C, and (K, D) be a log-explicit Liouvillian extension of $(C(x), d/dx)$ with $\mathrm{Const}_D(K) = C$. Write $K = C(x)(t_1, \dots, t_n)$ with each t_i either elementary or nonsimple primitive over $C(x)(t_1, \dots, t_{i-1})$, and let $E_{K/C(x)}$ and $L_{K/C(x)}$ be given by (9.6) and (9.7) respectively. If there are $v \in K$ and $u \in K^*$ such that $Dv = Du/u$, then there are $r_i \in \mathbb{Q}$ such that*

$$ v + \sum_{i \in L_{K/C(x)}} r_i t_i + \sum_{i \in E_{K/C(x)}} r_i a_i \in C $$

where $t_i = \exp(a_i)$ for $i \in E_{K/C(x)}$.

Proof. We proceed by induction on the number μ of nonsimple primitives among $t_1, \dots, t_n$. If $\mu = 0$, then K is elementary over $C(x)$ and the result follows by the Risch structure Theorem. Suppose that $\mu > 0$ and that the theorem holds for any log-explicit Liouvillian extension of $(C(x), d/dx)$ with constant field C and at most $\mu - 1$ nonsimple primitives. Let i_0 be the largest index such that t_{i_0} is nonsimple primitive over $C(x)(t_1, \dots, t_{i_0-1})$, $t = t_{i_0}$, $k = C(x)(t_1, \dots, t_{i_0-1})$ and $\theta_i = t_{i_0+i}$ for $i \in \{1, \dots, m\}$ where $m = n - i_0$. Then, k is a log-explicit Liouvillian extension of $(C(x), d/dx)$ with at most $\mu - 1$ nonsimple primitives, and $K = k(t)(\theta_1, \dots, \theta_m)$ is elementary over $k(t)$ by the maximality of i_0. Let $u_i = \theta_i$ and $v_i = a_i$ for $i \in E_{K/k(t)}$, $u_i = a_i$ and $v_i = \theta_i$ for $i \in L_{K/k(t)}$, $u_0 = 1$, $v_0 = t$, $I = E_{K/k(t)} \cup L_{K/k(t)}$, p be the cardinality of I, and $F = k(u_0, v_0)(u, v, \{u_i\}_{i \in I}, \{v_i\}_{i \in I})$. Since the degree of transcendence of K over $k(t) = k(u_0, v_0)$ is exactly p, and the degree of transcendence of $k(u_0, v_0)$ over k is 1, the degree of transcendence of F over k is at most $p+1$, hence strictly less than $p+2$. Since $Dv - Du/u = 0 \in k$, $Dv_0 - Du_0/u_0 = Dt \in k$ and $Dv_i - Du_i/u_i = 0 \in k$ for each $i \in I$, Theorem 9.2.1 implies that the elements $\omega = du/u - dv$, $\omega_0 = dt$ and $\omega_i = du_i/u_i - dv_i$ of $\Omega_{K/k}$ are linearly dependent over C, so let $c, c_0, \{c_i\}_{i \in I} \in C$ be not all zero such that $c\omega + c_0 dt + \sum_{i \in I} c_i \omega_i = 0$. Dividing by c if $c \neq 0$, we can assume that $c \in \{0, 1\}$. Since C is a vector space over $\mathbb{Q}$, there are $b_1, \dots, b_r \in C$ linearly independent over $\mathbb{Q}$, and m_{ij} such that $c_i = \sum_{j=1}^{r} m_{ij} b_j$. We can assume without loss of generality that $b_1 = 1$, hence that $c = cb_1$, since $c \in \{0, 1\}$. We then have

$$0 \quad = \quad c_0 dt + c\omega + \sum_{i \in I} c_i \omega_i \tag{9.17}$$

$$= \quad c_0 dt + cb_1 dv + \sum_{i \in I} \sum_{j=1}^{r} m_{ij} b_j dv_i - cb_1 \frac{du}{u} - \sum_{i \in I} \sum_{j=1}^{r} m_{ij} b_j \frac{du_i}{u_i}$$

$$= \quad d\left(c_0 t + \sum_{j=1}^{r} b_j y_j \right) - \sum_{j=1}^{r} b_j \frac{dz_j}{z_j}$$

where

$$y_j = c\delta_{1j} v + \sum_{i \in I} m_{ij} v_i, \quad z_j = u^{c\delta_{1j}} \prod_{i \in I} u_i^{m_{ij}}$$

and $\delta_{ij} = 1$ if $i = j$ and 0 if $i \neq j$. By Lemma 9.2.2, this implies that $w = c_0 t + \sum_{j=1}^{r} b_j y_j$ is algebraic over k, and that z_j is algebraic over k for each j. We also have

$$\frac{Dz_j}{z_j} = c\delta_{1j} \frac{Du}{u} + \sum_{i \in I} m_{ij} \frac{Du_i}{u_i} = c\delta_{1j} Dv + \sum_{i \in I} m_{ij} Dv_i = Dy_j \tag{9.18}$$

for each j, so

$$Dw - \sum_{j=1}^{r} b_j \frac{Dz_j}{z_j} = c_0 Dt + \sum_{j=1}^{r} b_j Dy_j - \sum_{j=1}^{r} b_j \frac{Dz_j}{z_j} = c_0 Dt.$$

Applying the trace from $E = k(w, z_1, \ldots, z_r)$ into k and Theorem 3.2.4, we get

$$c_0 [E : k] Dt = D(Tr(w)) - \sum_{j=1}^{r} b_j \frac{D(N(z_j))}{N(z_j)}$$

where Tr and N are the trace and norm maps respectively. Since Dt has no elementary integral over k by hypothesis, the above implies that $c_0 = 0$, hence that $w = \sum_{j=1}^{r} b_j y_j$ is algebraic over k. Let $F = E(y_2, \ldots, y_r)$. Since each y_j is either a logarithm or algebraic over E by (9.18), F is elementary over E, hence elementary over k, and therefore log-explicit Liouvillian over $(C(x), d/dx)$ with at most $\mu - 1$ nonsimple primitives. Furthermore, $\text{Const}_D(F) = C$ since $F \subseteq K$, so applying the induction hypothesis to $y_1 = w - \sum_{j=2}^{r} b_j y_j \in F$ and $z_1 \in F$, we get that there are $r_i \in \mathbb{Q}$ such that

$$y_1 + \sum_{i \in L_{F/C(x)}} r_i t_i + \sum_{i \in E_{F/C(x)}} r_i a_i \in C \tag{9.19}$$

where $t_i = \exp(a_i)$ for $i \in E_{F/C(x)}$. Since E is algebraic over k and each y_i is either a logarithm or algebraic over E by (9.18), $E_{F/C(x)} = E_{k/C(x)}$ and

$$\sum_{i \in L_{F/C(x)}} r_i t_i = \sum_{i \in L_{k/C(x)}} r_i t_i + \sum_{j=2}^{r} \overline{r_j} y_j = \sum_{i \in L_{k/C(x)}} r_i t_i + \sum_{j=2}^{r} \sum_{i \in I} m_{ij} \overline{r_j} v_i$$

where $\overline{r_j} = 0$ whenever y_j is not a logarithmic monomial, r_j otherwise. In addition, $y_1 = cv + \sum_{i \in I} m_{i1} v_i$, so (9.19) becomes

$$cv + \sum_{i \in I} m_{i1} v_i + \sum_{i \in L_{k/C(x)}} r_i t_i + \sum_{j=2}^{r} \sum_{i \in I} m_{ij} \overline{r_j} v_i + \sum_{i \in E_{k/C(x)}} r_i a_i$$

$$= cv + \sum_{i \in L_{k/C(x)}} r_i t_i + \sum_{i \in E_{k/C(x)}} r_i a_i + \sum_{i \in I} \widehat{r_i} v_i \quad \in \quad C$$

where $\widehat{r_i} = m_{i1} + \sum_{j=2}^{r} m_{ij} \overline{r_j} \in \mathbb{Q}$. Furthermore,

$$\sum_{i \in I} \widehat{r_i} v_i = \sum_{i \in L_{K/k(t)}} \widehat{r_i} \theta_i + \sum_{i \in E_{K/k(t)}} \widehat{r_i} \eta_i$$

where $\theta_i = \exp(\eta_i)$ for $i \in E_{K/k(t)}$, so

$$cv + \sum_{i \in L_{k/C(x)}} r_i t_i + \sum_{i \in E_{k/C(x)}} r_i a_i + \sum_{i \in L_{K/k(t)}} \widehat{r_i} \theta_i + \sum_{i \in E_{K/k(t)}} \widehat{r_i} \eta_i \quad \in C . \quad (9.20)$$

Since $\{\theta_i\}_{i \in I}$ is a transcendence basis for K over $k(t)$, $\{d\theta_i\}_{i \in I}$ is a basis for $\Omega_{K/k(t)}$ over K by Theorem 9.1.2. For $i \in E_{K/k(t)}$ we have $\omega_i = d\theta_i / \theta_i - d\eta_i$ where $\theta_i = \exp(\eta_i)$ and $\eta_i \in k(t)(\theta_1, \ldots, \theta_{i-1})$. For $i \in L_{K/k(t)}$ we have $\omega_i = d\eta_i / \eta_i - d\theta_i$ where $\theta_i = \log(\eta_i)$ and $\eta_i \in k(t)(\theta_1, \ldots, \theta_{i-1})$. In both cases, $\eta_i \in k(t)(\theta_1, \ldots, \theta_{i-1})$ implies that that $d\eta_i$ is the K-span over K of $\{d\theta_j\}_{j \in I, j < i}$ by Theorem 9.1.2. Hence, the matrix of $\{\omega_i\}_{i \in I}$ in the basis $\{d\theta_i\}_{i \in I}$ is a triangular matrix whose diagonal entries are either θ_i^{-1} or -1. This implies that $\{\omega_i\}_{i \in I}$ is linearly independent over K, hence that $c \neq 0$ in (9.17). Therefore we can assume that $c = 1$ in (9.20), and noting that $L_{K/C(x)} = L_{K/k(t)} \cup L_{k/C(x)}$ and $E_{K/C(x)} = E_{K/k(t)} \cup E_{k/C(x)}$ completes the proof.   □

The algorithms of Sect. 9.3 for determining whether new logarithms or exponentials over a differential field are monomials over that field having the same constants become now applicable to log-explicit Liouvillian extensions.

Corollary 9.4.1. *Let $C, x, K, E_{K/C(x)}$ and $L_{K/C(x)}$ be as in Theorem 9.4.1, $a \in K^*$ and $b \in K$. Then,*

(i) *Da/a is the derivative of an element of K if and only if there are $r_i \in \mathbb{Q}$ such that*

$$\sum_{i \in L_{K/C(x)}} r_i Dt_i + \sum_{i \in E_{K/C(x)}} r_i \frac{Dt_i}{t_i} = \frac{Da}{a} . \quad (9.21)$$

(ii) *Db is the logarithmic derivative of a K-radical if and only if there are $r_i \in \mathbb{Q}$ such that*

$$\sum_{i \in L_{K/C(x)}} r_i Dt_i + \sum_{i \in E_{K/C(x)}} r_i \frac{Dt_i}{t_i} = Db . \quad (9.22)$$

The proof and corresponding algorithms are exactly the same than for Corollary 9.3.1 and the algorithms following it. In the cases arising from the integration algorithm, we can always ensure that the differential field containing the integrand is a log-explicit Liouvillian extension of its constants by applying recursively the integration algorithm to primitives. For the general case, Rothstein and Caviness also proved that any Liouvillian extension of a differential field can be embedded in a log-explicit Liouvillian extension [69].

Exercises

Exercise 9.1. Let (k, D) be a differential field of characteristic 0 and a, b in k^* be such that $b^2 + 1 \neq 0$ and

$$\frac{Da}{a} = \frac{Db}{b^2 + 1}.$$

Show that $\sqrt{-1} \in k$.

References

1. S.K. Abdali, B.F. Caviness & A. Pridor (1977): Modular polynomial arithmetic in partial fraction decomposition, *Proceedings of the 1977 MACSYMA Users Conference*, NASA Pub. CP-2012, 253–261.

2. A. Akritas (1989): *Elements of Computer Algebra with Applications*, Wiley Interscience, New York.

3. E. Artin & O. Schreier (1926): Algebraische Konstruktion reeller Körper, *Abhandlungen aus dem Mathematischen Seminar der Universität Hamburg* 5, 83–115.

4. J. Ax (1971): On Schanuel's conjectures *Annals of Mathematics* 93, 252–268.

5. J. Baddoura (1994): *Integration in Finite Terms with Elementary Functions and Dilogarithms*, Ph.D. Thesis, MIT, Mathematics.

6. J. Baddoura (1994): A Conjecture on Integration in Finite Terms with Elementary Functions and Polylogarithms, *Proceedings of ISSAC'94*, ACM Press, 158–162.

7. T. Becker & V. Weispfenning (1993): *Gröbner bases: a computational approach to commutative algebra*, Graduate Texts in Mathematics 141, Springer–Verlag, New York.

8. L. Bertrand (1995): Computing a Hyperelliptic Integral using Arithmetic in the Jacobian of the Curve, *Applicable Algebra in Engineering, Communication and Computing* 6, 275–298.

9. L. Bertrand (1995): *Calcul Symbolique des Intégrales Hyperelliptiques*, Thèse de Doctorat de l'Université de Limoges, Mathématiques.

10. M. Bronstein (1989): Simplification of Real Elementary Functions, *Proceedings of ISSAC'89*, ACM Press, 207–211.

11. M. Bronstein (1990): A unification of Liouvillian extensions, *Applicable Algebra in Engineering, Communication and Computing* 1, 5–24.

12. M. Bronstein (1990): The transcendental Risch differential equation, *Journal of Symbolic Computation* 9, 49–60.

13. M. Bronstein (1990): On the integration of elementary functions, *Journal of Symbolic Computation* 9, 117–173.

14. M. Bronstein (1991): The algebraic Risch differential equation, *Proceedings of ISSAC'91*, ACM Press, 241–246.

15. M. Bronstein & B. Salvy (1993): Full Partial Fraction Decomposition of Rational Functions, *Proceedings of ISSAC'93*, ACM Press, 157–160.

16. W.S. Brown & J.F. Traub (1971): On Euclid's Algorithm and the Theory of Subresultants, *Journal of the ACM* 18, 505–514.

17. B.F. Caviness & M. Rothstein (1975): A Liouville Theorem on Integration in Finite Terms For Line Integrals, *Communications in Algebra* 3, 781–795.

18. P.L. Chebyshev (1857): Sur l'intégration des différentielles qui contiennent une racine carrée d'un polynôme du troisième ou du quatrième degré, *Journal de Mathématiques Pures et Appliquées (2eme série)* 2, 1–42.

19. P.L. Chebyshev (1864): Sur l'intégration de la différentielle $(x + A)dx/\sqrt{x^4 + ax^3 + bx^2 + cx + d}$, *Journal de Mathématiques Pures et Appliquées (2eme série)* **9**, 225–246.
20. G. Cherry (1985): Integration in Finite Terms with Special Functions: the Error Function, *Journal of Symbolic Computation* **1**, 283–302.
21. G. Cherry (1986): Integration in Finite Terms with Special Functions: the Logarithmic Integral, *SIAM Journal on Computing* **15**, 1–21.
22. G.E. Collins (1967): Subresultants and reduced polynomial remainder sequences, *Journal of the ACM* **14**, 128–142.
23. D. Cox, J. Little & D. O'Shea (1992): *Ideals, Varieties, and Algorithms*, Undergraduate Texts in Mathematics, Springer–Verlag, New York.
24. G. Czichowski (1995): A Note on Gröbner Bases and Integration of Rational Functions, *Journal of Symbolic Computation* **20**, 163–167.
25. G. Darboux (1878): Mémoire sur les équations différentielles algébriques du premier ordre et du premier degré, *Bulletin des sciences mathématiques*, 2eme série **2**, 60–96, 123–144, 151–200.
26. J.H. Davenport (1981): *On the integration of algebraic functions*, Lecture Notes in Computer Sciences 102, Springer–Verlag, Heidelberg.
27. J.H. Davenport (1983): *Intégration Formelle*, Rapport de Recherche RR375, IMAG Grenoble.
28. J.H. Davenport (1986): The Risch Differential Equation Problem, *SIAM Journal on Computing* **15**, 903–918.
29. J.H. Davenport, Y. Siret & E. Tournier (1988): *Computer Algebra*, Academic Press, London.
30. G.M. Fichtenholz (1959): *Kurs differencial'nogo i integral'nogo iscislenija*, Moscow.
31. K.O. Geddes, S.R. Czapor & G. Labahn (1992): *Algorithms for Computer Algebra*, Kluwer Academic Publishers, Boston.
32. P. Gianni & B.M. Trager (1996): Square-Free Algorithms in Positive Characteristic, *Applicable Algebra in Engineering, Communication and Computing* **7**, 1–14.
33. G.H. Hardy (1916): *The Integration of Functions of a Single Variable*, Cambridge University Press, Cambridge, England.
34. E. Hermite (1872): Sur l'intégration des fractions rationelles, *Nouvelles Annales de Mathématiques (2eme série)* **11**, 145–148.
35. E. Horowitz (1969): *Algorithms for Symbolic Integration of Rational Functions*, Ph.D. Thesis, University of Wisconsin, Madison.
36. E. Horowitz (1971): Algorithms for partial fraction decomposition and rational function integration, *Proceedings of SYMSAM'71*, ACM Press, 441–457.
37. I. Kaplansky (1976): *An Introduction to Differential Algebra*, 2nd Edition, Hermann, Paris.
38. P. Knowles (1992): Integration of a Class of Transcendental Liouvillian Functions with Error–Functions, Part I, *Journal of Symbolic Computation* **13**, 525–543.
39. P. Knowles (1993): Integration of a Class of Transcendental Liouvillian Functions with Error–Functions, Part II, *Journal of Symbolic Computation* **16**, 227–241.
40. S. Lang (1970): *Algebra*, Addison-Wesley, Reading, Massachusetts.
41. P.S. Laplace (1820): *Théorie Analytique des Probabilités*, 3eme Ed., Ve Courcier, Paris.
42. D. Lazard & R. Rioboo (1990): Integration of Rational Functions: Rational Computation of the Logarithmic Part, *Journal of Symbolic Computation* **9**, 113–116.

43. J. Liouville (1833): Premier mémoire sur la détermination des intégrales dont la valeur est algébrique, *Journal de l'Ecole Polytechnique* **14**, 124–148.

44. J. Liouville (1833): Second mémoire sur la détermination des intégrales dont la valeur est algébrique, *Journal de l'Ecole Polytechnique* **14**, 149–193.

45. J. Liouville (1835): Mémoire sur l'intégration d'une classe de fonctions transcendantes, *Journal für die reine und angewandte Mathematik* **13**, 93–118.

46. R. Loos (1982): Generalized polynomial remainder sequences, in: (Buchberger, Collins, Loos) *Computer Algebra, Symbolic and Algebraic Computation*, Springer–Verlag, New York. 115–137.

47. J. Luetzen (1990): *Joseph Liouville 1809–1882*, Studies in the history of mathematics and physical sciences 15, Springer–Verlag, New York.

48. D. Mack (1975): *On Rational Integration*, UCP-38, Dpt. of Computer Science, University of Utah.

49. J. Mařík (1991): A note on integration of rational functions, *Mathematica Bohemica* **116**, 405–411.

50. B. Mishra (1993): *Algorithmic Algebra*, Texts and Monographs in Computer Science, Springer–Verlag, New York.

51. J. Moses (1971): Symbolic Integration: the Stormy Decade, *Communications of the ACM* **14**, 548–560.

52. T. Mulders (1996): A Note on Subresultants and a Correction to the Lazard/Rioboo/Trager Formula in Rational Function Integration, submitted to the *Journal of Symbolic Computation*.

53. M.W. Ostrogradsky (1845): De l'intégration des fractions rationelles, *Bulletin de la Classe Physico-Mathématiques de l'Académie Impériale des Sciences de St. Pétersbourg* **IV**, 145–167, 286–300.

54. A. Ostrowski (1946): Sur l'intégrabilité élémentaire de quelques classes d'expressions, *Commentarii Mathematici Helvetici* **18**, 283–308.

55. J.C. Piquette (1991): A Method for Symbolic Evaluation of Indefinite Integrals Containing Special Functions or their Products, *Journal of Symbolic Computation* **11**, 231–249.

56. N.V. Rao (1994): A generalization of Liouville's theorem on elementary functions, Preprint, University of Toledo, Ohio.

57. R. Rioboo (1991): *Quelques aspects du calcul exact avec des nombres réels*, Thèse de Doctorat de l'Université de Paris 6, Informatique.

58. R. Risch (1968): On the Integration of Elementary Functions which are built up using Algebraic Operations, Report SP-2801/002/00, System Development Corp., Santa Monica.

59. R. Risch (1969): Further Results on Elementary Functions, Research Report RC-2402, IBM Research, Yorktown Heights.

60. R. Risch (1969): The Problem of Integration in Finite Terms, *Transactions of the American Mathematical Society* **139**, 167–189.

61. R. Risch (1970): The Solution of Problem of Integration in Finite Terms, *Bulletin of the American Mathematical Society* **76**, 605–608.

62. R. Risch (1979): Algebraic Properties of the Elementary Functions of Analysis, *American Journal of Mathematics* **101**, 743–759.

63. J.F. Ritt (1948): *Integration in Finite Terms*, Columbia University Press, New York.

64. M. Rosenlicht (1968): Liouville's Theorem on Functions with Elementary Integrals, *Pacific Journal of Mathematics* **24**, 153–161.

65. M. Rosenlicht (1972): Integration in Finite Terms, *American Mathematical Monthly* **79**, 963–972.

66. M. Rosenlicht (1976): On Liouville's Theory on Elementary Functions, *Pacific Journal of Mathematics* **65**, 485–492.

67. M. Rothstein (1976): *Aspects of Symbolic Integration and Simplification of Exponential and Primitive Functions*, Ph.D. Thesis, University of Wisconsin, Madison.
68. M. Rothstein (1977): A New Algorithm for the Integration of Exponential and Logarithmic Functions, *Proceedings of the 1977 MACSYMA Users Conference*, NASA Pub. CP-2012, 263–274.
69. M. Rothstein & B.F. Caviness (1979): A Structure Theorem for Exponential and Primitive Functions, *SIAM Journal on Computing* **8**, 357–367.
70. M.F. Singer (1992): Liouvillian first integrals of differential equations, *Transactions of the American Mathematical Society* **333**, 673–687.
71. M.F. Singer, B.D. Saunders & B.F. Caviness (1985): An extension of Liouville's Theorem on Integration in Finite Terms, *SIAM Journal on Computing* **14**, 965–990.
72. J. Slagle (1961): *A Heuristic Program that Solves Symbolic Integration Problems in Freshman Calculus*, Ph.D. Thesis, Harvard University.
73. R. Tobey (1967): *Algorithms for Antidifferentiation of Rational Functions*, Ph.D. Thesis, Harvard University.
74. B.M. Trager (1976): Algebraic Factoring and Rational Function Integration, *Proceedings of SYMSAC'76*, 219–226.
75. B.M. Trager (1984): Integration of Simple Radical Extensions, *Proceedings of EUROSAM'79*, Lecture Notes in Computer Science 72, 408–414.
76. B.M. Trager (1984): *On the integration of algebraic functions*, Ph.D. Thesis, MIT, Computer Science.
77. B.L. Van Der Waerden (1970): *Algebra*, vol.1, 7th edition, Frederic Ungar Publishing Co., New York.
78. J.A. Weil (1995): *Constantes et polynômes de Darboux en algèbre différentielle: application aux systèmes différentiels linéaires*, Thèse de Mathématiques, Ecole Polytechnique.
79. M. Weileider (1990): *Integration elementarer Funktionen durch elementare Funktionen, Errorfunktionen und logarithmische Integrale*, Doctoral Thesis, LMU München, Mathematik.
80. D.Y.Y. Yun (1976): On Square-free Decomposition Algorithms, *Proceedings of SYMSAC'76*, 26–35.
81. D.Y.Y. Yun (1977): Fast Algorithm for Rational Function Integration, *Proceedings of IFIP'77*, North Holland, 493–498.
82. R. Zippel (1993): *Effective Polynomial Computation*, Kluwer Academic Publishers, Boston.

Index

297

Computer language

R.E. Crandall
Topics in Advanced Scientific Computation

1996. XII, 340 pages.
Hardcover DM 78,–
ISBN 0-387-94473-7

The theme of this book is "hard computational problems made accessible". It is a collection of essays on modern computational problems whose solutions are difficult, if not impossible, to achieve. This publication covers hot topics of interest to a broad spectrum of professionals working with chaos and fractals, complexity, prime numbers and encryption, wavelets, fast Fourier transforms, signal processing, and much more. As a tour of modern algorithms it will be extremely useful to computational researchers and students for the following reasons: it focuses on solutions to problems, rather than problem posing; it contains explanations of the importance and origins of hard problems whose explanations are usually difficult to find in modern literature; it includes exhibitions of state-of-the-art algorithms for these difficult problems; and, it contains the actual code for most algorithms discussed. Enhancement files, program code and other data are available via the TELOS Web site: *www.telospub.com*.

W. Gander, J. Hřebíček

Solving Problems in Scientific Computing Using Maple and MATLAB

2nd, exp. ed. 1995. XV, 315 pages.
106 figures, 8 tables.
Softcover DM 68,–
ISBN 3-540-58746-2

Modern computing tools like *Maple* (symbolic computation) and *MATLAB* (a numeric computation and visualization program) make it possible to easily solve realistic nontrivial problems in scientific computing. In education, traditionally, complicated problems were avoided, since the amount of work for obtaining the solutions was not feasible for students. This situation has changed now, and students can be taught real-life problems that they can actually solve using the new powerful software. The reader will improve his knowledge through learning by examples and he will learn how both systems, *MATLAB* and *Maple,* may be used to solve problems interactively in an elegant way. This second edition has been expanded by two new chapters. All programs can be obtained from a server at ETH Zurich.

Please order by
Fax: +49 30 82787 301
e-mail: orders@springer.de
or through your bookseller

Springer

Springer-Verlag, P. O. Box 31 13 40, D-10643 Berlin, Germany.

Springer. Computer literacy

D. Cox, J.B. Little, D. O'Shea

Ideals, Varieties, and Algorithms

An Introduction to Algebraic Geometry and Commutative Algebra
2nd ed. 1996. Approx. 550 pages.
83 figures. (Undergraduate Texts in Mathematics)
Hardcover DM 68,-
ISBN 0-387-94680-2

This introduction presents the highly interesting but complex subject at a level understandable to the undergraduate. The authors approach the topic in a direct manner and prefer to explain rather than assume too much previous knowledge on the part of the readers. In this way, it provides the ideal lead-in for both computer scientists and mathematicians.
"I consider the book to be wonderful ... The exposition is very clear, there are many helpful pictures, and there are a great many instructive exercises, some quite challenging ... offers the heart and soul of modern commutative and algebraic geometry."
The American Mathematical Monthly

M.S. Malone

The Microprocessor

A Biography
1995. XIX, 333 pages.
Hardcover DM 48,-
ISBN 3-540-94342-0

Here, for the very first time, Michael S. Malone tells the complete story of this amazing invention in his well-known and witty style. However, this is anything but an electronics textbook. Rather, it is a riveting and incisive adventure story about extraordinary people and the legendary companies they have built. It is a tale of huge success and devastating failure, steadfast partnerships and bitter rivalries – plus a liberal sprinkling of greed and wealth. Malone closes with a tantalising look into the future: emerging technologies, new software, and even speculation about what might lie beyond the microprocessor era.
"Malone's account of the creation and historical development of the microprocessor is the closest account to the truth that I have seen. This book takes full advantage of a good opportunity to tell the story correctly."
Federico Faggin,
co-inventor of the microprocessor

Please order by
Fax: +49 30 82787 301
e-mail: orders@springer.de
or through your bookseller

Springer-Verlag, P. O. Box 31 13 40, D-10643 Berlin, Germany.

Springer
and the
environment

At Springer we firmly believe that an international science publisher has a special obligation to the environment, and our corporate policies consistently reflect this conviction.
We also expect our business partners – paper mills, printers, packaging manufacturers, etc. – to commit themselves to using materials and production processes that do not harm the environment. The paper in this book is made from low- or no-chlorine pulp and is acid free, in conformance with international standards for paper permanency.

Springer

Druck: STRAUSS OFFSETDRUCK, MÖRLENBACH
Verarbeitung: SCHÄFFER, GRÜNSTADT